CLIMATE CHANGE IN THE MIDDLE EAST AND NORTH AFRICA

Environmental factors in the Middle East and North Africa (MENA) have played a crucial role in the historical and social development of the region. The book delves into a broad set of historical literature from the past 15,000 years that neglected to consider environmental factors to their full effect.

Beyond the broad historic analysis, the chapters derive conclusions for today's debate on whether climate change leads to more social conflict and violence. Introducing a theoretical framework focused on adaptive cycling, this book probes and refines the role of climate in ancient and modern political-economic systems in the MENA region. It also underscores just how bad the 21st-century environment may become thanks to global warming. While the MENA region may not survive the latest onslaught of deteriorating climate, there is also some interest in how a region that once led the world in introducing all sorts of innovations thousands of years ago has evolved into a contemporary setting characterized by traditional conservatism, poverty, and incessant strife.

Emphasizing regional dynamics, the book's central question deals with the role of climate change in the rise and decline of the MENA region. The book will be a key resource to students and readers interested in global warming, including academics and policymakers.

William R. Thompson is Distinguished Professor and Rogers Chair of Political Science Emeritus, Indiana University; Editor-in-Chief of the *Oxford Research Encyclopedia of Politics*; a former President of the International Studies Association; and a former Editor-in-Chief of *International Studies Quarterly*.

Leila Zakhirova is Associate Professor of Political Science at Concordia College, Moorhead, MN, and a Co-Editor of *Asian Security*. She is currently researching the impact of climate change on human security.

CLIMATE CHANGE IN THE MIDDLE EAST AND NORTH AFRICA

15,000 Years of Crises, Setbacks, and Adaptation

William R. Thompson and Leila Zakhirova

Routledge
Taylor & Francis Group

LONDON AND NEW YORK

CONTENTS

FIGURES

TABLES

ACKNOWLEDGMENTS

We want to thank Robert Denmark for putting together a book panel for us at the Toronto, Canada, meeting of the International Studies Association. He was joined by Mark Boyer and Gabriella Kuetting in providing feedback on an early draft. We also thank Imad Mansour and Nizar Messari for their efforts to find feedback from Middle Eastern readers. We are also very grateful to Heather Boecker for her meticulous attention to detail and assisting us in preparing the manuscript for the publisher.

1

15,000 YEARS OF CLIMATE CHANGE IN THE MENA

Big history, big questions

Big questions and big history

One of the more fascinating questions in Big History is why do people in some parts of the world essentially get ahead of others who reside in other parts. It is argued, for instance, that the number of large mammals suitable for exploiting for transportation and other work made a major difference in which regions developed early. So, too, does the potential for East–West interaction patterns.[1] Similarly, to the extent that coal was critical to the advent of industrialization, easy access to large deposits of coal was at least instrumental to some regions becoming pioneers in industrialization.[2] Yet for every putative advantage there are also corresponding disadvantages. South America had few large mammals that could be exploited for much besides their wool. Sub-Saharan African inter-action patterns tended to work from north to south. Change, as a consequence, moved very slowly through deserts, jungles, and long distances. India had and has access to large deposits of coal but lacks the appropriate climate that might have encouraged using coal extensively for heating purposes when that was an important prerequisite to innovating novel approaches to generating energy else-where in the world. Now that it is ready to use its fossil fuel inheritance, coal has gone out of fashion.

From this perspective of auditing pluses and minuses, what can we say about the rise and demise of the Middle East and North Africa (MENA)? After all, many of the initial breakthroughs to greater complexity in human affairs had their origins in southern Mesopotamia. Writing, urbanization, wheels, the secondary products revolution, large-scale irrigation projects, armies, states, empires – to say nothing of beer (but not wine) – all can be traced to Mesopotamia. For all of its precocities, Sumer and its predecessors and immediate successors no longer exist and have not been around for thousands of years. Its early lead may have owed a great deal to environmental variety (Algaze, 2008) but its Achilles heel has long been its environmental fragility. In the MENA, abrupt climate changes periodically set back the early leaders in greater complexity innovations. Ubaid, Uruk, and Akkad represent early peaks of innovation of different kinds. Ubaid traits spread throughout the immediate region. In the Uruk period, an extensive regional network was established to funnel resources back to southern Mesopotamia, making it possibly the first economic empire.[3] Akkad constituted the first coercive empire buttressed by a standing army of 5,000 soldiers. Each movement toward greater complexity, however, was set back by climate change. Moreover, it turns out that this inherent fragility was not simply a southern Mesopotamian characteristic. It was and remains a liability characteristic of much of the MENA and one that has established significant parameters on what can be achieved and sustained in that part of the world. Contrary to what was mentioned earlier, South America no longer needs large mammals for transportation purposes. North–South interaction patterns in Africa can be overcome by planes, trains, trucks, and shipping. Coal has turned out to be more of a bane than a cure for generating energy. Middle Eastern environmental fragility, however, persists and could lead to the region, or large portions of it, becoming virtually uninhabitable in the current bout of climate deterioration.

J. R. McNeill (2013: 28) cautions against making such generalizations about the Middle East and North Africa without considerable temporal caveats:

> I do not argue that these eccentricities [referring to water, grass, and energy characteristics] as a whole either favored or disadvantaged the region. Such generalizations cannot be sustained across time because conditions change. At certain times, such as during the later nineteenth century age of coal and steam, it is probably safe to say that the MENA stood at a disadvantage with respect to many parts of the world because most of it lacked coal. Such specific moments – anchored in particular historical moments-are plausible. But generalizations made for all time are not…[4]

We think most analysts would agree with his prudence and assumption that the Middle Eastern environment has been fairly intermittent in its impact on behavior. Yet the question is whether standing liabilities and specific moments are incompatible. If a disadvantage such as environmental fragility persists through at least recorded history and even beyond, the specific moments may appear distinctive even while the nature of the underlying environmental problems are more uniform. Such is the case, we think, with recurring episodes of cold and dry climates that have marked the region's environmental fragility in the past – not just fluctuations in weather but severe and recurring onsets of unfavorable climate change for several centuries at a time. Life may go on but in a step forward, two steps backward pace.

Two pages later, McNeill (2013: 30) remarks that:

> The second eccentricity of water that shaped life and history in the MENA is more familiar: the sharply uneven distribution of fresh water, the prevalence of aridity, and the consequent ecological responsiveness to even modest climate change. That responsiveness took the form of florescence in times of plentiful rainfall … and of crisis in times of low rainfall.[5]

This last statement is precisely the long-term environmental fragility that we have in mind. It is one that encompasses at least the last 15,000 years and definitely continues today.[6]

Curiously, though, the full MENA climate story remains relatively unknown and under-appreciated. Historians of the Middle East have been reluctant to give much weight to climate change in their analyses, preferring instead to emphasize personalities, dynasties, and culture. It has only been fairly recently that some analysts have begun to move away from the assumption that the Middle Eastern climate has been relatively constant over the past 10–15 millennia. Historians tend also to compartmentalize their analyses temporally. Just as archaeologists tend to specialize in interpreting specific mounds of trash, historians specialize in the ancient world, Byzantium and early Islam, or more contemporary interactions. The possibility that certain continuities influence behavior across the entire time span of Homo sapiens occupation in the Middle East, as a consequence, has not been tested very often.

Another propensity is to instinctively explain what happens in terms of what people do and have done. Agency is paramount. Things do not simply happen. Humans must be responsible for whatever damage they have done. While this instinct is preferable to an older propensity to explain things in terms of the fickleness of the gods, it can get in the way of telling the story appropriately. For example, one school of thought emphasizes "overshoot." Decision-making and human activity exploit its resource base until the exploitation boomerangs. Deforestation, soil erosion, or soil salinization from over-irrigation undermine the possibility of agricultural productivity. Large populations, the beneficiaries of growth and exploitation, are at risk. Only Malthusian remedies will suffice.

Alternatively, another school of thought stresses the resilience of human populations. Collapses may come and go but people persist. The complexities they have developed at one point in time may need to be dispensed with as conditions dictate but people do not vanish. Instead, they simplify. If large cities cannot be sustained, they fall back on living in small villages until large cities can once again be constructed and maintained.

Still other, older arguments have emphasized barbarian attacks, trade disruptions, and the rise and fall of dynasties to explain what has transpired. Lean, mean savages from the mountains or the deserts periodically swarm down on defenseless cities and destroy civilization. Periods of increased economic integration fall apart when something intervenes – be it war, more barbarians, or disease

– to block the gains accruing from exchanges of information and commodities. The framework for summarizing ancient Egyptian history is most revealing. The Old Kingdom dynasties give way to the Middle Kingdom dynasties that in turn are replaced by the New Kingdom dynasties. Each dynastic cluster is separated by an "intermediate" period of turmoil, disaster, and disorder.

Our point is not that these interpretative emphases are wrong. Humans, individually and collectively, possess variable amounts of agency and resilience. Decision-makers zig when they might have zagged given the same operative environmental conditions. If people did not have some resilience, the human population would have withered away long ago as another failed experiment in species development. Barbarians from the mountains and deserts have roles to play in history. Periods of increased economic integration do give way to setbacks before more integration is accomplished. Governments do rise and fall. However, what is often missing is an important common denominator, and especially so in Middle Eastern stories. It is climate change that helps weave together these various strands of emphasis. Climate change does not trigger resource overexploitation but it exacerbates it. Climate change does not necessarily overwhelm resilience in every instance but it can if the deterioration of environmental conditions is sufficiently severe and prolonged. Barbarian attacks are not random. They are driven by pastoralists and nomads looking for water, pastures, and more benign climates. Some easy loot on the side would not be a bad thing either. These goals tend to be realized or at least attempted by moving toward cities. Thus, nomads come into increased contact and conflict with more sedentary populations. Interruptions of economic integration can often be traced to fundamental changes in climate that lead to breakdowns in exchange patterns. Governments fall, especially in an agrarian era, when they can no longer extract taxes from farmers who can no longer grow crops because it is too cold, hot, or dry. In other words, climate change is not deterministic but it is extremely influential in bringing together a number of problems that are sometimes too difficult to overcome or survive. It provides a context for when these types of problems are more or less likely to occur.

To be sure, the problems with attributing causal significance to climate change in history are daunting. The evidence is often sketchy. Historical climate data are measured in longer time units than are political and economic behaviors of interest. We have data measured in centuries. Yet governmental regimes often collapse in specific years. Can the longer-term data be correlated with the shorter-term behavior? Moreover, archaeologists revise their data periodically on the timing of dynastic rise and falls. Sometimes events once thought to be influenced by climate change turn out to have occurred prior to the climate change after the dates are revised. Moreover, rival hypotheses abound and frequently there is even less data to address them. Can we be confident that climate change is the key when it is difficult to assess the power of alternative explanations?

We seek to be reasonable in our undertaking. Sometimes the data support climate change playing a significant causal role. Other times, the evidence is lacking. If scholars allow themselves to take these data problems too seriously, analytical paralysis is the likely outcome. Instead, we need to do the best we can with what evidence we can develop. Ironically perhaps, one advantage that is found in looking at climate impacts is that we usually are not attempting to link the amount of rainfall in say one day, week, month, or even year to behavior in that same time period. The kind of climate change that we are examining in the present study tends to persist for long periods of time – even centuries. It is the prolonged change that is so powerful, and all the more so if it is a rather severe change. As Knapp and Manning (2016: 113) note:

> The minimum criterion needed to assess whether climate can be considered in any way directly associated with historical change is to establish a chronological linkage. Thus, if climate can be shown to be particularly positive, or negative, as relevant to a particular region or even throughout a hemisphere, during years X1 to Xn, and there is good archaeological or documentary evidence of historical impact and change (plausibly associated with such climate) in or immediately following those years, then it would be reasonable to assess whether there is a real linkage and a case of climate forcing, or affecting, history… Generally, significant change does not involve regular, high frequency, single-year 'blips,' whether good or bad: human societies are usually well adapted to overcome lean or bad years. Rather, it is longer, multi-year, even multi-decade climactic episodes that may undermine long-standing agrarian, economic, and/or political regimes and that might precipitate historical change.

We propose to take chronology one step further by looking at multiple cases over a very long period of time.[7] One way to underscore the role of climate in Middle Eastern affairs is to focus on salient cases in order to assess how much difference climate change might have made. Conventionally, authors stay within modest temporal parameters and look at, say, the fall of Akkad thousands of years ago or the Ottoman Empire in the 17th century CE. Less conventionally, another approach is to string together an overview of salient episodes of behavioral changes that occur within roughly predicted windows of climate change periods. We attempt the latter approach in this examination, beginning with the Younger Dryas era and the origins of agriculture question going back many millennia and ending our historical review with the Ottoman Empire that disintegrated in the 20th century. Throughout we rely heavily on other scholars' analysis. We do not have incredibly novel things to say about either the Younger Dryas era or the Ottoman Empire. But we think there is some definite value added by looking at and comparing episodes of rise and decline in the MENA that appear to be strongly influenced by climate change. We do so for the longitudinal advantages of examining the *longue duree* of 15,000 years. If we find recurring behavior when similar types of climate change prevail throughout history, it will help our case. We also do it because we think the previous history has value for anticipating what might happen to MENA in the next and ongoing bout of climate deterioration. That is why we are confining our examination to MENA climate changes and human behavior. Contemporary global warming is not totally similar to the repetitive bouts of cooling that we will be examining in the book. Yet the outcome may be similar in kind if both extreme cooling and warming lead to drought, famine, food insecurities, disease, and, possibly, increased conflict.

Climate change will have repercussions throughout the world but it seems set to transform the MENA region into something never experienced before. Should this come as a surprise? Climate change, after all, has been given a great deal of credit by some for once shaping socio-economics and politics in the ancient Middle East. But it was also there that the significance of climate was beaten down as grossly exaggerated as an important explanatory factor. Yet it is not so much a case of rampant determinism bestowed on an ancient single driver as it has been much easier to make a deterministic case in the ancient world than in subsequent years. That may change. If climate change becomes so horrendous that it is near deterministic in terms of what can and cannot be done in MENA, we will have returned to an earlier phase in the history of MENA.

One way to interpret this situation from a long-term perspective is that climate change was initially a major driver of Middle Eastern behavior but gradually lost its power thanks to greater complexity and the development of successful adaptive strategies. That is, until perhaps now. We think this is a plausible interpretation but that it is wrong. Climate change has been a very strong driver of Middle Eastern and North African behavior throughout the last 15,000 years. But this may appear to have been more characteristic of the ancient world than in more contemporary times in part because climate changes in this region became somewhat less volatile as the ancient world came to an end and because we have been slow to appreciate how climate change works with more subtlety in the more contemporary world. It is equally true that life has become more complex as time passes – a trend that gives more power to other processes without necessarily detracting from the potency of climate change. Perhaps it builds in some additional resilience as well because complexity brings more tools for coping with climate change. Yet resilience varies over time just as the extent of climate change does. The ultimate question is whether the degree of resilience extant at any given time suffices to withstand the extent of climate change. Sometimes it does. At other times, climate change is quite capable of overwhelming societal resilience.

A 15,000-year study, no doubt, seems audacious. However, the role of shaping human behavior in the Middle East is much longer. Evidence suggests that African migrants moved into and through the Middle East on multiple occasions. To survive passage through extremely dry landscapes, episodes of climate change (increasing the amount of rainfall (perhaps circa 75,000 and 45,000–50,000 years ago) would have been necessary. One suspected pathway connects what is now Sudan, Saudi Arabia, and the Levant but there are likely to have been others. People passed through and some stayed for a while. There may have been interactions between Neanderthal and homo sapiens. There may also have been movement back to Africa. While there is a flourishing sub-literature on questions pertaining to the Middle Eastern participation in the "Out-of-Africa" dispersal phenomena, we judged that there was not much to say all that concretely at this juncture and therefore passed on making our study a 75,000-year analysis.[8]

At the same time, we are becoming more attuned to how we have exaggerated the potency of climate change in the ancient world. Climate change was important to explaining what took place but far from deterministic. For instance, climate change made agriculture possible in the Middle East but the relapse to a colder/dryer climate in the Younger Dryas period – a fairly popular interpretation – does not appear to have the main responsibility for stimulating the domestication of plants and animals. At later points, a number of political systems were brought down by climate fluctuations in conjunction with other processes and yet these same processes do not seem to have had the same effects at the same time elsewhere in the region and not all that far away. For instance, the abandonment of cities in a period of intense drought may be matched by the survival and expansion of other cities in the same region.[9] If we focus on the abandonment, climate change seems omnipotent. In the face of variable reactions, its omnipotence is reduced considerably.

Yet climate fluctuations, we think, have been critical to understanding Middle Eastern and North African behavior. The region has prospered in more favorable climate regimes and suffered in less favorable settings. The more severe the climate deterioration, the greater has been the damage to political, economic, and demographic stability. For that matter, no other region has probably experienced the type of serial climate problems that the MENA has encountered and survived.[10] A long history of survival may be encouraging but it is no guarantee that persistent survival can be assumed.

Thus explanations about Middle Eastern behavior tend toward giving climate too much power or too little. The survival and intermittent salience of the region for some 15,000 years underscores the argument against accepting climate determinism. Arguments de-emphasizing climate, on the other hand, tend to promote the significance of human agency and complexity. But this is a false dichotomy if much of the observed human agency and complexity has in fact been influenced strongly by changes in climate, favorable or otherwise. That is, political and economic stresses and even land degradation as explanatory factors make more sense within the context of unfavorable climate changes.

Nonetheless, explanations must give way to contextual changes. In the past, two important factors facilitating resilience in the face of environmental deterioration have been population size and mobility opportunities. Resilience was facilitated by small population groups being able to move within the region. Impacted groups could migrate to less impacted areas. Urban dwellers could move to smaller villages. Farmers could move to less sedentary activities. Desert dwellers could move away from the desert. Yet 15,000 years of development mean that population sizes are now fairly large and definitely less mobile than they once were. The current reluctance of Europe and the United States to absorb Syrian refugees demonstrates the type of scalar restrictions on movements outside the region that have emerged in contemporary times. As a consequence, resilience in the face of severe climate deterioration associated with global warming over decades may be considerably less likely than it was in the past. Moreover, contemporary and immediate future climate deterioration may be as severe or worse than anything hitherto encountered. Significant portions of the region could cease to function altogether given sufficient increases in temperature. Yet few decision-makers either within or outside the region appear to be devoting resources or attention to this looming problem. Even if the worst possible catastrophes are somehow averted, further autocratization, impoverishment, and political violence should be anticipated in what is already the most conflict-prone region in the world system. Major humanitarian crises on a scale not yet seen may also be expected.

A minimalist framework/theory

We begin our examination some 15,000 years ago when glaciers were receding and creating more livable environments, in the Middle East among other places, if only temporarily. Telling a story that winds through thousands of years runs the risk of doing little more than generating a long-winded account of multiple millennia in which the reader suffers vertigo trying to keep up with the waxing and waning of rulers and empires with different names. That is not our goal. To the extent that we end up telling a story, it is not because that is our main objective. Rather, we are interested in regional dynamics. We know enough about the story already not to know it ends but how it has seemed to end temporarily on many occasions. Early communities vanish in the traces of archaeological rubbish. Empires collapse. Powerful states become less so. Yet new or revived communities, empires, and states, keep coming back. How is it that successive communities abruptly have ceased to exist only to come back in some other guise? What drives the collapses? Is it due to overwhelming exogenous forces that cannot be resisted? Is it due to internal weaknesses that make actors more vulnerable to collapse? Is it some combination of the two? If the exogenous or endogenous forces were so powerful, how is it possible, after some lag, to start all over again? Whether or not we can answer these questions precisely, we also need to ask if we should expect the regional dynamics to persist well into the future. Presumably, exogenous and endogenous forces vary in their impact over time. Can they be so overwhelming that there is no coming back?

From a social science perspective, we have an antidote to mere storytelling. We construct models, frameworks, and theories to guide our explorations. Instead of only telling a story, we try to assess these models, frameworks, and theories. Do they fit what we are observing? Do they help to package the story and make it more digestible? To cover 15 millennia, we have to keep the number of variables small. Therefore, we need what might be called a minimal analytical construction. There will be ample opportunities for more ambitious or elaborate models in more specific circumstances. For a 15-millennia arc, something less elaborate is desirable.

We think work on adaptive cycles and resilience (also known as panarchy) in ecology fits this bill.[11] The actual work is sometimes called a metaphor, a model, a framework, and a theory. In fact, it is all of these things simultaneously. The heart of the argument can be summarized by Figure 1.1's recumbent eight moving through four phases. Phase 1 (growth) is the beginning of a new regime or system in which resource potential is high and organization low. Expansion, growth, and innovation are strong. In phase 2 (conservation), these activities move slower as markets become saturated, profit margins diminish, and the overall system becomes more rigid and resistant to newcomers. Phase 3 (release) is a Schumpeterian period of creative destruction. More energy is allocated to maintaining the status quo than to exploring new frontiers. Phase 2 dynamics have made the system "an accident waiting to happen," which is to say that the system's rigidity makes it less resilient. Phase 3 then becomes (or can become) a phase of Schumpeterian creative destruction breaking down the high degree of organization and creating openings for innovation. The system is re-organized and/or renewed in phase 4 (reorganization) with a continuation of the recumbent eight dynamic flow.

Figure 1.1 can be drawn in various ways and the terminology has been subject to some flux. We prefer to think of the connectedness X-axis as one of organization. The "release" phrase is sometimes explicitly labeled as "creative destruction" and the main author of the conceptualization has acknowledged

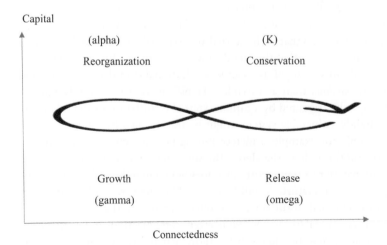

FIGURE 1.1   The adaptive cycle

borrowing from Schumpeter's (1939) economic argument (the main readings are Gunderson and Holling, 2002 and Gunderson, Allen, and Holling, 2010).[12]

The "conservation" label seems a euphemism for monopoly and resistance to innovation but a better phase title is not obvious. Nonetheless, what is important is to follow the flow from the "r" phase of growth to the "k" phase of conservation to the "omega" phase of release or creative destruction to the "alpha" phase of reorganization. However, as far as we can tell the argument emerged to explain how forest fires could be destructive in the short term while beneficial in the long term because fire removed some of the impediments to subsequent growth. It goes without saying that forest fires are sometimes so destructive that they can also foreclose subsequent growth. Climate change can also take on a double-edged impact.

There are a number of considerations to keep in mind vis-à-vis Figure 1.1. First, the phases should not be thought of as applicable to specific time periods or equal in duration. That is, it is not profitable to try to periodize 25 years of growth, followed by 25 years of conservation and then 25 years of creative destruction. One reason is that the pace of the dynamics is not constant through time.[13] Growth and conservation can last decades while release and reorganization can occur quickly. Alternatively, different systems can move through these phases at different gaits. A forest, for instance, can have a much different life cycle than a lake. So, too, might a community of hunter-gatherers versus a mighty empire with extensive regional reach.

A second consideration is that resilience tends to be stronger when connectedness or organization is weaker than when it is stronger. Mature and aging systems (phase 2: conservation) develop internal stress points that make them vulnerable to fragmentation and collapse. In this sense, there is a built-in response to the question of whether we should give more credit to external disturbances or internal processes. The model's assumption is that it is a combination of both. Systems made more vulnerable by internal processes tend to succumb to external disturbances when they are weakened as opposed to when they are stronger. That assumption does not mean that both internal and external factors must be equal in impact significance – only that they tend to work together. This attribute of the theory actually disguises one of its weaknesses. External disturbances are treated more as *deus ex machina* than as variables. Transformation in a highly vulnerable system might be triggered by a minor external disturbance or the other way around – a slightly vulnerable system might be transformed by a huge external disturbance (think, for example, a meteor hitting near the coast of Yucatan setting off firestorms throughout the globe). But since the external disturbance is not one of the variables in this framework, that does not seem to be a "burning" issue in the adaptive cycle literature. It will be for us. Therefore, we add the intensity or extent of the external disturbance as a fourth variable to the adaptive cycle's emphasis on three (capital/resources, connectivity/organization, and resilience).

The third feature of some interest is the term "panarchy," which is one of the main concepts in the adaptive cycle framework. The "pan" component refers to the god of nature while "archy" stands for "rules," thus the term is meant to encompass the rules of nature. But, of course, that is a rather wide compass. The panarchy argument in adaptive cycles is about there being a number of different variables/processes operating at different levels and at different speeds that govern change in the system. Slow-moving macro-level variables constrain the effects of faster-moving micro-level variables, or they can do so. Yet top-down control can be challenged by bottom-up revolts that move quickly and overwhelm slower-moving processes. Alternatively, slow and fast-moving processes can converge to bring about abrupt change and transformation. Social science has not given much explicit attention to the interaction of fast and slow-moving variables at different levels of analysis. We probably should pay more attention to this type of conceptualization.

Finally, the fourth feature of the adaptive cycle approach is that it is functional in nature but not necessarily conservatively functional. That is, the four phases are activities of a system that can lead to its renewal but that renewal or reorganization may not resemble the old system. This is not an argument about why structures persist or return to equilibrium after being pushed in one direction or another. It is an argument about why some aspects of the structure may persist while transformations are equally underway. At no point need it be assumed that renewal/reorganization must take place – only that there is some probability that it may, once processes of creative destruction are at work. What shape that renewal/organization might take is far less predictable than the probability that some reorganization is likely to be attempted. Graphically, adaptive cycle analysts demonstrate these types of movement as a string of boxes of "recumbent eight" dynamics residing within the interactions of resource potential, connectivity/organization, and resilience. A system moves through the dynamic phases in one era transiting into a new system and another set of phases, and then onto a third, and so on.

We adopt this model/framework/theory as one of our organizing devices knowing full well that it has limitations.[14] Perhaps its greatest limitation is that it appears difficult to develop hypotheses within its structure that go beyond the macro-probability of eventual crash and renewal.[15] We have already amended it by adding a fourth variable – the intensity of the external disturbance. There could well be other variables that need to be added for more complex and compelling explanation generations.[16] We leave that task to others and other venues. For the present, the adaptive cycle metaphor serves our immediate purposes and we will employ it as best as we can.

Road map

This book aims to probe and refine the role of climate in ancient and modern political-economic systems in the MENA region. It also seeks to underscore just how bad the 21st-century environment may become, thanks to global warming. While the MENA region may not survive the latest onslaught of deteriorating climate, there is also some interest in how a region that once led the world in introducing all sorts of innovations thousands of years ago has evolved into a contemporary setting characterized by traditional conservatism, poverty, and incessant strife. Our last empirical question, nonetheless, is how the MENA region compares to other parts of the world in terms of its future vulnerability to global warming.

The MENA has intermittently experienced major climate shifts since the retreat of the last glacial age. To establish a foundation for what may come in the 21st century, we first examine the impact of nine episodes of rapid climate change (RCC) beginning with the Younger Dryas event some 12,000 years ago, in Chapter 2. The RCC periodization is based initially on reading Arctic ice deposits with significant cooling and drying impacts in the eastern Mediterranean and other parts of Eurasia. We ask to what extent climate change has been a fundamental process within multiple political-economic and societal transformations in MENA. Our answer is that it has played a central role throughout the last 15,000 years and, not surprisingly, will likely continue to do so in the future.

Intermittent RCC bouts provide an armature for delineating transformations in MENA – an assertion we explore in Chapters 3 (focusing on the ancient Near East) and 6 (focusing on the period since the decline of the Roman Empire). In between these two chapters, Chapters 4 and 5 zero in on the centrality of climate change to political-economic transformations in ancient Mesopotamia and Egypt. Generating a series of hypotheses and data to test them makes it possible to delineate just how critical climate change has been in the heart of the ancient Near East. These findings reinforce our assertion in Chapter 6 that the environmental fragility of the MENA and the nature of responses to this long-term fragility have made the region increasingly less competitive with other regions in the decisive period leading up to the 19th century CE.

In Chapter 6, we return to applying the RCC framework to the post-ancient world – primarily the Late Antiquity and Early Modern eras. One of the underlying questions is whether the dynamics of climate change and related processes are transformed as we move toward the present day. There are differences. More people are involved due to population expansion and the rise in deadly epidemic diseases. Regimes do not necessarily collapse. Instead, they limp along subject to major restrictions on what resources they can mobilize and thus what they can do. But, more generally, the basic processes do not change all that much. Dryness leads to drought and famine. People are encouraged to move or die. This persistent vulnerability to environmental change has led to intermittent setbacks and a consistent check on the ability of the MENA region to flourish as it once did at the dawn of what we like to call civilization. This record goes a long way in explaining why the Near East was once the most innovative place in the world but no longer is.

A shift in orientation occurs in Chapter 7 where we move to contemporary problems with global warming in MENA. Global warming may or may not become an RCC but it is shaping up to have fairly similar effects to what occurred in past RCCs. Already underway, the buildup of carbon dioxide threatens to raise temperatures throughout the region to a level that will no longer sustain human life. In the past, climate refugees could move to other parts of the region that were not as strongly affected and, in effect, start over. That possibility was always critical to the functioning of the adaptive cycle. In the coming century, there will be climate refugees for sure but there may also be significant population casualties if the availability of water and food becomes more scarce. Given very large populations highly dependent on imports of food and likely to encounter even greater water scarcity than already prevails, the foundation is established for a major regional crisis in which the very survival of many of the region's inhabitants cannot be guaranteed.

That same heightened vulnerability to climate change is being manifested by global warming that is already underway and seems destined to become much worse. In Chapters 7 through 10, we establish the case for what global warming is doing to the MENA and what it might do or is likely to do later in the 21st century. In the most pessimistic scenarios with temperatures increasing as much as 6 degrees centigrade or more, conditions could become so deteriorated that human life will no longer be supportable in substantial segments of the MENA.[17] We try to pin down where the changes are likely to be most acute and in what ways. Three prominent cases of climate-induced conflict are specifically examined in Chapter 8. While we do not find support for climate change impacting Darfur and the two internationalized civil wars in Syria and Yemen, we still think they serve as harbingers of things to come – namely intense conflict amid deteriorating climates.

In Chapter 9, we return to the question of food and water deficits and assess where the MENA stands in comparison to other regions. There are other regions that appear likely to be hit harder even than the MENA. While that may seem to present a less dire picture for the MENA's future, it really means only that the number of people likely to be severely threatened by global warming is greater than the MENA's population. The sheer magnitude of the numbers involved will mean an even greater uphill battle for any policy responses that are forthcoming. In the second part of the chapter, we ask whether anything is being done to prepare for what is likely to be a widespread crisis, if not a catastrophe within the MENA. We find little evidence that attempts to adapt to climate deterioration are either underway or apt to prove sufficient given the kinds of change that may be coming. If not much more is done in preparation, the crisis/catastrophe, of course, will only be all the greater. Nor will it be confined or confinable to any one region. The migration issues associated with the Syrian civil war may well appear to be a tempest in a teapot compared to what might be coming in the next few decades.

Throughout our analysis, a primary concern is with the people who will attempt to find ways to cope with an increasingly malign environment. Even though we can demonstrate that, historically, the MENA has risen and declined on multiple occasions throughout the last 15,000 years, we are not optimistic that the next downturn will lead to an eventual renewal of organized life in the region. As depressing as that conclusion is, summarized in Chapter 10, it is one that it is difficult to escape in view of what could be coming down the pike in terms of climate change and how it has influenced the region in the past. Unfortunately, the fact that there are regions that are likely to be worse off may not make the likelihood of the outside world doing something about it more likely.

Thus, our book has multiple goals. We seek to give climate change a bigger explanatory role in impacting the political-economies of the Middle East and North Africa of the past 15,000 years. We want very much to draw attention to continuities in older forms of change and anticipated impacts in the near future associated with global warming.[18] Finally, we want to reinforce one answer to the question of how a leading region becomes something less than a leading region in the long term. If environmental variance encouraged the ancient Near East to take an early lead, its environmental fragility set it up to lose that lead and much more. Just how much will be lost because of its environmental fragility remains to be seen in this unfolding century.

Notes

1 Both the mammal number and the advantages of East–West interactions (versus North–South connections) have been advanced by Diamond (1997). Both point to definite advantages Eurasia possessed vis-à-vis Africa or the Americas. Yet underlying these advantages is perhaps the more basic Eurasian advantage of being the largest and most populated area. It was also the best connected land mass so that when experimentation in one part stagnated, other parts could forge ahead armed with information obtained in earlier innovative behavior.

2 See, for instance, Pomeranz (2000). There is no getting around the need for coal deposits in an era dominated by steam engines. Obviously, though, coal is found in a number of places. Easy and cheap access to coal deposits was hardly necessary and sufficient to modern economic growth.

3 In this respect, the Sumerians appear to have been imitating their Ubaid predecessors but the Sumerians took the ideas much farther.

4 Fortunately, that prudence does not inhibit McNeill from going on to make some interesting comparative environmental history generalizations covering millennia about the Middle East.

5 The first water eccentricity is that fish are less abundant and nutritious in warm water. Most of MENA's waters are warm, and the main exception is the Black Sea, which created a reserve pool of sailors that the Ottoman Empire was able to tap after its defeat in 1571.

6 McNeill (2013: 31) also adds that the nature of the aridity and lack of water meant that human settlement in the MENA was unlikely to resemble China or Europe in which dense and proximate settlements of people occurred. Instead, it was more like the South Pacific in which islands of settlement are surrounded by water or, as in the case of the Middle East, sand.

7 When mentioning that we had just completed a 15,000-year study of MENA climate change, a friend immediately queried whether one of us considered ourselves to be experts on the question. The flip retort was that now we were. However, the more accurate response is that a couple of non-experts were more likely to take on such a large chunk of history because if there are any 15,000-year MENA climate experts

out there, they would have known better. At the same time, we proceeded on the presumption that we need more ambitiously longitudinal studies to better understand the impact of climate. The audacity of the long duree perspective, moreover, is also lessened by the fact that a number of other scholars have investigated topics of interest with shorter temporal perspectives. Our task was to weave these shorter interpretations into a longer tapestry. If we had to weave the tapestry from scratch, it would never have been completed or perhaps even contemplated.

8 For more on this interesting question, see, among others, Shea (2008), Dennell and Petraglia (2012), Marks and Rosa (2012), Groucutt and Blinkhorn (2013), Petraglia et al (2015), Scerri et al (2014), Breeze et al (2016), Goder-Goldberger, Gubenko, and Hovers (2016), Langgut et al (2018), and Bar-Matthews, Keinan, and Ayalon (2019).

9 Two of the causal mechanisms that are often at work include population transfer (from cities in areas hit hard by climate change to cities in areas hit less hard) and different city endowments (some have more water, rain, or protection from attacks)

10 However, East Asia would seem to be a good candidate for regions that have endured serial bouts of climate change.

11 Applications of the framework to MENA historical topics can be found at Rosen and Rivera-Cullazo (2012) and Haldon and Rosen (2018).

12 However, the concept of creative destruction appears to be all that is borrowed from Schumpeter's argument. There is, of course, a long history of conceptual borrowing between biology and economics going back to the 19th century.

13 Nor is it necessarily uniform across space (Izdebski, Mordechai, and White, 2018).

14 One reviewer objected to the ahistorical nature of the framework. But, as social scientists, we view that as one of its virtues. We do not wish to ask different questions or make different assumptions for prehistorical, ancient, medieval, or modern times unless there is a compelling reason to do so. We think we are more likely to find the compelling reasons if we first look for them as opposed to constructing assumptions and answers to our questions ahead of time.

15 See Ollivier et al (2018) for an interesting critique of the applicability of the adaptive cycle argument to ecological transitions.

16 One of the subsequent developments in this research program is the recognition of a small number of traps that interfere with the smooth flow of the growth, crash, and renewal process. A rigidity trap was implicitly built into the k phase (conservation/monopoly) conceptualization but that phase might just as well be labeled the rigidity phase. There is also a poverty trap that can thwart renewal in the alpha phase. So we are not alone in wishing to see the introduction of more variables. See, for instance, Carpenter and Brock (2008).

17 These worst-case scenarios assume that, collectively, we have failed to make sufficient efforts to move toward a de-carbonized economy quickly enough to keep future temperature increases below the magic 2 degrees centigrade. Some observers believe that we still have a few years to turn things around if we accelerate decarbonization efforts before 2030 (UNEP, 2018). Yet, as the Global Commission on the Economy and Climate (2018) reminds us, what happens in the next 10–15 years hinges on investments made in the next 2–3 years. Sad to say, as we have expressed near the beginning of this chapter, there is little evidence of anything happening in the immediate future that would give cause for optimism. Worse-case scenarios become more probable as a consequence.

18 In the process, we think we will also lend credence to the securitizing of food and water deficiencies in the MENA. Conventional security concerns deal with war and threats of internal and external conflict. As argued in Swain and Jagerskog (2016), conventional concerns need to be broadened to take in a host of 21st century threats to human welfare in the Middle East and North Africa, as well as elsewhere.

2

CLIMATE CHANGE FROM THE YOUNGER DRYAS TO THE LITTLE ICE AGE

Until fairly recently, understanding of the climate history of the Middle East and North Africa (MENA) has been either one of unchanging constancy or a patchwork of information about climates becoming wetter and dryer or colder and warmer at various points in time. The conceptualization of RCCs – rapid climate changes – offers another path.[1] Alley (2000: 120) offers a useful introduction to the subject in this way:

> for most of the last 100,000 years, a crazily jumping climate has been the rule, not the exception. Slow cooling has been followed by abrupt cooling, centuries of cold and then abrupt warming, with the abrupt warmings generally about 1,500 years apart, although with much variability. At the abrupt jumps, the climate often flickered between warm and cold for a few years at a time before settling down. One can almost imagine a three-year-old who has just discovered a light switch, flicking it back and forth, losing interest for a while, and then returning to play with it again.

In the list from Younger Dryas to the Little Ice Age in Table 2.1, there have been many repetitive types of change. Is it possible that these periods of climate deterioration are characterized by a temporal pattern in which something similar reoccurs every so often? Moreover, it is conceivable that the pattern is global as opposed to regionally variable. Subject to continuing modification that involves further delimiting of the time periods involved, evidence for the advent of a global sequence of cooling periods that encompasses the known periods of climate problems has been generated. There are regional variations from time to time but the process appears to be generalizable. In the period with which we are most concerned, there have been nine intervals of global cooling sketched

TABLE 2.1 The RCC Sequence

No.	Name	BP	BCE
1	Younger Dryas	12700–11700	10700–9700
2	10.2 kya	10500–10000	8500–8000
3	8.2 kya	9000–8000	7000–6000
4	6.2/5.9–5.1/5.2 kya	6200–5000	4200–3000
5	4.2 kya	4200–3800	2200–1800
6	End of Bronze Age	3250–2800	1250–800
7	Late Antiquity Ice Age	1550–1300	450–700 CE
8	Medieval Climate Anomaly	1050–930	950–1070 CE
9	Little Ice Age	600–150	1400–1850 CE

TABLE 2.2 Brooke's Climate Favorability Schedule

Less Favorable	More Favorable
12000–9600 BCE	9600–7000
7000–6000	6000–4000
4000–3000	3000–1300
1300–700	700 BCE–300 CE
300–900 CE	900–1400
1400–1700	1700–2000
2000–present	

Source: Based on Thompson (2018).

in Table 2.1. The designated intervals are rough brackets of time and one can anticipate further refinements to their dating.[2]

The list put forward in Table 2.1 corresponds well to lists developed by anthropologists and archaeologists.[3] The difference is that the RCC list makes the timeline sketch somewhat less arbitrary. It is not just a crude appearance of repetitive processes dimly perceived by attempts to examine many thousands of years of behavior. If we can explain why RCCs work similarly over time, it is possible to focus more on the variations in behavioral response over time – as opposed to simply working out what seemed to take place in each episode separately. One still has to do the separate episode work but less attention needs to be given to deciphering the climate changes. Moreover, the search for similarities and differences over time is placed within a stronger framework of expectations. That is, we have better reasons to expect similarities and differences knowing that the climate changes are irregular manifestations of general and recurring processes originating elsewhere.

Table 2.1 also suggests a calendar of relatively malign and benign historical climate periods that represents an improvement on Brooke's (2014) innovative calendar, which is summarized in Table 2.2. The two calendars are not radically different but the one suggested by substituting RCCs for less favorable periods is more specific (fewer years are delineated) and more transparent about how good years (generally warm and wet) are differentiated from the bad ones (generally cool and dry). In time, a general explanation for RCCs, as opposed to various reasons on why climate might have changed here or there, may prevail as well.

Another major effort at establishing a climate change timeline is advanced by Issar and Zohar (2007). In this case, some of the same periods are identified as transitional in terms of climate change but there is some inherent disagreement registered with the RCC approach on whether cooling periods were also drying episodes (see Table 2.3). For instance, Issar and Zohar identify 3000–2300 BCE and 1500–1200 BCE as cool and humid periods followed by extreme drying (and warming in the 1200–850 BCE case). RCC events are generally cooler and drier. Since it is the dryness that is most critical to drought/famine outcomes, it may not make too much difference for application periods, cold or warm. At the same time, Issar and Zohar note that they disagree with previous claims about whether specific periods of time were cold or warm. One of the attractions of the RCC approach is that one does not have to determine the climate specifics of each temporal interval *sui generis*. RCCs are near cyclical in their evolution and regular in their general manifestation, if not in terms of duration or intensity.

All of this hinges on the ability to explain the RCC process in a generic fashion. Analyses are predicated on the examination of the evidence found in the indicators identified in Table 2.4. At this time, the general process can be traced to several alternative sources, including solar variation, volcanic activity, increased ice in the North Atlantic, and interruptions of the Atlantic circulation of cold and warm water.[4] These processes produce a colder than average North Atlantic which in turn alters the northern North American and Eurasian/Siberian climates. In Eurasia, very cold air moves through corridors (Weninger and Harper, 2015) to the Eastern Mediterranean (Pontic steppes down the Danube to the Aegean) and China (east of the Himalayas crossing China to the Pacific). The timing of these atmospheric changes tends to focus on the winter and early spring thereby altering the ability to grow crops. There is variation within these patterns of abrupt cooling. Every year the cooling period need not be the same but there tends to be at least one year of catastrophic cooling per generation (every 25 years or so) that makes it extremely difficult to continue as before.[5] In the northern hemisphere, it becomes extremely cold; further south, it becomes much more arid than before.

Nonetheless, it is critical to match global indicators, as in the case of ice cores, with local evidence of change to ascertain the degree to which possible global cases were truly global in effect. Alternatively, a "near" global event might have missed or passed over some patch of territory for a variety of reasons pertaining to countervailing influences. Put another way, we are not in a position to assume global climate change events. In our case, we need to be sure that global cooling and drying events influenced the MENA climate.[6] Table 2.5 summarizes specific attempts to match RCC events with MENA-centric climate indicators.

We had a choice to make in utilizing this material. Either we spent some time discussing the details for each case as we moved from one RCC event to another

TABLE 2.3 Issar and Zohar's Near Eastern Climate Change Timeline

Timing	Phase Description
Within the last 60,000 years	Climate fluctuations in the "Near Eastern bridge" made the passage from East Africa more feasible in a cyclical fashion
Some 12,000 years ago	Younger Dryas warm spell accompanied by sedentary population and deteriorating food supply, thereby encouraging the development of agriculture and irrigation
Some 8,000–7,000 years ago	Cold spell north of Black Sea and the Caucasus encourages migration southward but Mesopotamian population initially limited in size and therefore less affected
4000–3100 BCE	Proliferation of urban settlements and nomads
3500–3000 BCE	Warm and dry; semitic pastoral groups encouraged to move into northern Mesopotamia from the desert
3000–2300 BCE	Cool and humid
2400–2000 BCE	Increasing dryness encourages many groups to move toward areas with water; widespread abandonment/destruction of urban areas from the Balkans to Mesopotamia and Palestine; Akkad, Old Kingdom Egypt, Troy, and Indus collapsed
2000–1500 BCE	Climate deterioration (warm and dry) through 1800 and then improvement and gradual revival; however, increased precipitation less than optimal thereby continuing to apply pressure on nomads to migrate (e.g., Hyksos into Egypt and collapse of Old Kingdom Egypt)
1500–1200 BCE	Cold and humid
1200–850 BCE	Warm and dry, considerable pressure on Mesopotamian farming and herding; waves of people of Eurasian steppes move toward Anatolia; Sea Peoples reflect farmers marginalized by climate and turning to nomadism/ migration with movements into the Balkans, Greece, and Anatolia; Libyans moving east into Egypt
850–350 BCE	Climate improving toward 350 BCE; colder and more humid after 300 BCE but the degree of change remains moderate; Libyans, Nubians, Assyrians, Parni, Greeks, and Romans enter Egypt
350 BCE–150 CE	Cooling trend underway in Mediterranean area, with worsening climate in Eurasian steppes; nomads successfully resisted by Han Chinese military putting more pressure on the west; Parni move into Iran from the Caspian region; Nabateans attacking Petra 312 BCE; Alexander the Great benefited from improving climate as did the foundation of the Roman Empire; Germanic groups begin to be pushed into Roman empire by nomads from steppes; Roman climate highly favorable with above-average rainfall (50–250 CE) and then rainfall levels decline

(Continued)

TABLE 2.3 (*Continued*) Issar and Zohar's Near Eastern Climate Change Timeline

Timing	Phase Description
150–550 CE	In the 3rd century CE, rapid turnover of Roman rulers, along with mismanagement, heavy taxation, inflation, famines, climate improvement to 550 CE (with reinforcement of Byzantine Empire and then deteriorating until c. 900 CE); Goths moving into Roman territory 251 CE; Franks attack Gaul; some of the tribes responding to Hun pressures in the 5th century
550–900 CE	Desiccation from 600–1000 CE; Muslims move out of deserts; later, Vikings, Turks, Normans, and others move into Near East
900–1500/1600	Favorable climate at the turn of millennium; cooling peaks c. 1500 CE; Near East climate favorable despite Crusader invasion
1700–1800	Warm and dry; Ottoman Empire setbacks

Source: Based on the discussion in Issar and Zohar (2007: 39–225).

(Chapters 3 and 6) or we summarize the evidence and invite interested readers to examine the climate change evidence on their own. We opted for the latter approach because, frankly, discussing the various types of evidence, not all of which points in the same direction admittedly, is a major demand on readers' attention and interest.[7] Suffice it to say that there is evidence that RCC events have influenced the MENA climate. Is all of the evidence "ironclad"? The answer is no and the applications of the RCC interpretation will be subject to continued debate and testing. One thing is clear. Climate change rarely influenced all of the MENA evenly.[8] Yet, there appears to be enough evidence to proceed as if the RCC framework is appropriate for application to the MENA in general.

One of the advantages associated with working with recurring cooling periods is that a team of geographers appears to have cracked the code on how cooling periods work. It is more complicated than merely arguing that colder weather is not beneficial for growing crops. There are a number of interactions to consider. What Zhang et al. (2011a) have developed is a causal template of which variables interact with other variables in cooling crises. They first developed quantitative data for an array of 16 indicators that captured agricultural, economic, social, demographic, and war processes in early modern and pre-industrial Europe (1500–1800). This time period approximated the Little Ice Age. But they emphasize that the European Little Ice Age went through three phases: (1) mildly cool (1500–1559), (2) cold (1560–1660), and (3) mildly cool (1661–1800). The phasing of the period permits one to compare and contrast the mildly cool periods with the cold period. Using a variety of statistical techniques on their multiple time series, the modelers are able to construct the causality flow chart sketched in Figure 2.1. The question is which variable(s) influenced the subsequent values of which other variables?

TABLE 2.4 Paleoclimatic Proxies

Focus	Variables	Indicators
Ice	Atmospheric composition	Trapped bubbles
	Windiness	Dust grain size
	Source strength of wind-blown materials	Abundance of pollen
		Dust
		Sea salt
	Temperature	Ice isotopic ratios
		Borehole temperatures
		Gas isotopes
		Melt layers
	Snow accumulation rate	Thickness of annual layers
		In-situ radiocarbon
Ocean sediments and corals	Temperature	Species assemblages
		Shell geochemistry
		Alkenone thermometry
	Salinity	Shell isotopes after correction for temperature and ice volume
	Ice volume	Isotopic composition of pore waters
		Shell isotopes after correction for temperature and salinity
	pH	Boron isotopes in shells
	Ocean circulation	Cd/Ca in shells
		Carbon-isotopic data
	Corrosiveness/chemistry of ambient waters	Shell dissolution
Lake and bog sediments	Temperature	Species assemblages
		Shell geochemistry
	Atmospheric temperature and soil moisture	Washed- or blown-in materials including pollen and spores
		Macrofossils such as leaves, needles, beetles, and midge flies
	Water balance (precipitation minus evaporation	Species assemblages
		Shell geochemistry
Tree rings	Variations in the isotopic ratio of water related to temperature	Cellulose isotopic ratios
Speleothems/ cave formations	Moisture availability	Growth rate of formations
	Isotopic ratios of water related to temperature or precipitation rate	Oxygen isotopic composition
	Overlying vegetation	Carbon-isotopic composition

Source: Committee on Abrupt Climate Change, National Research Council (2002: 22).

TABLE 2.5 RCCs and MENA Climate Changes

No.	Name	BCE	Local Climate Evidence
1	Younger Dryas	10700–9700 BCE	Bottema (1995), Balter (2010), Makarewicz (2012), Rosen and Rivera–Collazo (2012), Carlson (2013); Contreras and Makarwicz (2016); Jones et al (2019)
2	10.2 kya	8500–8000	Weninger (2009); Weninger et al (2009); Berger et al (2016); Arranz–Otaegui et al (2017).
3	8.2 kya	7000–6000	Goodfriend (1991, 1999); Weninger et al (2009); Frumkin et al (1994); Lemcke and Sturm (1997); Bar–Matthews et al (1999); Weninger (2009); Weninger et al (2009); Flohr et al (2016); Berger et al (2016); Jones et al (2019); Palmisano et al (2019); Petraglia et al (2020).
4	6.2–5.2/5.1 kya	4200–3000	Lemcke and Sturm (1997); Bar–Matthews (1999); Cullen et al (2000); deMenocal et al (2000); Nicoll (2004); Kropelin (2005); Kuper and Kropelin (2006); Parker and Goudie (2008); Weninger (2009); Weninger et al (2009), Roberts et al (2011); Rambeau and Black (2011); Manning and Timpson (2014); Clare (2016); Lespez et al (2016); Petraglia et al (2020).
5	4.2 kya	2200–1800	Adams (1981); Weiss et al (1993); Frumkin et al (1994); Gasse and van Campo (1994); Johnson and Odada (1996); Ricketts and Johnson (1996); Bar–Matthews and Ayalon (1997); Bentalab et al (1997); Bottema (1997); Lemcke and Sturm (1997); Bar–Matthews et al (1998, 1999); Eastwood (1999); Von Rad et al (1999); Cullen et al (2000); Gasse (2000); Wick, Lemcke, and Sturm (2003); Phadtare (2000); Parker and Goudie (2008); Weninger et al (2009); Rambeau and Black (2011); Roberts et al (2011); Sharafi et al (2015); Sarker et al (2016); Weiner (2016); Bini et al (2019); Petraglia et al (2020)
6	End of Bronze Age (3.1/3.2 kya)	1250–800 BCE	Weiss (1982); Kaniewski et al (2008); Rohling et al (2009); Weninger (2009); Weninger et al (2009); Roberts et al (2011); Langgut, Finkelstein and Litt (2013); Langgut et al (2018); Sharafi et al (2015); Kaniewski et al (2019).
7	Late Antiquity Ice Age	450–700 CE	Rambeau and Black (2011); Dominguez–Castro et al (2012); Buntgen et al (2016); Maron et al (2019).
8	Medieval Climate Anomaly	950–1070 CE	Hassan (2011); Kaniewski, et al (2011); Roberts et al (2012); Toker et al (2012); Xoplaki et al (2016); Luming et al (2017).
	Black Death	1347–?	Dols (1979a, 1979b); Schmid et al (2015); Campbell (2016); Dobler (2018).
9	Little Ice Age	1400–1850 CE	Soon and Baliunas (2003), White (2011, 2013); Mikhail (2016); Xoplaki et al (2018).

Note: The episode (Black Death) between the 8th and 9th events does not represent a different RCC but rather a pandemic that emerged in the transition from a warm period to a cold period. We should also note that Fleitman et al (2008), Flohr et al (2016), Berger et al (2016), and Jones et al (2019) mention a 9.2 kya RCC but it is not clear how this event corresponds to the ice core evidence. Roberts et al (2018, see also Petraglia et al, 2020) find a 5300–5000 BCE aridification episode but that dating is neither corroborated nor challenged in other reports.

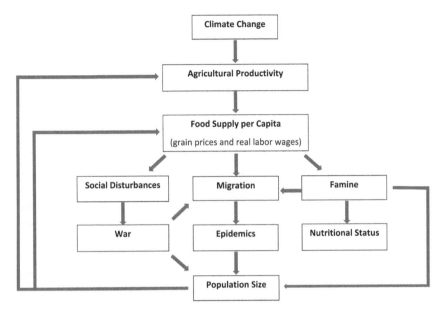

FIGURE 2.1   Cooling period dynamics

The nature of the model is a bit more complicated than it may look. The easiest way to convey the connections among the variables is via a 13-step recipe list:

1. Cooling shortens the plant growing season and shrinks the amount of cultivable land that is available for growing food.
2. The ratio of grain yield to seed decreases.
3. Even though food supply is decreased, population size continues to grow.
4. Grain prices climb and the real wages of labor decline.
5. High grain prices and low wages increase the frequency of famine.
6. Human nutritional status declines and, ultimately, human body height declines as well.
7. High grain prices and low wages intensify social problems.
8. Social conflicts (rebellions, revolutions, and political reforms) peak following decreases in temperature with variable lags of 1–15 years.
9. The number of wars increases substantially as does their lethality and duration.
10. The interaction of economic chaos, famine, social disturbances, and war encourages migration.
11. Migration and health deterioration encourage epidemics.
12. War fatalities and famine reduce the population growth rate and size.
13. After cooling dissipates, agrarian productivity returns and resets the interactions among the variables displayed in Figure 2.1.

Zhang et al. (2011a) contend that grain prices are the key to what happens because only grain prices manage to influence social disturbances, war, migration, epidemics, famine, and nutritional status. Their findings suggest that a lot of problems could be avoided by the massive import of food from somewhere with a more favorable climate. In the 16th and 17th centuries, it is not obvious where that external source of food supply might have been. Nor is it clear who would do the importing and pay the bill for the imports for a century of severe cold climate.

A second advantage associated with the model is that it suggests an alternative to the Malthusian model, which basically contends that a rapidly expanding population will eventually outstrip the region's carrying capacity. Food demand will exceed food supply and people will starve until there is a closer correspondence between supply and demand. The analysis by Zhang et al. (2011b) offers another interpretation. Food supply was less than what was demanded but it was due to climate-induced changes in how much food could be grown. Food supply shrank in this perspective as opposed to demand blooming.

We acknowledge that early modern Europe is not necessarily a universal prototype for cooling episodes anywhere or everywhere. But it seems to be our best approximation until more data becomes available on historical periods. The findings by Zhang et al. (2011b) are certainly plausible. As we traverse some 15,000 years of climate change–induced behavioral changes in the MENA region, the Zhang et al model seems to fit reasonably well across the millennia.[9]

Another work done by the Zhang team is also of great utility. The study by Zhang et al. (2011b) demonstrates that 80 percent of some 88 periods of negative population growth took place in cooling periods. Famines, wars, and epidemics served as the dominant triggers of population collapse. They argue that climate change diminished carrying capacity which led to Malthusian checks on population size.[10] Other studies by the same analysts have found that war frequency, dynastic changes, epidemics, macro-economic cycles, nomad migration, and nomadic-sedentary conflict are all linked to periods of climate cooling.[11] While these findings are tied to different times and places depending on data availability, they are all restricted to Eurasian pre-industrial periods and correspond to RCC's second millennium CE cooling episodes. Thus, we have some reason to anticipate that RCCs lead to drought, reduced food supply, negative population growth, social unrest, war, political change, disease, and migrations of agriculturalists and nomads – as well as their conflict over access to water. Exposed to cold, decreased precipitation, and consequently, poor nutrition, plants, animals, and people die in larger numbers than is the norm. Economies predicated on agrarian production falter and fragment. Population numbers are depleted. Some cities are abandoned completely. Survivors flee to more benign environments. Scarcities become more common as does human conflict over the distribution of territory and food. Political regimes collapse overwhelmed by the inability to collect tax revenues and manage the disorder that is encouraged by a deteriorating climate.

In the Rapid Climate Change approach, the major climate episodes are denoted by proper names – as in Younger Dryas, the 8.2 kya event, the 4.2 kya event, or the Little Ice Age (Table 2.1). The question then becomes one of correlation. Did behavioral changes follow changes in climate? If so, can the behavioral changes be attributed to the climate changes at that particular time? These questions will be addressed in the following chapters starting with the Near Eastern events and processes in Chapter 3. Pursued iteratively, we can map out the last 15,000 years of interactions between people and climate, along with the intervention of other processes.[12]

If this RCC conceptualization and set of behavioral expectations hold up to continued scrutiny, the recurring nature of these processes offers a powerful explanatory tool for what appears to be generally repetitive responses to deteriorating climate change subject to the variable intermediation of other factors.[13] It also facilitates greatly our analysis that stretches across 15 millennia of hot and cold alterations, culminating in the latest round of global warming. Global cooling has been the main atmospheric culprit in wreaking selective havoc in the past. The ongoing, human-generated, global warming may prove to have similarly disastrous effects at least in the MENA region and other parts of the Global South. And yet, humans did somehow survive the nine previous iterations of cooling by way of the adaptive cycle scheme (Figure 1.1). The open question, however, is whether the ongoing warming iteration will prove to be survivable in regions that will be hardest hit by the deteriorating climate.

Notes

1 RCC events are sometimes referred to as abrupt climate change events or ACCs. The ACC acronym may be gaining currency but note that the National Centers for Environmental Information define abrupt climate change as a change that occurs rapidly. RCCs are also sometimes called Riccies for short (Mayewski and White, 2002). It is hard to imagine Accies catching on. However, another name for these events are Dansgaard-Oeschger events which seems even less likely to catch on and are often applied to pre-Holocene occurrences, of which there have been some 24 prior to the Younger Dryas event in the past 100,000 years.

2 For instance, some archaeologists have already narrowed numbers 3, 4, and 6 to 8600–8000 BCE, 6000–5200 BCE, and 3700–2900 BCE, respectively, based largely on perceived behavioral changes. We approve of this narrowing but choose not to use the revised figures until they become subject to some consensus. It should also be noted that the Late Antiquity and Medieval events were not initially picked up in the ice studies around the turn of the 20th/21st centuries. They have emerged more recently and it is not clear at this time if they are one event with a hazy pause in the middle or two different events. Part of the problem is that ice needs to be interpreted and it is difficult to specify exactly when the evidence suggests a change episode began or ended. Applications to some parts of the world are also more difficult than to others. That is, we are unlikely to miss changes in Greenland when examining Greenland ice; other parts of the world may sometimes be more of a stretch.

3 This point will be demonstrated in Chapter 3.

4 It is possible that some events will be attributed primarily to seemingly idiosyncratic occurrences reflecting more general phenomena such as Lake Agassiz glacial water

entering the Atlantic as is often linked to the Younger Dryas episode and which mean that the sources of RCCs may be multiple. However, Wang et al (2013) suggest that several of the earlier Holocene events were caused by cold water entering less cold bodies of water while the most recent ones have been related to decreases in solar radiation. There is also a Gobekli Tepe variation. One of the megaliths has been interpreted as a column devoted to a number of prehistorical zodiac signs and might be read as suggesting a comet (or parts of it) striking the earth in the early 11th millennium just at the outset of the Younger Dryas period (Sweatman and Tsikritsis, 2017). Needless to say, this interpretation must be viewed as a minority hypothesis at this time but probably cannot be ruled out either.

5 Other types of variation include glaciers that advance in some but not all northern regions. On occasion, rains and floods increase in some tropical areas without necessarily offsetting the aridity tendencies. The duration of cold regimes varies as does the scale and tempo of the abrupt change with temperature decreases varying between 2 to 8 degrees centigrade.

6 One known liability of reading ice evidence is that it is likely to miss shorter intervals (Finne et al, 2019). Moreover, the ice readings are not always in agreement with other sources of evidence on old shifts in climate (Robinson et al, 2006).

7 We cannot over-emphasize this point – that the evidence is not overwhelmingly in favor of a close connection between climate change and political-economic and cultural changes. In the earliest cases, there is continuing debate about whether climate change precedes the claimed effects. See, for instance, Maher, Banning, and Chazan (2011). In both earlier and more recent cases, there are multiple drivers at work and it is difficult to assess the relative strength of one driver versus another. Yet we are also impressed by the number of cases over time that seem to work similarly. The repeated onset of cold and dry climates appears to have negative consequences of a similar ilk. The consequences are not manifested uniformly throughout MENA nor are they exactly the same over time. But they are also hard to miss. This only means that climate change deserves a piece of the explanatory action.

8 One reason for the unevenness is that the Eastern Mediterranean portion of MENA is influenced by Eurasian climate changes while North Africa tends to be subjected to Mediterranean climate influences. African climate influences are manifested most strongly in the Egyptian Nile area. That does not mean that the cool and dry alternations were only manifested in the eastern segment of the MENA. North Africa (Barich, 2014; Jaouade et al, 2016; Cheddadi et al, 2019; Roberts et al, 2019) also experienced periods of dryer/more moist alternations but not necessarily on the same schedule as in the east.

9 One obvious caveat is that markets for grain did not exist in prehistoric times. Still, the focus on food supply does apply even if grain "prices" only came much later.

10 The 88 collapses were restricted to the Northern Hemisphere between 800 CE and 1900 CE.

11 See Pei et al (2014, 2015, 2018); Pei and Zhang (2014); and Zhang et al (2006, 2007).

12 Our primary generalization would be that cool and dry periods are not especially conducive to innovation and breakthroughs to superior strategies. As with any generalization, there are exceptions. Early modern Northwest Europe appears to have been a beneficiary of the Little Ice Age although it might not have felt like it at the time. Another possible exception is linked to Stanley's (1996) thesis that cooling forced our species' predecessors to leave their African trees.

13 Note as well that we do not claim all political-economic/cultural setbacks are due to RCC episodes. An illustration is provided by the case of the Neo-Assyrian Empire which flourished in the first half of the last millennium BCE. A recent analysis (Sinha et al, 2019) of this empire which had its base in upper Mesopotamia shows that it was dependent on better than normal rainfall for agricultural prosperity. The life cycle of the empire appears to be highly correlated with more rain

in the expansionary phase of the empire and less rain in the second half of the empire's trajectory which was characterized by both decline and excessively brutal behavior as exemplified by forcing large numbers of conquered subjects to move far from their traditional homelands as control measures. Another instance appears to be domestic rebellion and constrained Ptolemic–Seleucid warfare in ancient Egypt due to lower levels of annual Nile River flooding (Manning et al, 2017). Climate change seems to have been at work in both cases but not of the RCC variety.

3

THE ORIGINS OF AGRICULTURE, DROUGHT, AND ANCIENT EMPIRES

In this chapter, our objective is to examine what sort of impacts a series of cold and dry climate changes had on human behavior in the ancient Near East. For our purposes, the ancient era encompasses several millennia prior to historical records and several millennia that have left stronger and more concrete traces of what transpired after the advent of the cold and dry regimes. As a consequence, there is a lot that we do not know about ancient times. Wherever possible we strive to demonstrate explicitly how our version of ancient dynamics is shaped by earlier studies.

We start with an image of ancient Near Eastern dynamics as one of intermittently rising and falling cities and states. While Jericho may have managed to persist for some nine millennia, the size of ancient villages and cities waxed, waned, and sometimes simply vanished. Powerful states and empires appeared and then eventually declined and were replaced by successor centers of power often somewhere else in the Near East. We have a schedule of periods of abrupt climate change that took place during the ancient Near East from the Younger Dryas some 12,000 years ago to the end of the Bronze Age around the end of the 2nd millennium BCE. We ask whether the sequence of rapid climate changes (RCCs) appears to have triggered episodes of creative destruction (Figure 1.1) that played a role in shifts from periods of societal organization to disorganization and back to reorganization. We do not assume that climate determined the disorganization. Our model only suggests that climate change could trigger or reinforce tendencies for things to fall apart. Of course, climate change is variable and long periods of drought would have been quite difficult to withstand so that the effects of climate change could certainly have been more than merely reinforcing. Just how far we can push the causal role of climate remains to be seen as we consider the historical evidence in this chapter and the three chapters that follow it.

Younger Dryas and the origins of agriculture

The origins of agriculture have a fascinating and checkered history.[1] The basic question involves discerning how the human species initiated the cultivation of domesticated plants and animals. Wild plants and animals coexisted with homo sapiens as long as the latter species has been around (about 200,000 years). But the deliberate cultivation of domesticated sources of food is another matter. The practice emerged independently in several parts of the world (the Middle East, China, western South America, and West Africa) at different times. To date, the Middle Eastern innovations have been granted the title of first even though the dating of Chinese domestication keeps getting pushed back in time.

The "to date" phrase is quite critical to understanding the development of explanations for the Middle Eastern onset of agriculture. The story changes over time as new evidence emerges – a normal characteristic of prehistory analysis. The basic progression starts with the ending of the last glacial age roughly between 20 and 15,000 years ago.[2] Most glaciers in western Eurasia were located north of the Mediterranean but the lands immediately south of the eastern Mediterranean were also affected quite strongly by the change in climate. Temperatures warmed and precipitation increased which meant more vegetation and animals could be supported. Small Natufian villages began to emerge in the eastern Mediterranean area almost 15,000 years ago. Despite the improving environment, these villages were always sited near the availability of abundant wild plants that could be consumed and dependable supplies of water.

Rather abruptly, climate change called the Younger Dryas era commenced. After a few thousand years of steady improvement and population expansion, the climate became colder and dryer. Vegetation retreated. Water supplies became less dependable. Sedentary villages were either abandoned or became more scarce as hunting and gathering activities were forced to become more pro-active and cover more territory. Evidence exists that the expansion of the human population had put pressure on the food supply even before the climate changed and drastically reduced the availability of food.[3] Thus, the Younger Dryas RCC at the very least dramatically reinforced an early crisis in population expansion and food supply shortages. The question is what did the Natufian population do in response?

In the early 20th century, the prevailing argument was the oasis theory (Pumpelly, 1908; Childe, 1936) which argued that food cultivation was the product of Middle Eastern desertification that followed the receding of the glaciers some 15 millennia ago. Populations clustered around sources of water for survival in the changing climate and were strongly encouraged to produce their own food as its general supply dwindled. However, early 20th century analysts had little reliable evidence on the Middle Eastern climate. In the mid-1960s, it was decided quite erroneously that no significant Middle Eastern climate change had followed the retreat of the glaciers. If the climate change focus could not be sustained, the alternative possibility was population increase exceeded the food supply. The eventual response was the innovation of cultivation to create greater human control over the food supply.

This interpretation flows logically from an implicit focus on the population-food supply balance. If the food supply is adequate for a given population, there is presumably little or no incentive to develop new procedures. If food supply falls below demand, as might be anticipated in an unfavorable climate, one of the responses could be innovations to increase supply. But there are other possible responses, including famine, population die-off, migration, and/or falling back on hunter-gathering. The other side of the equation puts the emphasis on population changes. If food supply remained roughly constant and the population expanded, there would still be an imbalance that encourages some form of response. Perhaps not surprisingly, then, agricultural origin explanations in the 1960s moved to an emphasis on population stresses outstripping the food supply.

There definitely is evidence of food scarcity problems during the Younger Dryas era. One example pertains to gazelle consumption. Natufians attempted to restrain their consumption of gazelles by focusing on males, which were needed in smaller numbers, compared to females, to maintain reproduction rates. But by selecting large males for eating purposes, the genetic outcome ultimately was smaller gazelles and less meat supply (Cope, 1991; Mithen, 2003: 47–48). One of the consequences, as previously noted, was shorter late Natufians compared to the heights of earlier Natufians.

By the 1980s, more evidence on climate change had become available. Bar-Yosef and other authors led the charge back to crediting climate deterioration in the Younger Dryas cold and dry interregnum (between the initial improving post-glacial climate and the return to warmer weather in the 8th millennium BCE) with a major role in human crop cultivation. Access to food diminished with the onset of the Younger Dryas period thereby encouraging some people in the Natufian culture to intensify ongoing sedentary cultivation activities around areas that had ample water supply while others pursued more mobile approaches to acquiring food along traditional hunter-gatherer lines. This argument also elevated the southern Levant area as the ground-zero of Neolithic agrarian innovation.

Table 3.1 helps situate these claims in time. Groups in the Levant are categorized by housing architecture, types of tools employed, and other cultural practices. The Epipaleolithic era gave way to the early Natufian culture which in turn was succeeded by pre-pottery Neolithic A and B (PPNA and PPNB) groups who are believed to be migrants into the southern Levant. Initially, the Natufian groups were concentrated along the Levantine coast. Gradually in the PPNA/PPNB/pre-pottery Neolithic C (PPNC) periods, groups emerged along the northern wing of the "Fertile Crescent" pattern (northern Syria and Mesopotamia) and eventually into parts of Anatolia (Figure 3.1).

While this argument that retains some elements of the earlier oasis explanation has not been abandoned, increasing problems have come to light in subsequent decades. Four problems seem most critical and all pertain to the discovery

TABLE 3.1 Time, Climate, and Successive Cultures in the Early Levant

Time Frame	Cultures	Climate Changes
20000 BCE	Epipaleolithic – last glacial maximum in effect	Cold and dry, limited lakes and rainfall
16500	Middle Epipaleolithic	Climate improvement, probably first in southern Levant – lake expansion and rising sea levels
12500	Early Natufian	Warm and wet phase
10800	Climate Crisis – Younger Dryas [RCC no. 1]	Abrupt deterioration – harsh cold and dry
9700	Climate Amelioration	Rapid return to warm and wet conditions – Mediterranean climate with hot summers and warm winters/seasonal rainfall emerges in that region as the norm
8550	PPNA	
7600	PPNB	
6200	Climate Crisis	Dry
5500	Climate Crisis Pottery Neolithic	Dry
4500	Chalcolithic	
3300	Early Bronze	
2100	Climate Crisis	Dry

Source: Based on integrating information from Verhoeven (2004) and Bar-Yosef (2011) with dating modifications.

of better data. The most important one is that it takes some considerable time to convert wild cereal grains to domesticated grain. Earlier evidence suggesting that domesticated grain had appeared during the Younger Dryas era has been refuted. The apparent ratio of domesticated seed to wild seed is too low to rule out the possibility that the apparently domesticated seeds were simply natural mutations and not evidence of domestication. Zeder (2011) argues, moreover, that the transition from wild to domesticated grain could have taken as long as 2–4,000 years. In other words, it was quite a protracted process – revolutionary in ultimate impact but hardly in terms of timing. Since experiments in domestication seem to have been ongoing even before the advent of an unfavorable climate period, there is no reason to give too much credit to climate deterioration for the initiation of new practices. It may very well have been the case that some greater incentive to experiment was experienced by small groups here and there.[4] But there is no evidence for a societal commitment to domestication strategies as a consequence of climate change. More likely, the unfavorable growing conditions at best only slowed the transition to agriculture which is now thought to have emerged in the PPNB era.

Two other types of findings suggest that animal cultivation, once thought to come later in time than plant cultivation, occurred more or less in synch

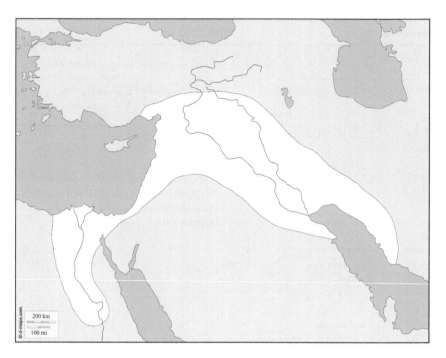

FIGURE 3.1   Map of the ancient Near East

with plant cultivation and was by no means restricted to the southern Levant. Innovations occurred throughout the Fertile Crescent era, with different areas specializing in different types of breakthroughs (a pig here, barley there, and cattle somewhere else sort of pattern) in expanding the food supply. Finally, little evidence of expanded population sizes at the right time (around the onset of the Younger Dryas) could be found either. Thus, in the past few years, the importance of climate change, a singular site of innovation, the possibility of population expansion, and plant versus animal domestication rates have all been reduced substantially in attractiveness.

Instead of an abrupt change in ways of doing things, the development of agriculture appears to have been a protracted process with quite early roots. Some evidence exists to suggest that human cultivation of plants was occurring in a small way in the Middle East as early as 23000 BCE. Essentially, these experiments continued at a very slow pace and only burst forth as clearly recognizable agriculture in the 8th millennium BCE when climate conditions, if nothing else, favored larger-scale agriculture. Table 3.2 offers a guideline to the pace of agrarian developments. The Epipaleolithic Kebaran culture were conventional hunter-gatherers who also harvested wild plants. The succeeding Natufian culture is considered a more complex form of hunting-gathering with more sedentary periods presumably to harvest/cultivate plants. The PPNA culture was even

TABLE 3.2 The Levantine Progression to Agriculture

Culture (BCE)	Architecture	Tools	Cultivation	Categorization
Kebaran (pre-12500 BCE)	Huts/tents made of branches/hides	Grinding tools appear	Wild grass/grain harvesting	Hunter-gatherer
Natufian (12500–9500)	Stone/caves in villages	Grinding tools common/sickles	Wild cereal harvesting – probably limited cultivation	Complex hunter-gatherer with sedentary episodes
PPNA (8500–7600)	Mud brick villages	Grinding tools common/sickles	Plant cultivation/wild fruit and seeds still collected	More sedentary complex hunter-gatherer
PPBN (7600–6000)	Stone/mud brick/plaster; Two-story structures	Grinding tools/sickles common	Cereal and animal domestication/wild seeds still collected	Hunting in decline
Pottery Neolithic (5500–4500)	Temporary occupation of small villages	Tools common Pottery introduced	Cereal and animal domestication	Farming and herding

Source: Based on Verhoeven (2004: 224–231). Time periodizations are approximate. Note that there is also an ill-defined PPNC culture thought to have followed PPNB and preceding the Pottery Neolithic.

more sedentary than the Natufians. So, too, were the PPBN group, which is now believed to have exceeded a ratio of wild to domesticated cereals that qualifies as genuine cultivation or farming. By the time the PPBN culture disappeared, farming and herding appear to be fully established.

Instead of a sharp takeoff during the Younger Dryas crisis, Figure 3.2 captures the more gradual movement toward full-blown agriculture in which the cooling crisis is viewed as a temporary retreat from further progress. A more supportive climate was needed to make more headway to the agrarian age.

The basic pattern that seems to have been repeated is one of each culture developing and expanding. Population growth led to environmental deterioration or less easy access to food as vegetation and wild animals were exterminated. Decreases in precipitation brought on by climate change increased the amount of turmoil associated with maintaining sedentary settlements. A mixture of village abandonment and cultivation experimentation resulted. It is certainly possible that agriculture was part of this experimentation but the pattern of relying on access to wild fruit, seeds, and animals persisted through PPNA and PPNB cultures. By the time the PPNC culture emerged toward the end of the 7th millennium (RCC 8.2kya), nomadic pastoralism, focusing on moving herds of goats, sheep, and cattle to greener pastures in an environment once again drying out could have been an explicit step toward more complex animal cultivation practices (Zarins, 1992; Rollefson and Kohler-Rollefson, 1993; Rollefson, Rowan, and Wasse, 2014).

Thus, the rise and fall of these early Near Eastern cultures seem to adhere closely to the expectations of the resilience theory's framework. Initial growth encounters problems linked to an expanding population and a dwindling supply of nearby food. Environmental deterioration was accelerated by cold and dry spells. Villages were abandoned and presumably surviving populations dispersed in the application of earlier hunting and foraging strategies. New cultures emerged with different tools, housing styles, and burial rituals. In some cases, they built on top of the older villages while in other cases new villages were constructed. The movement toward agriculture stumbled forward in a step forward, half a step backward fashion.

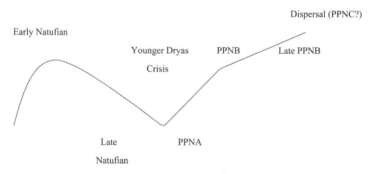

FIGURE 3.2 The movement toward agriculture in the Near East. Source: Based on Verhoeven (2004: S179).

None of this means of course that the Younger Dryas was without impact on the probability of farming emerging as a practice. Murphy (2007: 137) best captures a role for the Younger Dryas in the following way:

> The right climatic conditions for farming are twofold: first you need an adverse period to diminish returns from hunter-gathering, and second a prolonged favorable period to enable the fragile seedling of agriculture to take root. Such conditions were provided in some regions by the Younger Dryas episode at the Pleistocene/Holocene transition followed by an unprecedentedly long period of relative stability that persists to this day…
>
> Younger Dryas supplied the appropriate carrot and stick, where the stick was the steady decline in the availability of the majority of traditional food resources, and the carrot was the presence of high-yielding protodomesticants that could be stored for months or even years. The result was a gradual switch to farming by several societies in Asia and Africa soon after 11,000 BP.

At this time there is only one major caveat on this interpretation of the beginning of agriculture. Gobekli Tepe is an archaeological site located in southeastern Turkey (Anatolia) near the Syrian border. Its claim to fame is that a large number of carved megaliths were discovered there which may mean it was the world's first temple (Peters et al., 2014). No evidence has yet been uncovered that any permanent settlements are to be found there. The area served as a religious or spiritual site and burying ground from some time prior to PPNA and PPNB to about the end of PPNB at 8000 BCE. At that time it was not only abandoned but also filled in a deliberate and atypical burial process.

The reason why this case represents a major caveat is twofold. One, it reverses the usual expectation that these types of sites require an urban/agrarian foundation to support their development. It is not even clear how (or exactly why) hunter-gathers coordinated the great labor needed to construct the megalithic site over a long period of time. But, apparently, they did. The second dimension of the Gobekli Tepe mystery is that the original source of Einkorn wheat is located about 20 kilometers away from the site. The closeness raises the possibility that wheat cultivation (including transforming wheat into beer) came about in order to feed the labor applied to building the site. So far, there is no evidence that this occurred but it remains a speculative possibility. If it did happen this way it would probably not affect the gradual timing interpretation of the origins of agriculture but it would put an entirely different spin on it.

While these developments toward interpreting the earliest phases of the agrarian era have been forming since the late 20th century, a second thrust within paleoclimate circles has crystallized in the early 21st century that helps generalize the climate end of the equation. In other words, it is possible to move down a list of successive cultures/groups/empires/states and take note of climate changes to the extent that appropriate evidence can be found.

The agrarian ancient world

We have groups and regimes that virtually disappear or disintegrate in the Middle Eastern ancient world and a number of environmental cooling deteriorations. As highlighted in Table 3.3, they overlap a great deal. Not only do they overlap but the ends of the cultures or states take place toward the ends of the cooling intervals, suggesting that the people involved may have endured environmental deterioration for a time but ultimately could not persist. This does not imply that climate change determined what took place – only that it played a highly significant role in creating a context for multiple problems that ultimately defied human management. It also does not mean that humans really disappeared. Rather, their structures underwent extreme stress forcing people to scatter (assuming they survived the stress). In every case, re-grouping occurred sometimes in the same places as before after variable time gaps. At other times, parts of the Middle East were less affected by the stresses and therefore could have been favored for the re-emergence of concentrated structures.

The Natufian, PPNA, and PPNB cultures have already been introduced and summarized in Table 3.2. The point is that these groups began in the Levantine coastal area and gradually spread out in iterations in the Fertile Crescent pattern toward Jordan and Mesopotamia and into Anatolia. The archaeological traces of these cultures end abruptly and are supplanted by the next group after a period of time.[5] The presumption is that their settlements were abandoned until new groups with different cultural characteristics reestablished themselves sometimes in the same place and at other times in other places farther afield.

TABLE 3.3 Ancient World RCC Timeline

Groups, Episodes, and RCCs	Timing (all BCE)
Natufian culture	12500–10000/9500
RCC no. 1 [Younger Dryas]	10700–9700
PPNA culture	9500–8000
RCC no. 2 [10.2kya event]	8500–8000
PPNB culture	7000–6000
RCC no. 3 [8.2kya event]	7000–6000
Ubaid culture and the Sahara	6500–3800
Uruk Disruption	c. 3200
RCC no. 4 [5.9kya and 5.2–5.1kya events]	4000/3900–3000
Fall of Akkad, Mesopotamia	2200
Fall of Old Kingdom, Egypt	2150
Decline of Indus	2000–1800
RCC no. 5 [4.2kya event]	2200–1800
End of the Bronze Age	c. 1100
RCC no. 6 [End of Bronze Age]	1200–800

Fagan (2004: 127) captures the core processes at work in the following passage on the PPNB culture:

> The Mini Ice Age of 6200 to 5800 B.C. was a catastrophe for many farming communities between the Euxine Lake [Black Sea] and the Euphrates River. Month after month, a harsh sun baked soils that were no longer fertile. Dust cascaded out of a cloudless sky, lakes and rivers dried up, and the Dead Sea sank to record low levels. Farming societies shrank or evaporated in the face of unrelenting drought. Many turned to sheep farming as they resorted to the classic famine strategy of habitat tracking: moving to areas less affected by drying and cooling, where they could eke out a living off their herds.

The temporal correlations with the RCC periods do not mean that cold weather alone necessarily drove these early cultures from their places of habitation. It does mean that a cooler than average climate regime was in effect with presumed effects on the ability to cultivate and access a dependable food supply. That alone could provide the incentive to abandon one area for greener pastures. It is also an incentive that could be compounded by soil degradation and carrying capacity problems, dwindling numbers of animals to hunt, water shortages, flooding, conflicts with neighbors, and disease. Still, all of these other sources of incentive to move are hardly independent of a cooling regime. On the contrary, we might expect them to be associated with cooling episodes.[6] The available evidence allows us only to note that global cooling helps explain why these early cultures appear to have come to abrupt ends when they did. These same early cultures would have had few defenses against environmental deterioration. Ultimately, moving on to other places might have been the only realistic response as conditions worsened.

Ubaid and the Sahara

The relatively well-known Sumerian efflorescence in southern Mesopotamia was preceded by the Ubaid era (roughly 6500–4000 BCE). This period was named for southern Mesopotamians who influenced culture in a wide span of the Middle East from Arabia to Anatolia. One of the Ubaid era innovations was the pottery wheel, which led to the emergence of mass-produced pottery. As a consequence, one of the most prominent markers of their influence was the wide distribution of Ubaid pottery with a black-on-buff pattern. But there were other cultural traits, such as distinctive figurines, cemetery practices, and architectural styles that diffused broadly. Other innovations were the application of canal irrigation, initially developed farther north in central Mesopotamia, to the south, early urbanization (Uruk, Ur, Eridu), progress in animal domestication, and a network for the acquisition of commodities not available in the south. In many respects, the Ubaid might be said to have created a crude template for the subsequent Sumerian episode.

The Ubaid influence is thought to have been generated via trade and the inter-regional network established to bring in scarce commodities to the south in exchange for finished goods. It gradually became prevalent in northern Mesopotamia and then equally gradually disappears so that Ubaid traits survived only in the south by the end of the 5th millennium (Stein, 2010). Presumably, this cultural fade-out reflects a breakdown in the acquisition network. It is conceivable that increasing aridity in the 5th millennium contributed to this breakdown. But the evidence is by no means clear-cut given the little that is known about this period. With more time and fieldwork in a part of the world that is not easy to do said fieldwork, the demise of the Ubaid influence may be in some respects attributable to what is known as the 5.9 kya event (or 3900 BCE) which marks the high or low point, depending on how one looks at it, of another cooling and drying episode in the RCC schedule.[7]

However, Hole (1994) has a bold and attractive interpretation of this period. He notes that prior to 5000 BCE, villages were rather small (50–200 people), scattered, largely self-sufficient, and lacking in irrigation canals. After 5000 BCE, some villages were ten times as large as they had been, temples and storage facilities were prominent, as was elite stratification. Hole's explanation is that there were climate deterioration and a demographic crash beginning in the mid-6th millennium, which led to settlements either disappearing altogether or diminishing in size. Some of the survivors became nomads while others became refugees looking for more favorable environments. Some portion of the environmental refugees became a labor pool that could be utilized to build a new agrarian infrastructure based on irrigation canals. Larger, more centralized settlements were made possible as a consequence. This argument fits resilience theory's emphasis on creative destruction quite nicely.

The general timing of this process is not altogether clear. If applied to the 5th millennium instead of the 6th it would help explain how southern Mesopotamia was able to exploit northern refugee labor to construct a broader enterprise in Sumer than had been witnessed in the Ubaid era. But perhaps this process happened in both the Ubaid and Urukian eras. Increased aridity in the north drove refugees south where there was more standing water. Ubaid and Sumerian development thus proceeded by taking advantage of the lack of rainfall in the North.

Moving farther east, the Saharan Desert was "under construction" in this time period. For the past 20,000 years (Hamblin, 2006: 32), it had fluctuated between wet and dry regimes. Wet regimes supported animals and humans. Dry regimes drove both types of animals away. It had become much less inhabitable between the 4th and 3rd millennium BCE as it became more arid. Some migrants moved toward the Nile River and, as they became more sedentary, traded hunting and foraging as well as nomadic herding for agriculture.

Berking et al (2012) dryly describe what might be viewed as a general law for arid environments:

> The rise and fall of ancient cultures in drylands is mainly controlled by the availability of water. Where no perennial water sources are available (ancient) cultures in drylands often depend on the availability of water by

effective rainfall as a source for water-harvesting measures. These settlements are susceptible to climate changes....

If the critical rainfall declines sufficiently or ceases altogether, there are strong pressures to migrate toward perennial water sources. That is what happened eventually in the eastern Sahara Desert. Nomadic cattle herders began moving into the eastern Sahara after the 8th millennium BCE. By the 4th millennium BCE, they were increasingly moving toward the Nile and becoming part of the extended Egyptian society. Nicoll (2012) presents a calendar for these developments summarized in Table 3.4.

People moved into and stayed in the Sahara area until they could no longer survive easily. Aridification was a gradual process that, no doubt, was aided by accelerated periods of cool dryness. The major movement out of the Sahara toward Egypt in the 3300s took place shortly before the Urukian Disjunction circa 3200 BCE that is elaborated in the next section. In this case, the emphasis has to be placed on the causal effects of environmental deterioration and less on the rigidifications that come with population expansion and diminishing food supply – although the latter would certainly have been prevalent. This is a theoretical tension concerning the application of resilience theory that will reappear from time to time. Resilience theory tends to treat climate change as the straw that broke the camel's back but in some cases, its effect is far greater than the

TABLE 3.4 Sahara Becoming Dryer Timetable

Time (BCE)	Climate Conditions	Behavior	Egyptian Regimes
9000	Arid conditions		
8000		Opportunistic pastoralism	
7000	Transition to more rainfall	Use of wild millet, legumes; First cattle and pottery	
6000	Maximal vegetative cover; lakes filled		
5000	Intense drought conditions begin	First villages; sedentism; caprovids Nomadic pastoralism	Predynastic Egypt
4000	Water sources dry up – aridification	Migration from desert Nile Valley population boom	Pharaonic kingdoms
3000	Less surface water storage and vegetation Low Nile	Old Kingdom collapse	
2000	Arid		
1000			Graeco-Roman period
1	Pervasive droughts; hyperarid		

Source: Modified from Nicoll (2012: 159).

metaphorical last straw. No matter how people choose to adapt, the primary way to survive is migration to a different place with better prospects of facilitating survival.

The ascent and disruption of the Sumerians

The better known, post–Ubaid, Mesopotamian episode requires first establishing some background. In the first half of the 4th millennium, settlements in northern Mesopotamia were no less complex than those in the south (Figure 3.3). Something happened to change that initial homogeneity. One way or another it had a great deal to do with water. One of the main differences between northern and southern Mesopotamia was that northern agriculture depended on rainfall while southern Mesopotamia was distinctive in its dependence on irrigation. Not coincidentally, northern Mesopotamian topography was characterized by relatively arid plains while southern Mesopotamia was more varied and included a large marshy area that blended into the Gulf. Six thousand years ago, the marshy area and water table associated with it extended some 200 miles farther into the mainland than it does today. Moreover, the Tigris and Euphrates were then less separated and fed into the south as an integrated body of water.

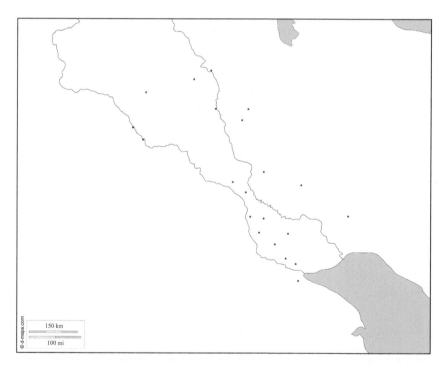

FIGURE 3.3 Map of settlements in ancient Mesopotamia (dots indicate their location)

These comparative advantages made it possible for southerners (in comparison with northerners) to:

1. Extract larger agricultural productivity
2. Do so with greater reliability and predictability
3. Support far larger herds of animals for food purposes
4. Exploit water transport to distribute southern products at lower costs
5. Exploit water transport to acquire raw materials from non-southern locations
6. Develop and maintain stronger ties to agricultural hinterlands
7. Create a large labor pool linked to maintaining water channels
8. Create stronger linkages between and among urbanized settlements to foster increased technological innovation[8]

As a consequence, it was possible for southern Mesopotamia or Sumer to develop into a sub-region of interactive and competitive cities that as a whole were much more technologically advanced than anywhere else in the world in the second half of the 4th millennium. Of course, the comparative geographical advantages did not make this outcome inevitable. But it seems far less likely to have taken place in a less conducive environment.

One of the things southern Mesopotamians did was to create a broad acquisition network that brought in raw materials from adjacent areas throughout southwest Asia and exchanged finished products constructed in southern Mesopotamia for them. The network encompassed adjacent Elam centered on Susa, northwestern Iran via Godin Tepe for lapis lazuli and tin from Afghanistan, Tigris–Euphrates routes for metals and timber, land and sea routes to Egypt, and a Gulf route to Dilmun (Hamblin, 2006: 41).

The nodes of the network were manned by three types of Sumerian outposts. One type consisted of large concentrations of Sumerian colonists occupying space in northern Mesopotamia that was relatively uninhabited. A second type consisted of traders who resided in non-Sumerian cities. A third type may have been settlements located in strategic areas that were taken by force (although there seems to be little support for this aside from one town that gives evidence of being under siege from another group). In part, it is this third possible arrangement that gives support to the notion that Uruk was the world's first coercive empire. The problem is that the evidence is stronger for the first two types, which might suffice to qualify Uruk as an empire according to definitions that do not require territorial expansion and control.

However that dispute is negotiated, Uruk may have initially been a beneficiary of dry times. In 3800 BCE, Uruk had a population of some 10,000. By 3200 BCE, its population had climbed to 50,000 (Burroughs, 2005: 242). It is presumed that rain-reliant northern farmers and others were driven south to places where irrigation supplied water for agriculture. Yet just as the city was expanding to what was a very large size for the times, its ability to acquire scarce commodities from its extended hinterland abruptly ended.

What is sometimes referred to as the Uruk Disjuncture is that the extended network for acquiring imported goods collapsed around 3200 BCE. Colonists withdrew presumably to the home cities. Trading outposts ceased to function. Sumerian overland trade with its southwest Asian interaction sphere came to an abrupt halt. The conflict between Sumerian cities seems to have increased. Exactly why this disjuncture occurred has never been clear. That it did happen has never been challenged. Nor does withdrawal from the acquisition network seem to have been voluntary on the part of Sumerian colonists and agents. If they were forced to leave, it means that their presence in the north was no longer tolerated. One clue may be the arrival of armed migrants from the Black Sea area who moved into Anatolia and down the Levantine coastline. Coercive pressure from the new arrivals, with or without the assistance of locals, might have sufficed to evict the transplanted Sumerians. While largely speculative, the motivation for southern migration from the Caucasus by pastoralists and nomads is an increasingly cold climate that drives people toward warmer areas. On the other hand, farming in northern Mesopotamia could also have collapsed in the context of an extended drought. Such a development would help account for the withdrawal of the colonists and increased hostility/desperation on the part of indigenous populations who had nowhere to go. Still, the Sumerians apparently ran into trade interruption problems in the direction of Iran as well which would not be explicable in terms of the same intrusions in the Anatolian and Syrian areas.[9] But Iran had its own pastoralists/nomads who were prepared to raid traders when times were hard. It may also be that trade was simply difficult to sustain in periods of extended environmental deterioration (RCC no. 4). Supply and demand are likely to suffer. Thus, a harsher climate could go some way to explaining the demise of the extended acquisition network. Later, in the next millennium, trade did resume but a new route, this time maritime connecting southern Mesopotamia and Indus via the Gulf, had to be forged to circumvent the hostile and/or barren overland routes.

Uruk thus represents an acquisition network breakdown for which the etiology is not altogether clear. Some of the breakdowns may be traceable to political-economic turmoil in the Urukian cities. Some of it may be due to the intrusion of new groups from the north or even a reaction to Urukian imperialism. Mesopotamia did not disappear. New approaches emerged, both in terms of new trade routes (a maritime route replaced the dubious overland approaches) and in terms of how Urukian cities interacted with one another. Centralization gave way to more independent cities at least for a time. Creative destruction appears to have worked once again, even if all the new features were not necessarily welcome advances. Political fragmentation, for instance, led to an increase in intercity warfare.

Akkad, the old kingdom, and Indus (end of the third millennium BCE)

After Uruk's centrality diminished, Mesopotamia came to be characterized by swings from political fragmentation to imperial centralization. Barjamovic (2013) captures this swing, which is summarized in Table 3.5:

TABLE 3.5 Imperial Centralization and Political Fragmentation in Mesopotamia

	Imperial Centralization	*Political Fragmentation*
2900–2350		Multiple city-states
2350–2215	Akkad	
2215–2100		Multiple city-states in S. Mesopotamia
2100–2000	Third Dynasty in Ur	
2000–1750		"hundreds of polities"
1750–1595	Hammurabi of Babylon in S. Mesopotamia gradually diminishing in size	
1650–1150	Hittite empire in Anatolia	
1600–1340	Mitanni empire in N. and W. Syria	
Mid-16th c.–1155	and N. Iraq	
	Kassite empire in S. Mesopotamia	
1050–900		Fragmentation throughout Near East
1360–609	Expanding Assyrian empire subject to temporary setbacks but unifying Near East by the first half of the 7th century	
626–539	Babylonian empire emerges from disintegrating Assyrian empire	

Source: Based on Barjamovic (2013: 124–126).

In southern Mesopotamia, roughly the first half of the 3rd millennium BCE was preoccupied with competition and conflict among a number of independent city-states. That all changed when Sargon created an Akkadian centralized empire in the 2300s. Sargon was a northern Mesopotamian Semite in the employ of the King of Kish, who somehow managed to overthrow the King in a palace coup and alter the ethnic composition of what had been a Sumerian monopoly. Sargon then used Kish military resources to coercively centralize control of the competing city-states and create the Akkadian empire in which Akkadian became the primary language. The empire lasted nearly 200 years and was characterized by external aggression in all directions (Anatolia, Syria, Elam, Dilmun, Oman, Canaan) and intermittent internal rebellions by dissatisfied cities. Toward the middle of the 2100s, Akkad was under pressure from several sources. Amorites and Gutians were moving into farming areas from the north and west. A major drought had reduced agricultural production in the north while the irrigation canals in the south were suffering from years of salinization.[10] Food became scarcer. The empire disintegrated into a mixture of rule by Gutian and Sumerian city kings in different parts of the former empire.

A contemporary explanation was that Naram-sin, the last Akkadian ruler, had offended the gods who had been responsible for the wealth and agricultural abundance of the early imperial era and was punished by them taking away what they had given before. The main punishment tool was the invasion of the Gutians, a mountain people that had been subordinated earlier by Akkad. In the Curse of Agade (written slightly later in the 3rd millennium BCE), the outcome is described in the way people who live in towns usually depict people who do not:

> [Enlil] lifted his gaze towards the Gubin mountains and made all the inhabitants of the broad mountain ranges descend (?). Enlil brought out of the mountains those who do not resemble other people, who are not reckoned as part of the Land, the Gutians, an unbridled people, with human intelligence but canine instincts and monkeys' features. Like small birds they swooped on the ground in great flocks. Because of Enlil, they stretched their arms out across the plain like a net for animals. Nothing escaped their clutches, no one left their grasp. Messengers no longer traveled the highways, the courier's boat no longer passed along the rivers. The Gutians drove the trusting (?) goats of Enlil out of their folds and compelled their herdsmen to follow them, they drove the cows out of their pens and compelled their cowherds to follow them. Prisoners manned the watch. Brigands occupied the highways. The doors of the city gates of the land lay dislodged in mud, and all the foreign lands uttered bitter cries from the walls of their cities. They established gardens for themselves within the cities, and not as usual on the wide plain outside. As if it had been before the time when cities were built and founded, the large arable tracts yielded no grain, the inundated tracts yielded no fish, the irrigated orchards yielded no syrup or wine, the thick clouds (?) did not rain, the macgurum plant did not grow.[11]

The Gutians are given primary credit in the above passage but note as well the emphasis on limited food supply, the absence of rain, the breakdown of the cities, and the prevalence of disorder – more specifically, in this case, in the form of civil war among contenders for the throne. These elements are all recognizable features of societal rigidities ala resilience theory.

While it is clear that the author of the Curse did not like Gutians, barbarian intrusions, themselves, tend to be driven by climate problems encouraging pastoralists and nomads to move toward water and grass for their herds. They are also less likely to invade when urban areas are capable of mounting strong defenses. Thus, a substantial portion of the credit for the downfall of the empire would seem to rest with the lack of rain, the breakdown of the Mesopotamian agrarian system first in the north and then in the south overwhelmed with northern refugees that, in turn, invited opportunistic attacks from inside (technically, the Gutians had been conquered by Akkad earlier but never fully integrated into the empire) and outside the weakened empire.[12]

Hallo and Simpson (1998: 63; Hamblin, 2006: 304) also stress that external threats and pressures were coming from multiple directions. The Gutians were coming in from the northwest. The Lullabi and Hurrians were growing stronger in northern Mesopotamia. Elamites were also gaining strength and power in the northeast. Amorites were pressing from the southwest. Akkad was being pressed from all sides suggesting, if nothing else, considerable weakness on the part of the Akkadians to defend themselves.

One of the Sumerian city-states was eventually able to recreate a more unified political system, the Third Dynasty of Ur (Ur III), that lasted about a century (2100–2000 BCE). It too, however, succumbed to a combination of environmental deterioration after a brief respite from drought and highly centralized intervention in food growing processes. A combination of famine and pressure from Amorites migrating in from the west and Elamites attacking from the east brought an end to Ur III.[13]

The general climate pattern prevailing in southern Mesopotamia is captured succinctly in the schedule reported in Table 3.6. Sumerian/Akkadian activities took place in generally optimal climate conditions except for around 3200 BCE and 2200–1900 BCE. Those years were also when the Sumerians and Akkadian encountered major setbacks in being able to grow or acquire food and developing greater complexity.

The problems that beset the Mesopotamians were not restricted to Mesopotamia. The 2200s are described (Hamblin, 2006: 288) as a time of "widespread destruction in western and southern Anatolia" as well. Three-fourths of the known city sites were leveled and depopulated. Surviving sites were diminished in size and populated by new groups. Luwians are believed to have been entering Anatolia from the Balkans at this time, along with the Hittites crossing the Caucasus Mountains. Most, if not all, cities in Canaan ceased to function and were replaced by nomadic herders. It may be that the cities had been destroyed by repeated Egyptian raids (Hamblin, 2006: 274–276). In Syria, the same time is characterized as "widespread urban collapse, population decline, decentralization, and devastation" (Hamblin, 2006: 251). Not every city was destroyed but most were affected in some fashion. Amorite rulers tended to emerge in the places that survived.[14]

TABLE 3.6 Climate Change Timeline in Southern Mesopotamia

Time Period (BCE)	Description
3500–3200	Warm and moist
3200	Drought
3000–2340	Warm and moist
2340–2180	Warm and moist to increasing drought
2200	Drought
2200–1900	Cool and dry
1900–1800	Warm and moist

Source: Based on Murphy (2007: 163–164).

Hamblin (2006: 250) also notes that

> The defining characteristic of the end of the Early Bronze Age is a region-wide crisis reflected in widespread ecological decay, dynastic collapse, urban disintegration, and tribal migration. Similar patterns of crisis are found in Egypt, Canaan, Syria, Cyprus, Anatolia, the Aegean, Mesopotamia, and Iran during this same period. [In the search] for region-wide factors contributing to this crisis, [o]ne theory posits that increasing aridity initiated a crisis for both farmers and herders. An extended drought reduced both annual crop yields and the overall carrying capacity of nomadic grazing grounds. An alternative, or perhaps complementary theory posits human-caused environmental deterioration, through over-grazing, soil depletion, salinization, deforestation, and to other forms of ecological degradation. The impact of either or both of these developments was decreasing productivity, increasing social stress, competition for resources, and population displacement through migration and war.... The cumulative result was a period of political disintegration, war, anarchy, social upheaval, and migration.

We do not really need to pick between the two theories given our employment of resilience theory which combines both emphases.[15] The climate change evidence is fairly firm and there is some evidence for human-caused ecological degradation, as in the case of the Mesopotamian salinization problem centered around its irrigation practices.[16] It seems likely that climate change served as a tipping point for unsustainable environmental practices that might otherwise have continued for a longer period of time. There is no need to award stronger causality to one or the other theory when both probably work better in tandem.

Egyptian dynasties

Between the 4th and 1st millennium BCE, the Egyptian government was associated with a long list of dynastic turnovers (Table 3.7) that are distinguished after a hazy beginning dealing with the unification of north and south by three clusters of more or less conventional rule (with rather unimaginative labels: Old Kingdom, Middle Kingdom, and the New Kingdom) that are interrupted by three Intermediate periods of disorder and turbulence.[17] The Intermediate periods also have kings and pharaohs but there tend to be multiple individuals ruling simultaneously in different segments of the Nile-configured Egypt and they do not have long tenures.[18] In later periods, some of the rulers are not Egyptians. Thus, the Intermediate periods can be summarized as periods of discernible political fragmentation and instability.

But what happened in these Intermediate periods in general? How do they contrast with the periods of centralization? In Table 3.8, Hallo and Simpson (1998: 191) give us a better fix on how the two columns encompass different types of behavior.

TABLE 3.7 Centralization and Fragmentation in Ancient Egypt

Timing (BCE)	Centralization	Fragmentation
2686–2160	Old Kingdom	
2160–2055		First Intermediate Period
2055–1650	Middle Kingdom	
1650–1550		Second Intermediate Period
1550–1069	New Kingdom	
1069–664		Third Intermediate Period
664–525	Saite Restoration	
525–404, 343–332	Persian rule	
332–30	Ptolemaic rule	
	Roman rule	

Source: Based primarily on Manning's (2013) treatment.

TABLE 3.8 Egyptian Variations in Cultural Styles

Old/Middle/New Kingdoms	Intermediate Periods
Strong monarchies ruling entire land	Weak governments
Stable bureaucracy with effective means of taxation	Interference from foreigners
Firm control of the borders	Decrease in monument building
Exploitation of mines and quarries	Rival contemporary dynasties
Building of temples and funerary structures	Societal restructuring
Achievements in architecture, sculpture, and painting	Opportunities for changes in religion and state structure

Source: Hallo and Simpson (1998: 191–192).

We will return to the Third Intermediate period in the section below at the end of the Bronze Age. The First Intermediate period (2150–1980 BCE using our earlier schedule rather than the one in Table 3.7) took place roughly at the same time as the collapse of the Akkadian Empire.[19] Egyptologists emphasize this period as one of an unraveling central government with provincial governors (nomarchs) assuming increasing powers of governance after the very long rule of Pepy II. The central government was also experiencing increasing fiscal and administrative problems due to exemptions from taxation and bureaucratic employment being controlled by a small number of families. There may also have been intrusions from non-Egyptian groups in the northeast. But political entropy was paired with significant declines in Nile river flow and increased aridification which meant sand was encroaching on agricultural areas. A tomb inscription for Ankhtifi, the nomarch of the third nome of Upper Egypt during the First Intermediate period, gives some taste of the nature of the crisis.

> I gave bread to the hungry and clothing to the naked: I anointed those who had no cosmetic oil: I gave sandals to the barefooted; I gave a wife to him

who had no wife. I took care of the towns of Hefat and Hor-mer in every [situation of crisis, when] the sky was clouded and the earth [was parched(?) and when everybody died] of hunger on this sandbank of Apophis…

The whole of Upper Egypt died of hunger and each individual had reached such a state of hunger that he ate his own children. But I refused to see anyone die of hunger and gave to the north grain of Upper Egypt…

Aridification, crop failure, famine, and even cannibalism should sound familiar and might have sufficed to disrupt central government whether or not trends toward decentralization were already in progress. It seems safe to say that Egypt was experiencing some of the same problems as those prevalent in Mesopotamia. Whether the cooling in Mesopotamia influenced the Nile flow problems directly or only indirectly thanks to some common root (Nile flow is governed by fluctuations in northeast African monsoon rains) is less clear. Yet the Old Kingdom seems to have fallen apart roughly at the same time as the Akkadian empire in roughly similar contexts. When climate conditions improved, the possibility of central governance returned in both areas.

The adaptive cycle model that we adopted in chapter 1 (Figure 1.1) is premised in part on things going wrong (stresses and strains) as systems age and elites attempt to monopolize resources. This period of "conservation" is brought to an end by some type of trigger that breaks up the monopolies and permits renewal and reorganization of resources in the alpha period. Karl Butzer (2012) developed a very similar model (Figure 3.4) that he applied to several cases but focused mainly on Egyptian developments, and principally but not exclusively aimed at the demise of the Old Kingdom.

Butzer's take on the decline of the Old Kingdom stresses a concatenation of events that conspired to bring the old regime down. The last major authority figure of the Old Kingdom, Pepy II, had ruled for nearly a century. His death catalyzed civil wars among provincial warlords for control over a state that had allowed its fiscal resources to diminish over time by exempting prime land from

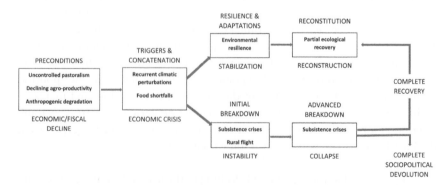

FIGURE 3.4 Butzer's collapse model. Source: Based on Butzer (2012: 3636).

taxation. Trade with the outside world had been centered on the Byblos entrepôt in the Levant and this critical outlet had been destroyed about the same time by Mesopotamian attacks. Egypt was in an intense economic crisis that was made all the worse (see triggers and concatenations in Figure 3.4) by low Nile floods and famine.

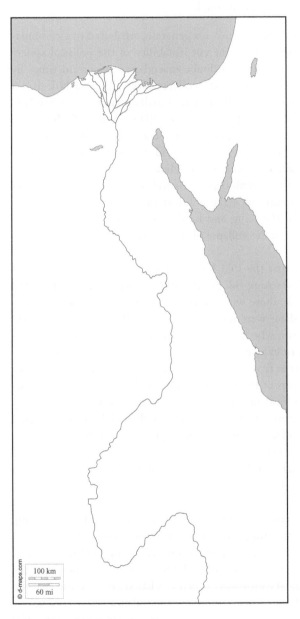

FIGURE 3.5 Map of ancient Egypt

In his interpretation, Egyptians demonstrated resilience (and military coercive ruthlessness) in reconstituting the Middle Kingdom, thereby avoiding full collapse. Things worked out much differently in the Third Intermediate crisis and the collapse of the New Kingdom.

Butzer acknowledges climate problems but assigns them to a trigger category. There is ample scholarly reluctance to concede a strong role for climate change. Manning (2013: 75) provides a good example:

> The collapse of the Old Kingdom ... is generally attributed to a combination of factors ... stresses the inherent instability of the political system itself. Political instability was no doubt exacerbated at certain times by other factors that led to central state collapse inter alia poor Nile River flooding and external threats. Scholars have usually sought a change in Nile flooding patterns as the main reason for the collapse of the Old Kingdom state ... The end of the Old Kingdom and the rise of the First Intermediate Period, lasting about a century, was a period during which drier, more arid conditions prevailed throughout the Middle East..., and indeed there is good evidence, both in tomb depictions and in literary descriptions, of famine .... But the failure to make political adjustments in the evolving relationships between the King and local power bases were perhaps the most important factor in the collapse of the central state....

There can be little doubt that the Old Kingdom had developed a number of political fault lines. The last strong ruler of this dynasty had somehow persisted for 94 years. With his death, there were a plethora of candidates to replace him but the most senior candidates were quite senior in age. Pepy II was succeeded immediately by his oldest son who lasted only a year. His wife then broke precedent with male succession rules by governing for five years. But she was then succeeded by possibly as many as five rulers in the next nine years (Hamblin, 2006: 351). The turmoil at the top of the power pyramid was paralleled by regional power fragmentation with local authorities (nomarchs) increasingly acting independently of the central throne. This fragmentation followed decades of gradual shifting control of once-central resources to the regional level as rewards, political payoffs, and religious subventions.

The conjunction of disarray at the top and power diffusion at the middle was accentuated by suffering at the base.

> this period seems to have been one of extended drought and low flooding levels of the Nile, decreasing food production and creating the potential for famine, while increasing competition for decreasing food resources.... drought and food shortages undercut both the real economic power of the king and his ideological legitimacy, which was based on the claim that the king was the divinely sanctioned sustainer of moral and natural order.
>
> *(Hamblin, 2006: 351)*

Thus, it is certainly possible to pick and choose which structural factor was paramount but it is less clear how to rank order the combination of factors coming together at this time. The easy way out is to describe the end of the Old Kingdom as an outcome brought about by a perfect storm of factors contributing to the breakdown of the central regime. Perhaps any one or two of the three factors described above would not have sufficed to bring down the Old Kingdom. Each one increased the probability of something changing; all three likely made regime dissolution highly probable.

The Second Intermediate period (1630–1520 BCE) falls outside of the fifth RCC event (see Table 3.3) by a few hundred years. Again, there are problems with Nile river level declines but in this case, the decline followed unusually high levels causing flooding so that the effects were less problematic and perhaps even appreciated. The fragmentation of central governance is highlighted by the conflict between non-Egyptian rule over parts of the north and a more typical Egyptian government in the South. The political origins of the Hyksos rulers in the north are murky but it is believed that they came from the northern Levant area, either by sea or land and had access to weaponry that was more advanced than Egyptian arms. What is less clear is whether they used their military superiority to gain control over parts of northern Egypt or instead were simply enabled to stay in power until Egyptians in the south caught up technologically. Egypt was reunified after fighting on a north–south axis was won by the south and the Hyksos were expelled.

The Second Intermediate period may be one of those political episodes that have a climate angle but are not necessarily linked directly to cooling dynamics. Still, there might be a down-the-line "knock on" effect that does not seem to have been discussed much in the traditional literature.[20] In Mesopotamia, the rulers that followed the downfall of the Sumerians and the Akkadians, like Hammurabi, had their origins in the semitic tribes coming in from the Syrian desert. Might this same ethnic movement have contributed in some way to the Hyksos showing up in northern Egypt?

The murky circumstances are actually even more complicated. The alien Hyksos (so-called "rulers of foreign lands" by the Egyptians) were once depicted as an Asiatic horde that descended on Egypt at a time of great vulnerability. But the "Asiatic" term was used rather loosely to refer to non-Egyptian populations to the east of Egypt. Since there is little evidence of fighting between the presumed arrival of the Hyksos and Egyptians, it is not clear that their arrival resembled a military invasion. On the contrary, there had been migration into the Egyptian Delta by the Canaanites and others from the east for some time. So they were probably already there before the Second Intermediate. It is quite possible then that the Hyksos takeover of part of northern Egypt was a more indigenous movement within a fragmenting Egypt than an outright invasion from outside the country. Moreover, the main impetus for eastern migration into Egypt seems to have been climate deterioration. This inducement also encompasses Indo-European and Indo-Aryan migrations out of the Central Asian steppes toward

South Asia, Iran, and the Near East that began around the end of the 3rd millennium. One hallmark of these migrations is that steppe migrants brought chariots and other superior military technology and established elite control over various groups that they encountered on that basis. As a consequence, migrants to the east, and this probably includes the Hyksos, tended to be mixtures of multiple ethnicities. To the extent that these characterizations hold up to further scrutiny, it is conceivable that the Second Intermediate, while in part a reaction to local Nile fluctuation problems, is also linked fairly directly to cooling in the east.

In any event, and in marked contrast to the haziness of Egyptian Old and Middle Kingdom events, the demise of the Indus Valley in this era is quite stark.[21] While not usually considered part of the Middle East, Indus, centered in an arc above the Thar Desert that helps demarcate one of the current borders between Pakistan and India, interacted with Mesopotamia through maritime trade fairly intensely in the 3rd millennium BCE, especially after the Uruk Disruption. Encompassing an area larger than Mesopotamia and Egypt combined, Indus agrarian productivity, centered on the major rivers of the area, led to the development of a number of fairly large cities that compared favorably to the urbanization ongoing in Mesopotamia at roughly the same time. Still, the Indus area was highly dependent on a monsoon rain system that determined river levels which had generally been weakening since the 5th millennium BCE.[22] Farmers initially benefited from the weakening because it meant less flooding and more silt to enrich farming soil. Later, the slow pace of change allowed adaptation to the gradual weakening by changing the types of crops that were grown (switching from wheat and barley to millet and rice). What they could not adapt to readily was a shift in the monsoons in which the traditionally rainy months began to resemble the non-rainy months in terms of the amount of precipitation that fell.[23] Interestingly, the Indus area had already experienced at least one severe decrease in monsoon rain between 4400 and 3760 BCE according to Petrie et al, (2017) without an outcome in which the population was forced to leave. A second round between 2200 and 2000 BCE was shorter but more severe.[24]

The second half of the 3rd millennium was warm and wet – so wet that flooding was a recurring problem. Then a severe and prolonged drought began around 2000 BCE. By 1800 most of the major cities had completely lost their populations and, unlike Mesopotamia, agriculture had come to a halt. Again, unlike Mesopotamia, urbanization did not reemerge in the Indus area for another millennium.

Murphy (2007) argues that the difference between what transpired in Mesopotamia and Indus can be traced to the drying up of the Saraswati River and the diversion of the Indus River. One might also add the enlargement of the Thar Desert that also took place at the same time. In both of the river cases, major Indus cities rather abruptly lost their access to water as did the adjacent agrarian hinterland. There was little time to prepare for the prolonged crisis and few reserves to cope with the hungry refugees that probably descended on the cities. In Mesopotamia, the Tigris and Euphrates continued to flow albeit at lower levels. Southern irrigation could also ameliorate somewhat the food crisis there. Refugees could flock to the south in Mesopotamia and the north could be revived when conditions improved.[25] In Indus, the surviving populations first moved to smaller villages and then had to leave the area entirely and move to the east from which they never returned.

There are of course rival hypotheses to the climate change explanation. Invasions from the north, trade interruptions with Mesopotamia, and domestic political unrest have been advanced. As usual, some of these processes may have been in play. Prolonged famine easily leads to political instability. The trade interruption with Mesopotamia presumably followed the Indus decline as opposed to setting it off. Evidence for foreign invasions has never held up to close scrutiny. None of these, however, would readily account for massive migrations to other parts of South Asia. Water shortages and food scarcity do fit what happened.

The end of the Bronze Age (late 2nd millennium BCE)

One of the more remarkable events in the climate-induced behavioral annals is what is termed the end of the Bronze Age.[26] The labeling is deceptive in part because it references the transition from bronze to iron metal usage which actually took some time to accomplish. So, bronze did not disappear abruptly but the eastern Mediterranean societies that used bronze encountered serious damage and entered a period of little growth of any kind entirely deserving the appellation of a Dark Age that took several hundred years for demographic and economic growth to reemerge.[27]

The several hundred years prior to the end of the Bronze Age (circa some point in the 12th century BCE) were years of economic growth and expansion in the Eastern Mediterranean. Trade boomed as never before. Warfare between Egypt and the Hittites had ended. Yet between 1250 and 1100 BCE, the eastern Mediterranean world had been destroyed by violence around the maritime rim from Greece to Egypt. The Egyptians claimed to have defeated two Sea Peoples attacks in 1207 and 1177 and two major Libyan attacks in 1179 and 1173. However, in a rolling discussion of what cities were destroyed in the 12th century, Cline (2014: 108–136) lists 62.[28] The first six of Cline's 62 cities were located in northern Syria, the next nine were in southern Syria/Canaan, Babylon was in Mesopotamia, the eight cities that follow Babylon were located in Anatolia, then come 29 Greek sites, followed by six Cypriot towns. This list may not be entirely comprehensive. They are sites that have been found by archaeologists to have been destroyed by fire, often with an abundance of arrowheads in walls, or abandoned in the late 13th and early 12th centuries BCE. Some of the sites were never re-occupied.

Knapp and Manning (2016) note that the reasons for this sudden collapse are many: migrations, external predations, internal political struggles, center–periphery inequalities, climatic change, natural disasters, and disease. They note that there has "never been an overarching explanation and [we] should not expect one." And yet a few pages later they almost provide one through a hypothetical statement:

> One or more factors – for example, prolonged negative climatic change – undoubtedly precipitated others to create a potentially disastrous cycle that undermined existing social, economic, and political structures.
>
> *(Knapp and Manning, 2016: 113)*

Only climate change, dry and cold weather, and protracted famine could have set large numbers of refugees, many of them accompanied by whole families, in motion on land and at sea. One envisions large-scale migrations of people uprooted from areas as far away as the Black Sea moving west and south, primarily along the coastlines but not exclusively so. We know from the Egyptian evidence that some of the attackers were the famous Sea Peoples, many of whom may have been based in Greece. But some of the attackers were almost sure to be local, non-urbanized tribes such as the Libyans on the fringe of Egypt or the Kashka enemies of the Hittites in Anatolia. Famine may well have exacerbated domestic political struggles here and there. Earthquakes could have caused some destruction but they are not usually accompanied by large numbers of arrowheads. Disease kills people but it does not destroy buildings. But only a widespread undermining of the agrarian economy could have caused similar outcomes in multiple sites around the same time. None of this means that climate change and famine are the sole answer to how the Bronze Age ended but it does mean that climate change and famine are very likely central to the answer. Singer (1999: 717 and discussed in Cline, 2014: 144) has it right when he concluded that the messages exchanged between the Hittites, Egyptians, and Ugarit about famine and the most urgent need for grain in conjunction with the archaeological evidence suggests "climatological cataclysms affected the entire eastern Mediterranean region towards the end of the second millennium BCE." It would not have been the first time nor, as we will see, was it the last time that something on this order occurred.

Despite all of the destruction, there was still variation in the response. Langgut, Finkelstein, and Litt (2013: 167), whose particular interest is the Levant, help clarify some of the variations in a chart (Figure 3.6) with three channels stemming from food supply problems. In Greece and Anatolia (Mycenaen and Hatti), the food supply problems led to famine, migration, and destruction. The Egyptian Pharaoh, perhaps because the Nile tends to be independent of the eastern Mediterranean climate fluctuations, can at least claim to have fended off groups seeking to do some damage in Egypt, although we only have his word for it. Canaan, in comparison, was characterized by instability, nomadic attacks (Apiru), destruction, and decline. Non-identical multiple outcomes do not necessarily invalidate a shared root cause. Indeed, we should expect variations no matter how devastating the climate-induced food supply crisis was. Different circumstances,

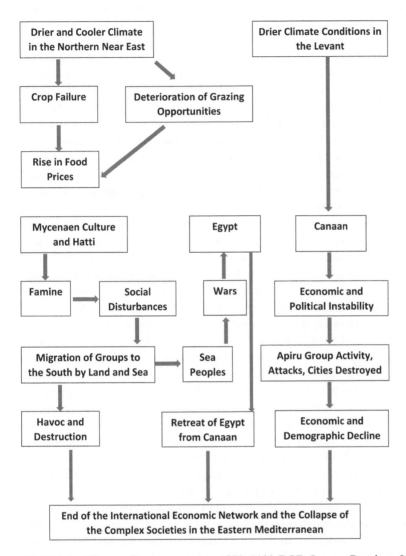

FIGURE 3.6 Climate change outcomes, 1250–1100 BCE. Source: Based on Langgut, Finkelstein, and Litt (2013: 167).

different actors, and perhaps even variations in the degree of climate change are likely to interact in different ways, thereby generating variable outcomes.

We have seen how each iteration in the cycling between favorable and unfavorable climates is associated with a variety of crises leading to shortages of food, famine, de-urbanization, negative or limited population growth, societal conflict, dynastic overthrows, alternations in political centralization and fragmentation, and attacks emanating from hinterland groups moving in from the mountains or deserts. In a number of cases, cities have been abandoned completely for centuries. In other cases, new groups have moved in and taken over political and societal control.

In Egypt in particular, though, there were considerable domestic problems that compete with the eastern drought and its consequences as an explanation of what transpired in Egypt, which was nominally outside the eastern drought zone. The end of the New Kingdom was marked by turmoil over local food scarcity linked to low Nile flooding and resulting in food price inflation and unpaid state wages which, in turn, led to increased strikes and rioting. The destruction of cities throughout the eastern Mediterranean meant foreign commerce had come to a standstill. Libyan marauders from the west were also taking advantage of governmental preoccupation and weakness. Finally, a schism emerged between the high priest and the Pharaoh on how to cope with these problems that led to civil war between the two bureaucratic wings of the old regime and divided territorial control and, ultimately to Nubians and Libyans who entered Egypt to install their own dynasties for the first time. In Butzer's (2012) view, this outcome manifested a failure to demonstrate Egyptian resilience and bring about reconstitution (see Figure 3.4). But while that judgment is hard to fault, it still leaves open how much credit to give to climate change in fomenting the breakdown of the New Kingdom. As for the end of the Bronze Age, climate change seems to loom quite large in the overall inventory of causation.

There seems to be ample support for the adaptive cycle (Figure 1.1) flow from exploitation or growth to "conservation" to creative destruction to reorganization (even if it took centuries as in the case of the end of the Bronze Age) and back to exploitation. Frequently, there is evidence that societies were becoming more rigid as they aged. Manifestations came in various forms ranging from political decentralization to increasing center-periphery frictions. Abrupt climate change made things worse. Invading groups, either from the nearby hills or marauders from far away gradually became more significant intrusive agents of destruction, driven presumably by water and food shortages. Societal organizations collapsed not everywhere but in prominent places. Reorganization occurred after a suitable lag, sometimes in different places (as in PPNA and PPNB emerging south of the Natufians or the switch from the Levant focus to a Mesopotamian one in the 7th millennium), sometimes in the same places (as in the Old, Middle, and New Egyptian kingdom transitions). The evidence for the adaptive cycle sequences in the ancient Near East is summarized in Table 3.9.

The ancient Near East thus experienced repeated rise and decline sequences. They appear to be closely related to abrupt climate change. Long periods of cool and dry weather made drought and famine more probable. Cooler and dryer climates thereby made agrarian exploitation and political order more difficult. If there were already political-economic strains, the changes in climate became overwhelming in areas that were already relatively arid.

While there is certainly anecdotal evidence for climate change-induced ramifications, Chapter 3 still cannot speak to the extent to which climate change and political-economic and societal changes are systematic. That is, are the behavioral

TABLE 3.9 The Ancient Near Eastern Adaptive Cycle

Organization/ Exploitation/Conservation	Creative Destruction	Reorganization and Renewed Exploitation/Growth
Late Natufian	RCC 1 Younger Dryas	PPNA
PPNA	RCC 2 – 10.2 kya	PPNB
PPNB	RCC 3 – 8.2 kya event	Ubaid
Ubaid	RCC 4 – 5.9 kya event	Uruk
Uruk	Continuing RCC 4 – 5.2/5.2kya event	Akkad
Akkad/Old Kingdom/ Indus	RCC 5 – 4.2kya event	Third Dynasty in Mesopotamia and subsequent fragmentation there and elsewhere
Eastern Mediterranean/ New Kingdom	RCC 6 – End of Bronze Age	Widespread decline in population growth into the 9th–8th centuries CE

changes that are linked to abrupt onsets of cold and dry climates due regularly to climate change per se? Or, do we find much the same sorts of behavioral changes in periods not characterized by climate change? We would also like to know just how much of the behavioral change is due to changes in climate: 10%, 25%, 50%, 75%? The odds are that the effects of climate changes are not deterministic. But that only tells us that the causal role of climate change is less than 100%.

There are strong restrictions on how well we can delineate the proportion of change in what people do is brought about by ancient episodes of climate change. Yet we can move away from presenting anecdotal evidence to looking at the serial correspondence of climate change periods and changes in things such as fluctuations in population sizes, the collapse of trade routes, the decline of ruling regimes, the major attacks on sedentary settlements by hinterland groups, and instabilities in the availability of water. This is the task taken on in the next two chapters.

Notes

1 See, for instance, Bar-Yosef and Belfer-Cohen (1989); Wright (1994); Pringle (1998); Nesbitt (1999); Verhoeven (2004); Bellwood (2005); Sherratt (2007); Zeder (2008, 2011); Brown et al (2009); Abbo, Lev-Yadin and Gopher (2010); Conolly et al (2010); Fuller, Wilcos, and Allaby (2011); Price and Bar-Yosef (2011); Henry (2013); and Wilcox (2013).

2 More specifically, Roberts et al (2018) indicate that the warming transition which ushered in the movement away from a glacial age occurred circa 14.5 Ka BP.

3 One of the gold standards of assessing the health of a population is the trend in the heights of skeletons. People deposited in sedentary burial grounds were becoming shorter in response to a deteriorating diet.

4 Zimmer (2016) stresses that farming developed independently in several parts of the eastern Middle East. Thus, it is possible that different innovators were exposed to different climatological settings.

5 There is a growing literature on the 8.2kya event. See Simmons (2007); Asouti (2009); Rosen (2011); Beihl et al (2012); and Akkermans et al (2015), in addition to a conference held in Leiden in 2010 that resulted in a number of papers: Renssen (2010); Biehl et al (2010); Ryan and Rosen (2010); Marcinick and Czerniak (2010); Niuwenhuyse and Akkermans (2010); Russell et al (2010); and Zerboni and di Lernia (2010). Cooling was evident, the event lasted some 500 years, Catalhoyuk was apparently abandoned and then rebuilt nearby, and increased movement on the part of pastoral groups was found.

6 Water shortages and flooding might seem contradictory but devastating flash floods in arid lands are hardly unknown. Nor are tendencies toward increased instabilities in weather patterns in periods of climate change.

7 Some scholars push the end of the Ubaid era back to 4200 BCE.

8 These advantages are taken from Algaze (2008: 63, 119–121).

9 The Sumerian contact with Egypt was also severed which suggests the extensiveness of the trade acquisition disjuncture.

10 On the salinization argument, see Jacobson and Adams (1958); Powell (1985);, and Artzy and Hillel (1988). Other useful sources on Akkad are Yoffee (1991, 2010); Weiss et al (1993); Van De Mieroop (2007); Ur (2010); Oates (2014); Weiss (2014, 2016); and Foster (2016). A cluster of critical pieces were published in 2017 (see Burke, 2017; Dornauer, 2017; Genz, 2017; Hoflmayer, 2017; Reihl, 2017; Schloen, 2017; and Schwartz, 2017) but none deny that Akkad collapsed. Their criticisms are more about whether other parts of Mesopotamia and Syria were hit equally hard by climate and Gutians. By and large, they argue that the evidence suggests that they were not. As Ur (2015: 69) puts it and demonstrates with an analysis of three Northern Mesopotamian cities at the end of the 3rd millennium, climate-induced responses customarily tend to be nonuniform. Of the three cities, one was abandoned in 2200 BCE, another one suffered the same fate sometime between 2100 and 2000, while the third was never abandoned but was substantially downsized.

11 Enlil is a reference to one of the Mesopotamian gods. Question marks in the text indicate the presence of untranslatable words.

12 Burroughs (2005: 256) suggests that about this time the pig, which had been domesticated for a fairly long time in the Near East, was demonized in Mesopotamia and Egypt as something to avoid consuming. Why only the pig was singled out is less clear.

13 A rather formidable wall (named "Repeller of the Amorites") was built to keep the Amorites out but, perhaps not surprisingly, it failed to do the job just as walls elsewhere have not sufficed to stop migrations and attacks.

14 Elsewhere, Hamblin (2006: 159) compares the role of Amorites in this period to similar roles played by Germanic people in the fall of Rome or the in-migration of Turkic people into the medieval Islamic empire. Initially used as mercenaries, they gradually increased their political and military power to the point of supplanting their former employers.

15 Danti (2010: 141) refers to them as the "Catastrophic Collapse" and the "Brittle Economy" models and nominates Weiss et al (1993) and Wilkinson (1994, 1997), respectively, as leading examples. Interestingly, he also notes that the tendencies stressed by the two models tend to emerge in discussions about the effects of 6.2kya, 5.2kya, and 4.2kya climate change events. In the first model, climate and invaders abruptly overrun urban areas and in the second urbanization and agricultural practices lower resilience to climate change. Part of the reason for the two opposing models is that Mesopotamia is or was a highly heterogeneous landscape that provides evidence for both in different regions. That is, both sets of processes could have been ongoing in different places. At the same time, one has to wonder to what extent are they diametrically opposed to interpretations? Both processes could have been ongoing in the same places as well. In some places one set of processes could have been predominant while the other set had more effect elsewhere. The eventual outcome, nonetheless, seems more uniform than

heterogeneous in the sense that de-urbanization and increased abandonment of dry farming practices were widespread in Mesopotamia after the 4.2kya event.

16 Periodically flooding land allows mineral deposits to accumulate which eventually interferes with agrarian growth.

17 See Bell (1971, 1975); Butzer (1976); Bard (1993); Hassan (1997, 2007); Seidlmayer (2000); Moeller (2005); Wilkinson (2010); Van De Mieroop (2011); Muller-Wollerman (2014); Ilin-Tomich (2016); Adams (2017;, and Dee (2017).

18 Egyptian rulers were called kings until the New Kingdom.

19 This statement must be made tentatively since the exact periodicity of Egyptian dynasties is controversial and frequently subject to some revision as better or different information becomes available.

20 See, however, Van Seters (2010); Sayce (2014); and Charles River Editors (2016) for a variety of related interpretations.

21 On the demise of Indus, see, among others, Possehl (1997a, 1997b); Staubwasser et al (2003); Madella and Fuller (2006); Lawler (2008); Berkelhammer et al (2012); Glosan et al (2012); Kathayat et al (2017); and Petrie et al (2017).

22 The strength of monsoons are governed by land sea temperature contrasts and wind patterns, both of which are influenced by temperature. Hence, cold periods are linked to weak monsoons and warm periods with strong monsoons (see Kathayat et al, 2017).

23 Kathayat et al (2017) note that there seems to be a general pattern of periods of stronger monsoons facilitating the construction of new regimes, as in the case of the Indo-Aryan migrations and the beginning of the Mauryan Empire. Weak monsoons are correlated with the end of the migrations and the collapse of the Mauryan Empire.

24 Note that both of these drought periods correspond to RCC timing.

25 Giosan et al (2012) suggest that the Indus civilization had the option of moving east unlike the cases of Mesopotamia and Egypt which were surrounded by arid lands. But as we have seen that is not quite the case in Mesopotamia.

26 The end of the Bronze Age literature is extensive. For a variety of more recent interpretations, see, for instance, Liverani (1987); Drews (1993); Finkelstein (1998, 2000, 2007); Zuckerman (2007); Monroe (2009); Rohling et al (2009); Kaniewski et al (2010, 2013); Langutt, Finkelstein, and Litt (2013); and Knapp and Manning (2016).

27 On the issue of Dark Ages, see Chew (2006).

28 Ugarit, Emar, Ras Bassit, Ras Ibn Hani, Gibala, Sumur, Deir 'Alla, Akko, Beth Shan, Megiddo, Hazor, Lachish, Ekron, Ashdod. Ashkelon, Babylon, Hattusa, Troy, Alaca Hoyok, Alishar, Masat Hoyuk, Mersin, Tarsus, Karaoglan, Mycenae, Tiryns, Katsingri, Korakou, Iria, Lakonia at the Menelaion, Pylos, Teikhos Dymaion, Boeotia, Phokis, Orchomenos, Gla, Krisa, Argolid, Corinthia, Berbati, Prosymna, Zygouries, Gonia, Tsoungiza, Lakonia, Ayios Stephanos, Messenia, Nichoria, Attica, Brauron, Eutrisis, Lefkandi, Kynos, Kition, Enkomi, Maa-Paeleokastro, Kalavasos-Ayios Dhimitrios, Sinda, and Maroni.

4

DROUGHT AND POLITICAL-ECONOMIC TRANSFORMATIONS IN THE ANCIENT NEAR EAST[1]

Despite widespread concerns about contemporary global warming, Old World archaeologists remain by and large reluctant to embrace climate considerations. A reluctance to welcome a return to anything resembling climate determinism is quite understandable. Being cautious about weak and missing climate data is also easy to understand. Nonetheless, the current ambivalence toward climate as an explanatory variable may be sacrificing an important key to ways in which ancient systems behaved, both similarly and differently. For instance, the first two areas to develop in the Old World, Mesopotamia and Egypt, differed from one another tremendously. Mesopotamia was often politically polycentric, early to urbanize, and taken over by successive waves of people who had moved into Mesopotamia from adjacent hinterlands. Egypt became politically unicentric fairly early on, was slow to urbanize, and was comparatively more successful at resisting outsider takeover bids through the 2nd millennium BCE. Despite some early Mesopotamian influences on Egypt, their cultures, religions, political ideologies, and languages were quite different. While unlike in many ways, the two earliest systems did share something that might be termed "political-economic rhythm." The parallels in the timing of the changes of successive regimes and periods of greater and lesser turmoil in the two ancient Near Eastern centers are quite remarkable. Cultural and ideological differences cannot account for these similarities. To the extent that climate change can contribute to explaining these similarities, climate would appear to have been a particularly significant parameter in the functioning of the ancient southwest Asian or Near Eastern world.

It is not difficult to find possible linkages between specific events and changes in the climate. But, even if it is safe to say that environmental change can be linked without much doubt to a few well-known examples, the question remains whether we can relate climatic deterioration systematically (as opposed to episodically as in the previous chapter) to processes such as regime transitions, center-hinterland conflict, and trade collapses? Moreover, is it possible to demonstrate even further that similar climatic deterioration experiences in Mesopotamia and Egypt led to similar and acute societal problems at roughly the same time? If both questions can be answered positively and successfully, it should be obvious that invoking a climatic factor need not lead to environmental determinism. Even if the two systems faced similar environmental problems at roughly the same time, their general responses were not identical. Otherwise, their differences in organizational strategies and belief systems would have not been so pronounced. Climate changes can create opportunities and challenges, but they need not dictate the outcome.

Hypotheses (and indicators) on climate change, river levels, regime transitions, trade collapse and reorientations, and center-hinterland conflict are derived from a simple model of general processes and then applied to Mesopotamia and Egypt for the 4000–1000 BCE period. A strong case can be made for linking water scarcity problems systematically to conflict, the fall of governmental regimes, and the collapse of trading regimes. Various Mesopotamian and Egyptian innovations may also be traceable to environmental change. Thus, the effect of climate change was not merely that of an occasional catastrophe or gradual drying. Climate effects persisted throughout the duration of the ancient world. They helped foster the emergence of an ancient world in the first place and then played a key role in its demise. The influence of climate change between the origins and the end of the ancient world also appears to be highly significant and persistent.

The argument

The argument to be developed here is certainly not that all problems in a system can be blamed on the climate. Nor is the argument really reducible to a straightforward climate-problems type of explanation. In the ancient world, climate was a critical contextual factor that arguably influenced a large number of other important processes, such as agricultural prosperity, state making, governmental legitimacy, population size, urban labor force, religion, trade, warfare, and so forth. Climate was particularly critical to the dynamics of the ancient world carrying capacities. The easiest generalization to make is that turmoil of various sorts was more likely to develop when the carrying capacity was threatened or exceeded. Climate, therefore, can be linked fairly easily and fundamentally in the abstract to intra-elite and center-hinterland conflict in systems in which economic survival is vulnerable to environmental deterioration. However, fringed by Saharan deserts in the west, Arabian deserts and the Indian Ocean in the south, the Mediterranean and Syrian deserts in the north, and mountains in the west, the ancient Near East region appears especially vulnerable to periods of extensive drought. In that respect, the fragility of the system to environmental deterioration was quite pronounced but it was not unique. Comparable situations can be found in, but are not limited to, the American southwest (LeBlanc, 1999) or even Polynesian islands (Kirch, 1984).

Climate emphases are hardly new. Three items are needed to add something new to the debate. First, we need a model, which locates climate change within the nexus of a network of interactive societal processes. Some effects of climatic deterioration are direct while others are not. If we are not promoting a simple, bivariate relationship between climate and conflict, how might we expect its manifestations to be best revealed and how extensive (or superficial) are these manifestations thought to be? Second, we need data on the processes and relationships that the model suggests are most worth examining empirically. Finally, we need explicit tests of hypotheses derived from the model (Watson et al., 1984). Granted, hypotheses about ancient system processes are not easily tested. The data are incredibly recalcitrant when they exist in the first place. But the effort should be made. Otherwise, it will be difficult to evaluate the relative accuracy of the model's hypotheses in terms other than face validity or logical rigor. Given the academic unpopularity of climatic explanations, attempts at empirical substantiation are especially indispensable.

In this context, Joseph Tainter's (1988) book has become something of a standard reference point in the studies of ancient rise and fall and collapse phenomena. The reasons for this centrality have to do with the author's logical dissection and critique of earlier studies, his analysis of, and references to, a number of different cases, and a general explanation put forward to encompass a long time frame and wide geographical scope. Tainter is especially critical of arguments that rely on resource depletion – whether gradual or abrupt, a category which includes climate change, catastrophes – viewed as a special case of dramatically abrupt resource depletion and intruders. He asks, how can economic weaknesses account for collapse when societies and governments are designed to deal with problems in fluctuating productivity? If the economic problems are so overwhelming, where is the evidence? If less than overwhelming, what accounts for the failure to develop appropriate counter-measures? Moreover, how is it possible that resource stress can sometimes lead to collapse while at other times it can lead to increased innovation?

If catastrophe explanations are a special case of resource depletion arguments, barbarian intruders are a special case of catastrophes. Intruder arguments bother Tainter in part because they are said to be a random accident used to explain a recurrent collapse phenomenon. Then, too, there frequently is no supplementary explanation provided as to how a less complex group is capable of defeating a more complex society. A third issue is why the intruders should seek to destroy the wealthier societies that presumably generate the incentive to intrude in the first place? These sorts of problems prompt Tainter to dub such explanations as "poltergeist models" that invoke inexplicable, mysterious troublemakers whose presence is difficult to show as a *deus ex machina* of collapse.

Tainter's own argument focuses on marginal returns. Societies solve problems but require energy to maintain their solutions. As societies become more complex, the associated costs rise and the marginal returns from continued investments in complexity may begin to decline. Thus, collapse may be a rational

response to an evaluation of prospective costs and benefits from further investments in complexification. Putting aside the awkwardness of imagining human
actors rationally choosing collapse over order as a least costly, relatively beneficial
decision, one has to wonder if resource depletion, climate change, catastrophe,
and intruders are not likely to be critical to the acceleration of marginal returns
and the development of collapse – if not by choice, then by default? This would
seem to be all the more the case if, in fact, the resource depletion/intruder problems are overwhelming. For that matter, whether such problems are random
accidents or recurring themselves depends on how one looks at what is to be
explained. If one tries to explain the demise of Egypt's Old, Middle, and New
Kingdoms as three separate cases, the factors involved are more likely to seem
idiosyncratic than if one attempts to account for the intermittent, but serial,
departures from centralized, dynastic order in Egypt.

The argument here, therefore, is that an emphasis on resource depletion, climate changes, catastrophes, and barbarian intruders can lead to unsatisfactory
explanations of societal and governmental collapse. Ignoring them as accidental,
misleading, or only part of the general problem, however, runs the risk of missing some significant elements of what transpired. We need to be able to answer
Tainter's questions. We may need to fit in his emphasis on marginal returns. But
we need not necessarily throw out resource depletion, climate change, catastrophe, or intruders in historical examinations of collapse – at least not if they help
make sense of widespread periods of crisis. To do so is to throw out the baby
with the bathwater.[2]

The model

The episodes centered on the six RCCs reviewed in Chapter 3 suggest that
diminished food supply should be the central factor in explaining rise and fall
phenomena. The inability to feed a population, either urban or rural, under-
mines the whole scheme of societal interactions. People need food for energy and
to survive. Cold and dry climates reduce the amount of water that is available for
growing crops and maintaining herds. Trade depends on continued supply and
demand, which assumes functioning societies and, to some extent, concentrated
urban dwellers. Governments need continued tax revenues from farmers, traders, and city dwellers to maintain their own type of energy fuel. Diminished
food crops, famine, and de-urbanization pull the material rug out from under
these relationships. As a consequence, population levels decline, cities de-popu-
late, and political regimes collapse.

However, food supply is difficult to measure serially in ancient contexts. We
know when things go badly wrong but we do not have much in the way of flows
of grain and rice to be able to assess fluctuations in how much food might be
available in any given year. Yet, food supply is conditioned on water availability,
either in terms of direct precipitation or river flow suitable for irrigation. River
flow is something that we can estimate. A prime mover in the model (depicted

in Figure 4.1), therefore, is water availability, which is predicated, in turn, on climate change and precipitation levels.[3] These two variables were especially critical for ancient systems, it is argued, because they were highly vulnerable agrarian economies and therefore quite sensitive to changes in precipitation and temperature. A second reason is that these prominent early societies emerged in river valleys and, not coincidentally, were highly dependent on the predictable flow of water in the rivers around which their agrarian activities were organized.[4] Climatic deterioration impacted the predictability and reliability of the amount and even the location of water in these central rivers. A third reason is that the earliest Old World civilizations were first enabled by more favorable climate conditions that preceded many of the markers that we associate with the genesis of societal complexity (urbanization, writing, organized religion, governmental coordination) and then strongly influenced by climatic deterioration. The presumption is that the packages of societal complexities that did emerge might have developed quite differently in the absence of climatic deterioration. Similarly, in the absence of benign climate, the complexities might have been less likely to have emerged in the first place.

The abundance or scarcity of water in the central rivers (e.g., the Tigris–Euphrates and the Nile) influenced how much land could be cultivated. Less scarcity meant more land could be cultivated. More scarcity led to a contraction in the amount of land under cultivation, crop failures, famine, and increased interest in irrigation practices. As river levels dropped, several things happened. Human settlements that had been, and could be, widespread in more moist times were abandoned as villages that were now too far from a water supply became untenable. Some of their populations were dislocated toward areas that still had access to water. Population density increased, as a consequence. In southern Mesopotamia, the surviving population centers grew into cities that attracted refugees from more arid northern Mesopotamia. In southern Egypt, population concentration near the Nile increased. The size of the labor force, swollen

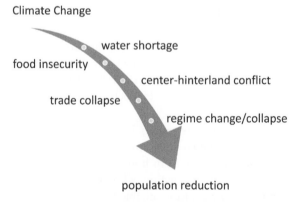

FIGURE 4.1 The model: climate change and its correlates

with economic refugees from dryer areas, expanded. Economic specialization became both more possible and more necessary. So, too, did agricultural intensification, the construction and maintenance of irrigation infrastructure, and the building of temples and monumental architecture (pyramids, ziggurats). More complex divisions of labor, greater density, and public goods increased the need for coordination, thereby expanding the role of bureaucratic and governmental managers.

Not everyone chose to move to more densely populated centers. A different type of increased economic specialization occurred as some people chose to focus on raising herds of animals in less populated zones. Finding food for these animals meant seasonal migrations. As a result, the distinctions between sedentary and nomadic populations became greater.

Urbanization, nomadism, religion, government, writing, and trade intensification were outcomes of these processes. The urbanization proceeded, particularly in Mesopotamia, from the concentration of population around a few nuclei where once the population had been more widely dispersed. Millennia-old trade and trading networks expanded to meet the increased demands of larger, denser, and richer communities that invariably developed in areas critically lacking important basic commodities such as wood and copper. Religion became more critical as the environment became less predictable. Some way to communicate with, and to appease, the gods in hopes of obtaining better weather became an increasingly organized activity. A variety of other activities needed coordination. Agricultural labor, irrigation management, surplus food storage, dispute adjudication, and defense were some of the most prominent. Bureaucrats were needed to count the number of laborers and slaves, the number of bowls necessary to feed them, taxes, and commodities exchanged between communities. Writing developed from these mundane accounting practices. Governments developed from coordination and supervision practices designed to enhance economic productivity, suppress domestic disorder, expand the community's access to resources, and protect the community from external attack.

Governments then, as now, were evaluated on their ability to perform their coordination/protection responsibilities. They were also blamed for climate-induced economic deterioration even if there was not much that could be done about it. That too has not changed much. In periods of severe economic deterioration, governmental legitimacy could be expected to suffer. Hungry populations were more likely to protest and riot. Unpaid armies were more likely to rebel. Provincial governors were more likely to act more autonomously from a weakening central government.[5] Political organization, in general, cycled back and forth from relatively centralized to decentralized processes. One implication of decentralization was a weakened resistance to incursions from tribes inhabiting the hinterland who were attracted to the river valleys (and the cities located in them) by the survival problems associated with climatic deterioration. To add to their already expanding problems, the probability of hinterland attacks on cities increased as the central cities became more vulnerable to attack.

Several different types of warfare developed in the ancient systems. On the one hand, cities needed protection from bandits, nomads, and unwanted refugees. Walled cities developed in response. Ambitious and vigorous political leaders would also go on the offensive to demonstrate their vigor and right to command by attacking threats outside the city walls. The threats were often real in the sense that the attacks from the center were responding to tribal pressure on outlying farms or extended trade routes. Many of these offensive raids could be expected in the early years of a new reigning monarch, as both internal and external demonstrations of fitness to rule. The soldiers used for these purposes were recruited from the urban labor surplus and from the hinterland populations brought under central control.

As cities became states, inter- and intra-state conflict also became more probable. Intra-state conflict was inherent to the cycle of political centralization-decentralization mentioned above. The first new states, as did those that were to follow, also needed cultivable land, access to raw materials not available at home, and secure trading networks. Expansion of the state to acquire these goals increased the probability of conflict with other states that were also expanding in the same direction or with states whose territory or spheres of influence were being encroached upon. A deteriorating environment would make these interstate frictions all the more threatening. Contrary to contemporary myths, state sovereignty issues did not first emerge in the 1648 (CE) Treaty of Westphalia. They emerged thousands of years before in the Sumerian city-state system, the conflicts between Upper and Lower Egypt and their neighbors and, subsequently, in the conflicts of successive empires in the Near East.

Climate

One survey of the ancient Near East (Van De Mieroop, 2004: 9) is representative of what prevailing professional attitudes on the role of climate in older southwest Asian processes has been until fairly recently:

> Long periods of drought could easily have occurred in the time span [3000-323 B.C.E.] we study here, however. While we can assume that over the last 10,000 years the climate in the Near East has not substantially changed, it is certain that even marginal variations could have had serious consequences for the inhabitants. The question arises as to whether the so-called Dark Ages resulted from a drying of the climate, which made rainfed agriculture impossible in zones usually relying on it, and which lowered the rivers to such an extent that irrigated areas were substantially reduced. Or should we focus on human factors in trying to explain such periods? So far, insufficient data on the ancient climate are available to serve as a historical explanation for the drastic political and economic changes we observe.

This extended passage is revealing in a variety of ways. Dark Ages are problematic for historians because few written records survive. Hence these eras are "dark." Yet concentrations of power and patterns of interaction are different before and after these historical blackouts. In the intervening periods, we are left with vague impressions of de-urbanization, economic decline, and increased turmoil. The conservative approach, which is certainly understandable given the lack of documentation, is to treat these eras very gingerly – if at all. Yet these Dark Ages recur and it is difficult to escape the feeling that the Dark Ages were instrumental (or might have been instrumental), if not crucial, to bringing about long-term changes in the areas in which they take place. There are also fairly strong clues that these Dark Ages are often preceded by drought, famine, and migrations induced by climate change.[6] But while we may have evidence of drought, famine, and migrations, the evidence on climate change is often weak.

Central to this argument is the idea that water was more readily available immediately prior to the 4th millennium. The ready availability of water initially encouraged the expansion of the human population in the Near East, particularly in the river valleys.[7] When precipitation and river levels declined, subsequent water scarcities forced or enabled the development of new adaptive strategies leading to the acceleration in the processes of nomadic-sedentary divergence, urbanization, writing, government, religion, and state making. Climate change did not determine what transpired. There were a host of possible responses, all of which were pursued.[8] The most successful strategies, however, involved the relatively novel development of cities and states. Climate change may not even have been necessary for these developments to occur. Some early concentrations of population such as Catalhuyuk and Jericho emerged prehistorically and prior to the historical period designated as the ancient world. Yet even these early urban settlements experienced their own significant climate change problems. In general, from our perspective, relatively favorable climates at the very least accelerated the initial development of multiple, interactive processes to new thresholds of activity. Where once populations had been more dispersed, increasingly large and concentrated population centers began to emerge and to persist, despite and, in some respects, because of a less favorable climate, with a number of implications for other societal processes. Subsequent climate changes continued to shape behavior in the evolving societies, with particular emphasis on conflict among elites, between elites and masses, and between center and hinterland.

Systematic records were not kept on weather and river levels in ancient times. Partly as a consequence, the traditional opinion emerged that "over the last 10,000 years the climate in the Near East has not substantially changed" (see above). If that was the case, why even bother exploring climate as an explanatory factor? Thus, the study of climate is plagued at the outset by a combination of bad data and limited professional incentives. A third problem is the ambivalence hinted at above as to whether to stress climate effects or human reactions to climate change. If too much emphasis is placed on the climate effects, the explanation easily begins to sound deterministic. The weather changed. People died from insufficient food and water supplies or moved elsewhere. But we know that not everyone died and not everyone moved. Obviously, different strategies were employed to cope with environmental changes, if that is what occurred. We also know or suspect that other things besides climate change (e.g., corruption, internal and external violence, governmental entropy, local interventions in trade, declining productivity) were ongoing. Are all of these processes also effects of climate? Should they share explanatory credit? Or is it possible that climate changes are spurious factors and the long-term changes we observe are due to "on-the-ground" processes related exclusively to complex human interactions?

At the same time, though, the recurrence and probable significance of Dark Ages are difficult to evade. We also have increasing reason to challenge the generalization that southwest Asian climate has not changed in the past 10,000 years. Moreover, in studying the recurring phenomena of Dark Ages in the long term, it is impossible to avoid the likelihood of some substantial connection to the onset of climate problems.

It should also be recognized that there are dissenting views on climate among scholars of the ancient Near East. Bell (1971, 1975) had earlier drawn attention to the importance of Nile fluctuations and their implications.[9] She also suggested that these periodic interactions with deteriorating environments should be anticipated in other parts of the ancient world, not merely Egypt. Weiss and Courty (1993), Weiss et al. (1993), and Weiss (1997) have stressed environmental change as critical to the turmoil of the 2200–2000 era. Elsewhere, Weiss (2000: 77) has also suggested that climate changes might have been responsible for problems at the end of the 5th (the Ubaid-Uruk transition) and the 4th millennium Uruk urbanization and its trading network retrenchment.[10] Butzer (1995: 138) goes one step further by drawing attention to what he calls "first-order anomalies" of general atmospheric circulation. Three pan-Near Eastern "dry shifts" are identified as having occurred around 3000, 2200, and 1300 BCE. He also notes that these dry shifts were not manifested exactly the same in Egypt and Mesopotamia. Fairly abrupt shifts in Nile flood levels were registered around 3000, 2250, 1850 (high floods as opposed to low floods), and 1200.[11] Protracted periods of low Mesopotamian rainfall, as registered in Lakes Van and Zeribar sediment layers and pollen traces, occurred around 3200–2900, 2350–2000, and 1300–1200 – all of which overlap with RCC periods.[12]

While Butzer (1995) did not pursue the political-economic implications of the first-order anomalies,[13] Matthews (2003: 100–01) criticizes an analysis of cycles of political consolidation and fragmentation (Marcus, 1998) for ignoring the "central role of climate and environment" in bringing about the breakdown of Mesopotamian states toward the end of the 4th, 3rd, and 2nd millennia. Unfortunately, Matthews stops short of pursuing this issue beyond advancing the criticism.[14] Bell, Weiss et al., Butzer, and Matthews are most suggestive and encouraging but all stop short of fully developing systematic linkages to the political-economic consequences of environmental deterioration. Issar and Zohar (2007) and Brooke (2014) have been exemplary in emphasizing environmental connections to the ancient political economy but they too are not particularly interested in establishing systematic linkages.

In a generally arid area, the first weather dimension that comes to mind is rainfall. One problem with generalizing about precipitation levels in the Middle East is that rainfall for the entire region may be averaged for any given time period but different areas will still be affected differently. If one area averages 100 millimeters a year and another 500 millimeters and they both experience 100 millimeters less in a dry year, the first area will become unlivable while the second one may experience some mild drought. There are varying precipitation zones in southwest Asia (for instance, precipitation currently tends to increase as one moves away from Arabia in a northeasterly direction – see Nissen, 1988: 59).

Temperature fluctuation is a second obvious dimension of weather. On this score, there does not appear to be a great deal of year-to-year, let alone day-to-day, information. We have some limited information on relatively short periods of time (e.g., less than a millennium) but only a few series encompass our full "ancient" period.[15] Accordingly, we will fall back on our cold-dry rapid climate change episodes that are enumerated in Table 2.1 to generate our six hypotheses. For the ancient Near East, RCCs 4 (4200–3000 BCE), 5 (2200–1800 BCE), and 6 (1250–800 BCE) are the most germane episodes of deteriorating climate change. Our first hypothesis, therefore, is that

H1: Significant shortfalls in water availability in the ancient world were more likely in periods of cool-dry climate deterioration.

One might also wish for more information on climate such as wind forces and directions or even dew levels, which can make some difference in dry-zone agriculture, but it is doubtful that too much more detailed information is likely to be forthcoming. One exception, and it is one that will prove to be a rather major one, is river level data. Both Mesopotamia and Egypt were constructed around river systems and much of their agrarian success and transportation opportunities were governed by the volume of water flowing through the Tigris–Euphrates and Nile rivers, respectively. As it happens, we have some reconstructed information on river volume (presented in Chapter 5, Figures 5.1 and 5.2) that is easier to interpret than general episodes of increasing and decreasing temperature and precipitation.[16]

There was a sharp descent in river levels beginning in the mid-4th millennium to about 3200 BCE. Some improvement in water availability was registered after that before another marked decline began in the mid-3rd millennium that continued to the end of the millennium. In the 2nd millennium, river levels ratcheted upward in 1300 BCE before declining once again in 1200–1100 BCE. In Egypt, there was a short descent in river level at the beginning of the 4th millennium in 3700–3600 BCE. Levels improved in 3300 BCE only to descend again to a low in 2900 BCE. An upward ratcheting that is similar to the Mesopotamian pattern is exhibited in about 1500 BCE before another decline sets in around 1200–1100 BCE.

H2: Significant center-hinterland conflict in the ancient world was associated systematically with periods of deteriorating climate and diminished water supply.

"Barbarian" attacks on cities have been held responsible for the downfall of whole civilizations. Yet barbarians were simply what people who lived in cities called those who did not live in cities. Center-hinterland conflict in the ancient world system was neither rare nor all-pervasive. Vigorous city-based states and empires sent periodic expeditions into the hills or deserts to pacify tribes causing problems, and especially problems related to interference with trade routes or attacks on outlying farming communities. Tribal attacks were usually emboldened by declining sedentary capabilities to punish hinterland incursions. Urban centers that did not or could not fight back must have been especially appealing. Sometimes, the attacks were also encouraged by pressures on the tribes themselves brought about by the in-migration of new and more powerful tribes into their traditional sphere of influence. In any event, it is often difficult to tell just how much causal agency to bestow on hinterland attacks (Bronson, 1988). Did they genuinely contribute to the organizational decline of the centers, or were they simply symptoms of general turmoil set in motion by other forces? Were they cause or effect – or both, simultaneously? Some authors (Hallo and Simpson, 1971: 72) go even further and suggest that the alternation between imperial centralization in the Tigris–Euphrates and Nile river areas and Bedouin-nomadic-led decentralization was almost paradigmatic for the ancient world system. If the ancient world system did alternate systematically, major hinterland attacks on the center would have played fairly significant roles in this process even if the attackers were merely responding to changes in the environment.

That is, of course, one of the questions. Were the hinterland attacks on the center prime movers or were they propelled by other factors? There is no question that some alternating tendency toward greater and lesser political order prevailed in both Mesopotamia and Egypt. In Mesopotamia, early Uruk predominance was followed by a period of inter-city-state competition and then intermittent attempts at imperial centralization (Akkad, Ur III, and Old Babylon). The circumstances are even more clear in the Egyptian case with Old, Middle, and New Kingdoms interrupted by intermediate periods of disorder. Both ends of the ancient world system shared similar security problems with desert and mountain tribes along the eastern and western frontiers. The Egyptians also had problems in the south and occasionally in the north. When the Egyptians were undergoing their intermediate phases and fighting with Bedouins, Nubians, and Libyans, the Mesopotamians were attempting to cope with Gutians, Amorites, Kassites, and Aramaeans, among others.

H3: Regime transitions in the ancient world were associated systematically with periods of deteriorating climate and diminished water supply.

Was the first major Egyptian regime transition, as the prehistorical era was being transcended, associated with climate deterioration? Did the south attack the north because it needed cultivable land that was becoming scarcer in the south, with the width of the Nile contracting due to receding water levels, as Hoffman (1979) has suggested? Did southern control of the north expand because southern elites wanted greater direct control over northeastern trade routes that were threatened in some way (Wilkinson, 1996)? We may never know the full story but we can at least check the political timing of regime transitions against the timing of the environmental change.

The major regime transitions that took place after about 3100 BCE are not particularly controversial, even though the precise dating will probably always remain debatable. The transitions in Mesopotamia that are of most interest are summarized in Table 4.1.

These regime transitions are thought to be significant because they represented discontinuities in politics, economics, and culture. One type of way of life gave way to a discernibly different type. Pottery styles may change, burial practices may be revised, and/or new technology may be introduced. The new regimes can also be reduced in part to new people seizing elite status. Sometimes they involved new people moving into Mesopotamia or Egypt but, most of the time, they were about a circulation of elites within the two main nodes of the ancient Near East. Established elites do not usually surrender political power peacefully. They also have to be rendered more vulnerable than is the case normally for new elites to be able to usurp successfully. Climatic deterioration offers one way in which these transitions might have come about. Environmental problems could have weakened the ability of established elites to resist challengers. Their material resources became diminished and their perceived legitimacy was also likely to come under attack. Environmental problems, therefore, can help create windows of opportunity for successful challenges. The existence of a window of opportunity does not mean that anyone will choose to jump through it. But it could very well increase the probability of someone doing so.

TABLE 4.1 Mesopotamian and Egyptian Regime Timing

Mesopotamia	BCE	Egypt	BCE
Ubaid	5000–4000	Merimda–Baderi	5000–4000
Uruk	4000–3100	Naqada I–II	4000–3200
Jemdet Nasr	3100–2900	Naqada III	3200–3050
Early Dynastic	2900–2300	Early Dynastic	3050–2575
Akkad	2350–2150	Old Kingdom	2575–2150
Third Dynasty Ur III	2100–2000	First Intermediate	2150–1980
Old Babylonian	2000–1600	Middle Kingdom	1980–1630
		Second Intermediate	1630–1520
Kassite	1590–1150	New Kingdom	1540–1070
"Various Dynasties in Babylon	1150–730	Third Intermediate	1070–715

Sources: Based primarily on Baines and Yoffee (1998: 202), with some modifications based on Wilkinson (1996: 15).

H4: Diminished water availability and climate deterioration discouraged population growth and urbanization in the ancient world.

If deteriorating circumstances led people to move away from sedentary settlements for survival purposes, city sizes in the ancient world should contract. Some cities might cease to exist altogether while a few cities might have grown larger by taking in climate refugees. But if the population is faced with drought and famine, not everyone will manage to migrate successfully. Increased mortality in the face of water scarcity and climate deterioration should lead to reductions in the size of the total population as well.

H5: Periods of ancient world trade collapse were associated systematically with periods of deteriorating climate and diminished water supply.

Ancient trade depended on urban development and thriving demand. Tendencies toward de-urbanization and reduced demand for products in urban areas make it increasingly difficult to sustain external acquisition networks. Suppliers respond to the reduced demand and increasing inability to pay for trade goods by making fewer goods available. Suppliers that are highly dependent on markets that deteriorate may not be able to find alternative consumers and, essentially, go out of business. To the extent that periods of economic depression are associated with declines in governmental order maintenance, transaction costs will also increase as trade routes are increasingly intercepted by groups who find less organized opposition to their raiding.

Long periods of economic deterioration and depression can lead to a complete collapse of commercial interactions across long and intermediate distances.

H6: The conjunction of significant political-economic crises in the ancient world was associated systematically with periods of deteriorating climate and diminished water supply.

If some or all of the first five hypotheses are supported by the empirical data, it stands to reason that political-economic crises involving economic and political collapse and turmoil are likely to overlap. To the extent that these crises do overlap, societies and governments are likely to be overwhelmed by a deteriorating environment over which they have decreasing control. If this hypothesis is borne out, analytical choices between emphasizing climate, marauding barbarians from the hills or desert, trade collapse, or political turmoil may often constitute false choices in ancient world analyses. Variations on the compounded onslaught of related stressors should be expected to predominate in the ancient world. The problem was not that individual crises could not be surmounted. Multiple shocks coming more or less at the same time, however, tended to be overwhelming in response to protracted deterioration and other problems – with or without the probability of land degradation.

Exploring the applicability of these hypotheses requires systematic information on relevant fluctuations in Mesopotamia and Egypt. Chapter 5 describes these data and uses them to test the six hypotheses as best as we can, given the obvious limitations on capturing ancient world dynamics systematically.

Notes

1 Chapters 4 and 5 are based on Thompson (2004); Thompson (2006a); Thompson (2006b); and Thompson (2006c).

2 In this and the next chapter, we are focusing on climate change's relationship to processes such as regime transformation or center-periphery conflict. This emphasis does not imply that we are ignoring the kinds of problems that emerge in the resilience theory's conservation phase – namely, diminishing resources and efforts to monopolize them. We are merely simplifying the modeling effort for Chapters 4 and 5.

3 The sketched model is based on arguments found primarily in Bell (1971, 1975) and Hole (1994). Bell explicitly states that her argument is also applicable to Mesopotamia even though she confines her own analyses exclusively to Egypt. Hole's argument is explicitly restricted to the Ubaid–Uruk transition in Mesopotamia prior to 4000 BCE.

4 Butzer (1976: 108) has described Egyptian economic history as "primarily one of continuous ecological readjustment to a variable water supply."

5 Vercoutter (1967: 280) argues that since Egypt was 35 times as long as it was wide, natural conditions made centralized authority desirable while, in practice, geography made intermittent dismemberment probable.

6 On the general relationship between Dark Ages and environmental problems, see Chew (1999, 2001).

7 The Egyptian population, according to Butzer (1976: 86) quadrupled in the 1500 years preceding the establishment of the Old Kingdom.

8 Hole (1994: 137) suggests seven possible "fight or flee" strategies in response to persistent crop failures. Variations on fighting include focusing on alternative subsistence tactics (such as fishing), intensification of cultivation combined with surplus storage, expanded religious activity, and external war for desired resources. Variations on fleeing include expanded nomadism, marginal subsistence in isolated villages, and migration to more desirable locales. His main point is that all seven were pursued in Mesopotamia between 5500 and 3500 BCE.

9 For Egypt, see, as well, Hassan (1994, 1997, 2011).

10 Interestingly, Weiss (2000: 84) stresses what he calls the incursions of southerners (Sumerian and Akkadian imperialism) into north Mesopotamia as a response to ecological problems. This emphasis reverses the focus in this chapter on hinterland incursions against the center, but is consonant with the examination of governmental regime transitions. At the same time, Weiss (2000: 88–89) acknowledges increased pastoralism and nomadic incursions against sedentary areas in periods of aridification.

11 Butzer (1995: 136) also observes that Nile flood levels did not improve until 800–1000 CE, ending two millennia of weak floods.

12 At the same time, these "anomalies" suggest causation at a level greater than the local or even regional weather systems. Whereas Egypt and Indus shared an African monsoonal common denominator (Weiss, 2000), Mesopotamian river levels are predicated on Anatolian precipitation. Therefore, similar climate problems in Mesopotamia and Egypt at roughly the same time, especially in conjunction with similar problems outside southwest Asia, suggest world-level climate change was at work.

13 Butzer (1997) focuses on Egyptian militarism in Palestine toward the end of the third millennium and backtracks from attributing much importance to environmental problems of that time.

14 Butzer (1995) is cited as the source for the timing of "episodes of climatic adversity and aridification." However Matthews curiously places the second millennium episode in 1400, as opposed to Butzer's 1300–1200 identification.

15 Sources on ancient southwest Asian climate include Diester-Haass (1973); Butzer (1976, 1995); Oates and Oates (1977); Van Ziest and Bottema (1977); Erinc (1978); Neumann and Sigrist (1978;, Schoell (1978); Van Ziest and Woldring (1978); Hoffmann (1979); Kay and Johnson (1981); Nissen (1988); Fairservis (1992); Bottema (1997); Fairbridge et al (1997); Lemcke and Sturm (1997); Potts (1997); Kerr (1998); and Weiss (2000).

16 The source for the river level data is Butzer (1995: 133) who refers to his series as "inferred" volume flows but, unfortunately, does not discuss his specific approach to inference. The Nile plot, however, resembles Butzer's (1976: 31) plot of East African lake levels which feed into the Nile, that suggests that the Tigris–Euphrates reconstruction is based on Anatolian lake data. Other analyses of reconstructed Tigris–Euphrates data are reported in Kay and Johnson (1981) and Bowden et al (1981). Obviously, the data are not as hard as we might all prefer, but it is unlikely that superior alternatives for river level estimations will soon be forthcoming. A related question is whether it might be possible to push the time frame back even further. Hole's (1994; see also Stein, 1999) comment that an Ubaid peak was attained in 5200 BCE, followed by a decline, the abandonment of some parts of Mesopotamia by 4800 BCE, and an increased interest in religion suggest that the Tigris–Euphrates water levels may have been influencing behavior even farther back in time. See Algaze (2000) for an argument that the attributes of the southern Mesopotamian ecological niche enabled that area to be the first to take the lead in developing "complex civilization." Butzer (1976: 23) notes as well that the 7000–4000 BCE rise in Mediterranean levels, due to melting glaciers and precipitation, transformed the northern third of the Egyptian Delta into swamp and lagoon.

5

COOLING, WATER SCARCITY, AND SOCIETAL CRISES IN ANCIENT MESOPOTAMIA AND EGYPT

Chapter 4 laid out a model relating the interactions among climate change, water supply, center–hinterland conflict, population/urbanization size, trade crises, and regime transitions in the ancient world. To assess the utility of the model, in this chapter, we assemble information on the scale and timing of their fluctuations. We begin our analysis with a discussion of these indicators of relevant fluctuations in Mesopotamia and Egypt. We then turn our attention in the second half of the chapter to test the six hypotheses outlined in the previous chapter.

Water scarcity

Eyeballing the reconstructed figures available for Tigris–Euphrates and Nile river flow presented in Figures 5.1 and 5.2, yield the following schedules.[1] In the case of the Tigris–Euphrates rivers, the pattern is one of precipitous declines in the level from 3500 to 2900 BCE, 2500 to 2000 BCE, and 1300 to 1100 BCE. In between these phases of declining water levels are periods of increasing water levels although they never return to the levels of 4000 BC until early in the first millennium.

The inferred Nile pattern (Butzer, 1995: 133) is not identical to the Tigris–Euphrates history. Nile levels cascade downwards from 4000 BCE to intermediate troughs in 3800, 2900, and very low troughs in 2000 and 1200–1100 BCE. The river level was falling between 4000 and 3800, 3300 and 2900, 2500 and 2000, and 1600 to 1200, although one might argue that the repercussions of each successive fall may have been more serious than the preceding one(s) because the levels were becoming successively lower.

Thus, we have the pattern summarized in Table 5.1 for water availability in Mesopotamia and Egypt. The patterns are not identical but they certainly overlap to a considerable extent. Butzer (1995: 137) commented presciently that

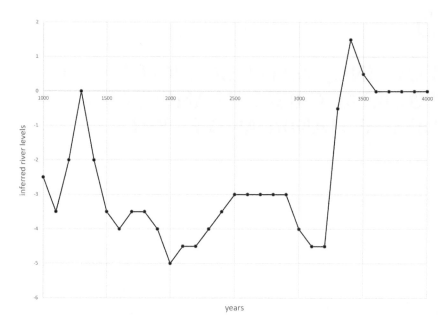

FIGURE 5.1   Tigris–Euphrates river levels

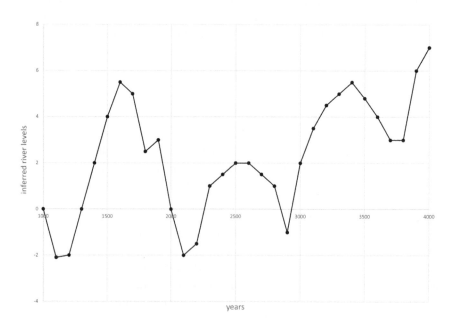

FIGURE 5.2   Nile river levels

TABLE 5.1 River Levels in Mesopotamia and Egypt

Declining in Mesopotamia	Declining in Egypt
	4000–3600
3500–3200	3300–2900
2500–2000	2400–2000
1300–1100	1500–1100

the parallels in the most dramatic reductions (around 3000, 2200, and 1300 BCE) across some 40 degrees of latitude implied "major anomalies of the general atmosphere, potentially of a global scale."

Center–hinterland conflict

Fluctuating amounts of conflict between center and hinterland persisted for at least the three millennia period (4000–1000) with which we are most concerned. What might explain an alternating pattern in center–hinterland relations? Given the less than benign ecology of the Near East, changes in climate offer an extremely attractive explanation. The model in Figure 4.1 (see the previous chapter) suggests that both push and pull factors might be at work. Climatic deterioration and the diminishment of water supply could cause problems for nomads and sedentary folk alike. Hinterland tribes needed access to water, grass, and other commodities that the cities controlled. Their perceived needs were apt to be more acute in years of sustained drought that were felt in the lowlands, the highlands, and the deserts. Urban resistance to hinterland attacks should also be hampered just as they were becoming more probable. Thus, even if hinterland groups were not being driven away from deserts and mountains by climate problems, the pull of weakened resistance to their raiding could suffice to escalate center–hinterland conflict levels.[2] An economic and governmental collapse could also greatly facilitate the large-scale movement of hinterland peoples into areas hitherto controlled by sedentary peoples – at least until the sedentary peoples manage to reorganize themselves and evict the interlopers.

Data on incursions/migrations are intended to capture some sense of the timing of conflict between centers and their hinterlands. In the ancient world system, there is little likelihood of capturing this rhythm with any great precision but it should be possible to develop a sense for its basic beat. By combing references in the historical literature to center–hinterland clashes and movements of peoples, the idea is to systematize as best as possible what we know at this time, subject to the inherent limitations of the subject matter.

Most typical of the literature are fairly vague references to some ruler fighting hill or desert tribes early or late in his reign.[3] Which hill tribes are not always noted. Exactly where they clashed or who initiated the encounter is frequently absent. The size of the fighting forces is almost never noted (or known). What they are fighting about is simply a question that only rarely surfaces. Consistent with these approaches to the subject matter is what might be called a professional bipolarity. Most historians are not too interested in the clashes between hinterland and centers. Granted, the information that is available is quite poor and thus there is a very good reason for professional caution. But, it also seems fair to say that most historians treat hinterland peoples much like their favorite subjects did – that is as intermittent nuisances of the general landscape – or they casually bestow great causal agency on hinterland depredations as significant factors in the downfall of various centers without really pondering too much about the implications.

An enumeration of ancient southwest Asian center–hinterland conflict (Thompson, 2002) lists little activity prior to the end of the 4th millennium. An important evolutionary development helps explain this slow-to-emerge phenomenon. All groups were initially mobile, moving from campsite to campsite, subject to natural catastrophes and the exhaustion of local food sources. Only gradually did cities or concentrations of sedentary people emerge. Equally gradual was the emerging division of labor between sedentary farmers and city dwellers and nonsedentary herders. Only after these societal dualities emerged was it possible for the center–hinterland conflict to emerge fully in the late 4th and 3rd millennia.

McNeill (2013: 34–40) argues that the potential for sedentary-pastoral conflict was higher in the Middle East than elsewhere. He notes that in Eurasia, Africa, and North America, grasslands tended to be distant from large cities. In the Middle East and North Africa (MENA), however, grasslands tended to be relatively close to large cities. Around the Eurasian steppe, nomads could retreat to places that land armies could not follow without large amounts of food and water. When nomads were less aggressive, very large empires could be constructed in places such as China and India. In the Middle East, though, the hostile interactions between farmers and nomads would come sooner, more frequently, and in a way that made large empires last for shorter durations. The interspersed pattern of grassland and desert (as opposed to clear and distinctive zones favoring one type of economic production over the other) limited the upward scale of imperial efforts prior to the advent of the Ottoman Empire. As a consequence, we should anticipate regular and significant interactions of a hostile nature between nomads and sedentary political systems in the MENA that parallel the history of the binary but relatively proximate split in agrarian styles.[4]

Scholars do not fully agree on the timing of the emergence of a division of labor between sedentary and nonsedentary economic activities. For instance, Sherratt (1996) argues for the late 2nd millennium BCE in terms of the emergence of classic nomads. Hole (1994) stresses that the bifurcation in economic strategies was roughly simultaneous with the first emergence of cites, which would place it in the 4th millennium BCE. But even if the division of labor emerged early, it is quite likely that some time was needed for the emergence of meaningful center–hinterland conflict. Sedentary people initially had the advantage in manufacturing weapons and constructing large armies. If nothing else, the development of a strong hinterland military mobility capability awaited the domestication of horses and camels (Butzer, 1997: 266), to say nothing of compound bows. One would also expect that the initial contacts would come in the form of center attempts to either protect vulnerable trade routes and/or acquire resources needed for urban life. Based on the better known East Asian case, raiding and trading should also have preceded any hinterland attempts to take over the center. In Mesopotamia and, to a lesser extent, Egypt, hinterland infiltration of the center over hundreds of years also preceded any takeover attempts. Moreover, the center forces initially should have had some capability advantages over hinterland tribes that gradually eroded as the highland and desert tribes learned how to cope with center forces, especially when those center forces grew weaker. It was only later (early to mid-2nd millennium) that hinterland forces emerged with superior weaponry in the form of chariots and compound bows.[5]

The incursions/migrations data, unfortunately, do not lend themselves readily to treatments as time series. It is better to regard them as approximations of fluctuations in center–hinterland conflict. The most major periods of conflict can be reduced by interpretation of the following short list. In the Mesopotamian case, it was first the movement of Trans-Caucasians and Amorites toward the end of the 4th millennium. Beginning around 2200 BCE, Gutians and Lullubi were causing increasing problems in the east. In the west, Subartu, Hurrians, and Amorites in general were in motion, attempting to move into Mesopotamia. From about 2000–1750, there was still activity but it seems much less than the amount that was associated with the 2200–2000 period. The pace picked up again in the early 1700s with Gutians, Kassites (in the east), and Sutae (from the west) on the attack. Few entries are recorded for Mesopotamia in the 1600s but that suggests more center disintegration than hinterland quiescence. In the late 1590s, Kassites took over Babylon, apparently by default, in the aftermath of a Hittite assault on that city.[6] Some Assyrian resurgence was demonstrated in the 1300s and early 1200s but toward the end of the period of primary interest, the 12th century BCE and later, Mesopotamia was characterized by a large-scale in-migration of Aramaeans from the western deserts.

The ancient Egyptian record works neatly in the sense that the major incursions were associated with the end of the Middle and New Kingdoms, most definitely in periods of crisis that encompassed deteriorating climates. Egypt initially enjoyed some insularity from external attacks ("Asiatics" from an Egyptian perspective came from the east and Libyans from the west). Prior to the mid-3rd millennium, there does not appear to be much evidence of a threat of serious attack. The Egyptians were sending expeditions into the Sinai for hinterland control purposes as early as the end of the 4th millennium, with the Egyptians initially projecting their coercive power into the hinterland, rather than the other way around. Infiltration was taking place or being attempted from both eastern and western deserts but little in the way of serious threats is noted in the record prior to the discernible increase of hinterland activity in the 2300–2000 BCE period.

There were three major bouts of hinterland attack – the three intermediate periods and the Sea Peoples attacks that began around 1200 BCE, in conjunction with Libyan attacks. In the first intermediate period (2150–1980 BCE), the incursions came from the west and the east. The threat in the second intermediate period (1630–1520 BCE) led to a Hyksos takeover of parts of Egypt from the west. The Sea Peoples came from the north and northeast, along with Libyan allied attacks in the late 1200s and early 1100s. The Libyan pressure continued into the third intermediate phase (1070 BCE on) and led ultimately to Libyan Pharaonic dynasties. Since the Sea Peoples attacks and the third intermediate period blend into one another, the third cluster translates into a 1200–1000 period for our purposes. Eventually, Egypt simply lost its autonomy completely to expanding states (e.g., Persians, Greeks, Romans). Table 5.2 converts these Mesopotamian and Egyptian center–hinterland conflict clusters into centuries of especially significant hostility.

Regime transitions

The historical sequence of ancient regimes in Mesopotamia and Egypt is characterized by a fair amount of consensus; however, there are still differences of opinion about the precise number of years to assign to each period. Yet the differences do not seem so great that any obvious validity threats are posed by adopting one chronology over another. More threatening perhaps is the tendency to revise the chronological schedules every so many years as new information comes to light. Table 5.3 lists the regime periodization scheme that will be employed in this investigation. It is based on fairly recent periodization principles, circa the mid-1990s, but it is still not quite the last or even the latest word on the subject. Joffe (2000) discusses some of the implications of the most recent revisions in the Mesopotamian schedule, which are not fully reflected in Table 5.3. Essentially, these revisions have the effect of extending the Uruk phase farther back in time and, as such, address an earlier period than the one that will receive most of the attention in this analysis. Prehistorical dating, no doubt, will continue to be debated precisely because it is about prehistorical periods of time and the temporal evidence is often based on indirect measurements. Dating regime transitions

TABLE 5.2 Years of Most Significant Hinterland Incursions against the Centers

Mesopotamia	Egypt
3200–3000	
2200–2000	2200–2000
1600–1500	1700–1500
1200–1000	1200–1100

TABLE 5.3 Chronologies of Regime Transitions for Egypt and Mesopotamia

Egypt	Years	Mesopotamia	Years
Merimda-Baderi	5000–4000	Ubaid	5000–4000
Naqada I–II	4000–3200	Uruk	4000–3100
Naqada III	3200–3050	Jemdet Nasr	3100–2900
Early Dynastic	3050–2575	Early Dynastic	2900–2300
Old Kingdom	2575–2150	Akkad	2350–2150
1st Intermediate	2150–1980	3rd Dynasty of Ur III	2100–2000
Middle Kingdom	1980–1630	Old Babylonian	2000–1600
2nd Intermediate	1630–1520	Kassite	1590–1150
New Kingdom	1540–1070		
3rd Intermediate	1070–715	"Various dynasties in Babylon"	1150–730

Source: Based primarily on Baines and Yoffee (1998: 202), with some Egyptian modifications based on Wilkinson (1996: 15).

in the 3rd and 2nd millennium BCE – the period about which the hypotheses are most oriented – tends to be more stable.

Is "regime" the right word for these chronological markers? Perhaps not, given all the connotations associated with the regime concept. It is employed here in the context suggested by Spier (1996: 14) as "a more or less regular but ultimately unstable pattern that has a certain temporal permanence." In short, these are sets of recognizable patterns that give way to other recognizable pattern packages and can be applied to multiple types of human activity.[7]

De-urbanization/population growth problems

With the de-urbanization hypothesis, the premise is that political regimes were increasingly built around major cities that fluctuated in size in part due to how well the political regime was doing. One might expect some de-urbanization as regimes deteriorated or collapsed. The causality can easily be reversed as well. De-populated cities suggest turmoil and fewer taxes for government coffers. Regimes are more likely to end when they demonstrate pronounced inabilities to address societal problems. Challengers are apt to arise from within a fragmenting center or the periphery. Table 5.4 offers a quick scan of trends in ancient urbanization as best as we can grasp the movement over time.[8] However, Figure 5.3 is probably easier to read.

The data summarized in Table 5.4 are unquestionably rough. These are estimates of the number of cities with some minimal size (10,000 people). No doubt, some information is or has been lost along the way. For instance, if populations were moving back and forth from city to rural village, these data would not necessarily capture the total population – only the city part, presumably, is estimable. Nevertheless, the data do show increases and decreases in specific cities. The data also demonstrate the life cycles of a number of places. In one period, they are there and in the next period, they seem to have vanished. We might expect this

TABLE 5.4 Population Estimates Based on Urbanization in Ancient Mesopotamia and Egypt (in Thousands – All Years Are BCE)

	3700	3500	3300	3000	2800	2500	2400	2300	2200
Mesopotamia									
Eridu	6–10	10	10	10					
Uruk		14	40	40	80	40	30	30	
Larak		10	10	10					
Nippur				10					
Shuruppak					30	30	10		
Suheri				10	10	10	10		10
Anshan			10	10	10	10	10	10	10
Nagar				20					
Ur					12	10	10	20	40
Zabalam					10	10	10	10	10
Kish					30	20			
Lagash						60	30		
Girsu							40	80	50
Umma					20	40	40	40	10
Kesh					40	10	10	10	
Adab					10	20	10	10	30
Akshak						10	20		
Akkad									30
Larsa					16	10			
Total	6–10	34	70	110	268	280	230	210	190
Egypt									
Memphis						30	30	30	30
Heliopolis							10	10	
Total						30	40	40	30

	2100	2000	1900	1800	1700	1600	1500	1400	1300
Mesopotamia									
Eridu									
Uruk	30	30	30					30	30
Larak									
Nippur			20	20				20	30
Shuruppak									
Suheri	10	10							
Anshan	10	10		10					
Nagar									
Ur	100	20	10						
Zabalam	10	10	10	10					
Kish									
Lagash									
Girsu	80	40							
Umma	20	25		40					

TABLE 5.4 (*Continued*) Population Estimates Based on Urbanization in Ancient Mesopotamia and Egypt (in Thousands – All Years Are BCE)

	2100	2000	1900	1800	1700	1600	1500	1400	1300
Kesh									
Adab	10	10	10	10					
Akshak									
Akkad									
Larsa		40	40	40					
Nina	10	10	10						
Isin		40							
Badtibira				10					
Masham-Sha			10	15					
Eshnunna				10					
Babylon					60	60			
Nineveh						10	10	10	10
Assur			10	15	10	10	10	10	15
Total	280	245	150	180	70	80	20	70	85
Egypt									
Memphis				30					
Heliopolis								30	30
Avaris				20	20	50–100			
Hermopolis									30
Thebes				10+	10+	10–20	60	80	80
Elephantine			16	16	16				
Total			16	76+	46+	60–120	60	110	140

	1200	1100	1000
Mesopotamia			
Eridu			
Uruk			
Larak			
Nippur	20		
Shuruppak			
Suheri			
Anshan			
Nagar			
Ur			
Zabalam			
Kish			
Lagash			
Girsu			
Umma			
Kesh			
Adab			
Akshak			
Akkad			

(*Continued*)

TABLE 5.4 (*Continued*) Population Estimates Based on Urbanization in Ancient Mesopotamia and Egypt (in Thousands – All Years Are BCE)

	1200	*1100*	*1000*
Larsa			
Nina			
Isin	40		
Badtibira			
Masham-Sha			
Eshnunna			
Babylon	75	75	100
Nineveh	10	10	10
Assur	20	12	12
Total	165	97	122
Egypt			
Memphis	75	100	100
Heliopolis	30	20	20
Avaris			
Hermopolis			
Thebes	150	100	120
Elephantine			
Pi-Ramses	160	120	
Tanis			35
Total	415	340	275

Source: Information based on Modelski (2003).

in the 4th millennium when larger towns were just beginning to emerge. The tendency continued through the next two millennia as well.

In the aggregate, though, the 4th millennium is not characterized by de-urbanization in Mesopotamia. The number of people living in cities of more than 10,000 people expand until the middle of the 3rd millennium. The locations varied but urbanization proceeded even if one city tended to be the main focus, just as the early, if short-lived predominance of Uruk is validated. There is a downward turn after 2500 BCE that continues to 2200 BCE that suggests the success of the Akkadian takeover may have profited from southern de-urbanization. A brief upward tick is centered on Ur around 2100 BCE before a long downward spiral to 1500 BCE reflecting, among other things, the gradual ascent of Amorites to power in Mesopotamia.

In contrast, the famous tendency of Egypt to avoid large cities is visible prior to the 3rd millennium. Yet there is a marked upward trend to 1200 BCE and then a reversal in the next century. Mesopotamian urbanization, on the other hand, exhibits a much less pronounced response to the End of the Bronze Age crisis. In this instance, it helped to be farther removed from the eastern Mediterranean meltdown.

Thus, Mesopotamian de-urbanization and possibly population down-sizing appears to have been going on in the 2400s, 2300s, 2200s, 2000s, 1900s, 1700s,

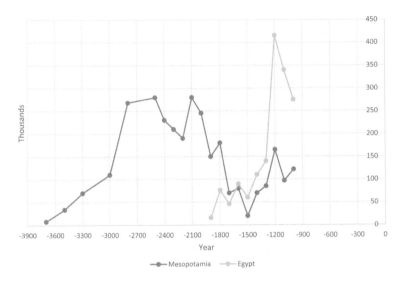

FIGURE 5.3 Estimated urbanization in ancient Mesopotamia and Egypt

and 1500s. In Egypt, years of de-urbanization/possible population down-sizing are the 1700s, 1500s, 1100s, and 1000s.

Trade networks and crises

Thompson (2001a, 2001b, 2006a) has surveyed the emergence of southwest Asian trading networks from about the 8th millennium BCE to the end of the Bronze Age. This summary narrative of ancient southwest Asian trade (presented in Table 5.5) can serve as something akin to a very crude time series. These generalizations are hardly carved in stone. They simply represent an attempt to interpret the flow of trade in one part of the world over a long period of time.

Evaluating the trade data

An even more abridged version of the ancient world trade panorama sketched above might read as follows (see Table 5.6). A southwest Asian trade network focused on commodities such as obsidian and lapis lazuli was in place considerably prior to the emergence of urbanization and states. The early leaders in urbanization, the Sumerians, created in the 4th millennium an ambitious acquisition network for raw materials and luxury goods that were absent from southern Mesopotamia. The network encompassed Mesopotamia, Syria, parts of Iran and Anatolia, and Egypt until toward the end of the 4th millennium when this network was forced to retrench by new, less cooperative people moving

TABLE 5.5 Reconstructing the Ancient Southwest Asian Trade

8th–6th Millennia BCE	A long-distance exchange network for tool-making obsidian linked eastern Anatolia and Mesopotamia (but not Egypt) in a limited, indirect, down-the-line, overland network through the 4th millennium (Culican 1966: 11–12; Zarins 1989: 342; Moorey 1994: 12; Stein 1999: 84–9).
Late 6th and 5th Millennia BCE	Initially, northern and southern Mesopotamia were not in direct contact but Halafian penetration from the north into the Zagros overlapped in timing with the expansion of Ubaid influence into northern Mesopotamia and southeast Anatolia (Marfoe 1987: 28; Hole 1994: 135; Lupton 1996: 1). Ubaid long-distance networks linked southern Anatolia, the Zagros, northern Syria, Mesopotamia, and the Arabian coast, may have reached southern Syria, Palestine, and Egypt. It remains unclear the extent to which the Ubaid network constituted a cohesive trade network, the diffusion of a cultural style, or both (Wright 1968: 248, 252; Marfoe 1987: 28; Porada *et al.* 1992: 88; Oates 1993: 408–409; Roaf and Galbraith 1994; Mark 1998: 6–7 122–123; Stein 1999: 84– 85). Peaking roughly mid-5th millennium, there is a retraction in Ubaid influence that resembles the Urukian retrenchment at the end of the 4th millennium (Wright 1968; Oates 1993: 410; Kuhrt 1995: 22). A Palestinian network (4500–3300 BCE) linked the Sinai, the Red Sea, southern Syria, Anatolia, and possibly Egypt (Mark 1998: 8, 122–128). Egypt was obtaining material from northeast Africa via the Red Sea and possibly Nubia, Palestine, and Syria (Zarins 1989: 342; Kantor 1992: 12; Mark 1998: 12, 122–128).
4th Millennium BCE	Urukian expansion began on a small scale in 3700 BCE and increased toward 3100. Based on increased demand for raw materials in southern Mesopotamia, the acquisition network was composed of large settlements in the north where local population density was limited and trading enclaves situated along the principal, river-centered trade routes (Algaze 1993: 46, 117; Hole 1994: 136; Lupton 1996: 41; Mark 1998: 122–128; Stein 1999: 91–92). By the middle of the millennium, the initially northern thrust was augmented by links from the eastern

Mediterranean to the Gulf (Moorey 1994: 6). Egyptian overland connections with Syria, Palestine, the Sinai, and Nubia increased. Maritime links probably via Byblos connected Egypt to Anatolia and Mesopotamia. Sumerian influences in Egypt ended with the Urukian retrenchment and collapse of the Syrian colonies (Adams 1977: 165–166; Trigger 1983: 39; Marfoe 1987: 26; Moorey 1987, 1994: 235; Zarins 1989: 368; Kantor 1992: 3–4, 13–14; Stager 1992: 41; Potts 1994: 74, 161; Mark 1998: 16, 19–20, 122–128, 130; Joffe 2000: 116, 119.

Late 4th–Early 3rd Millennia BCE

North-south Mesopotamian trade broke down around 3100 BCE due to a combination of problems, including social disruptions in southern Mesopotamia, a trans-Caucasian influx from the east, and immigration into Syria by Semitic groups (Algaze 1986: 289; Postgate 1986: 96; Kelly-Buccelati 1990: 122; Steinkeller 1993: 115–116; Lupton 1996: 72, 105). The Urukian contraction made space for local substitutes to continue some levels of overland trade flow (Chavalas 1992; Crawford 1992; Algaze 1995: 96) even while Sumerian contacts were increasingly focused on Iran and the Persian Gulf areas and Dilmun (Potts 1990: 90–92, 1994: 225; Moorey 1994: 245; Mark 1998: 10, 12).

Mid-3rd Millennium BCE

Exchange between southern Mesopotamia and the north may have become extensive again in the Early Dynastic I period, with cities in northeastern Syria operating as gateways and the southern cities competing for access to trade routes (Kelly-Buccelati 1990: 123; Algaze 1993: 107–108; Moorey 1994: 8). By the second quarter of the 3rd millennium, Mesopotamian trade was increasingly coming from the east via south Asia, Dilmun, and the Gulf, peaking between 2600 BCE and 2230 BCE and leading to the collapse of the overland network linking Central Asia, Afghanistan, Iran, Baluchistan and Indus around 2500 (Muhly 1969: 593, 595–596; Dales 1976; Rattenagar 1981: 208–209, 238–239; Frank 1993: 392–393; Potts 1993: 396; Potts 1994: 159, 281). Akkade attempted to reconstruct the Urukian expansion formula and managed to link the Gulf to southern Anatolia and the Levant, but encountered more resistance than earlier had been the case (Edens 1993: 125–126; Algaze 1993: 2–3, 108). Dilmun's entrepot functions between the Indus and Mesopotamia had been relocated away from the Arabian mainland to Bahrain by the 24th century (Potts 1990: 182, 189).

(Continued)

TABLE 5.5 (*Continued*) Reconstructing the Ancient Southwest Asian Trade

	Central Asian raw materials began appearing in the Aegean via the Gulf route during the Akkadian period (Cline 1994: 25). Cypriot trade mediation with the Asian mainland began in the last quarter of the 3rd millennium (Muhly 1969: 84).
	Byblos appears to have provided Egypt with its principal Mediterranean connection before and after the trade interruptions associated with the First Intermediate Period (Culican 1966: 19; Kantor 1992: 21; Stager 1992: 41; Butzer 1997: 259; Watrous 1998: 19). A Cretan connection was also established by 2600 (Warren 1995: 12). Stronger maritime connections implied less overland trade via Palestine (Marfoe 1987: 27; Stager 1992: 41).
Late 3rd Millennium BCE	Mesopotamian trade increased its reliance on Dilmun's Gulf middleman role, even though Indus merchants were established in Sumerian cities as trading agents (Muhly 1969: 72; Porada *et al*, 1992: 78; Tosi 1993: 369). Urban collapse was experienced throughout Central Asia between 2200 BCE and 2000 BCE (Kohl 1987; Tossi 1993: 373). In response to Akkadian attacks on Egypt's Syrian trade monopoly, Egypt appears to have attempted to hold on to its position in Palestine by force, leading to widespread urban destruction in Palestine, thereby further weakening the Egyptian regional economy (Kantor 1992: 21; Butzer 1997: 281–282). The Egyptian Delta was infiltrated by Bedouin in the First Intermediate Period, interfering with overland trade between Egypt and Palestine–Syria, but not maritime trade (Muhly 1969: 617; Ward 1971: 49). Minoan trade with the Levant and Egypt may have begun to expand after 2050 BCE (Watrous 1998: 20–1).
2nd Millennium BCE	Overland routes into Mesopotamia had been reopened by the beginning of the millennium (Muhly 1969: 603; Potts 1990: 258; Algaze 1993: 116). By the end of the millennium's first quarter, trade via the Gulf seems either to have come to a halt or declined substantially in volume, and roughly at the same time as the Indus collapse and a marked decline in Mesopotamian productivity (Muhly 1969: 582, 585; Rattenagar 1981: 235–237; Potts 1990: 258; Edens 1993: 130–132; Shaffer 1992: 450; Moorey 1994: 246). The subsequent Kassite control of trade moving through southern Mesopotamia is not well documented but they probably attempted to maintain the Old Babylonian flow of eastern goods into the Mediterranean economy (Leemans 1960: 120–121, 136; Piesinger 1983; Potts 1990: 298; Kuhrt 1995: 340).

Cretan–Syrian and Egyptian–Cretan contacts increased early in the millennium (Culican 1966: 27; Muhly 1969: 84; Kantor 1992: 4; Klengl 1992: 42; Cline 1994: 32; Merrillees 1998: 154; Watrous 1998: 21). Egyptian–Aegean exchange, via Syrian entrepôts, actually increased in the Second Intermediate Period, with Egyptian goods increasingly becoming the central focus of exchange in the eastern

Mediterranean (Culican 1966: 10; Redford 1992: 130; Cline 1994; Knapp and Cherry 1994: 128–134; Kuhrt 1995: 180; Warren 1995: 11). Cyrus became an increasingly important trading actor after 1700 BCE. Minoan Crete's role declined after the Mycenaean takeover of eastern Mediterranean trade routes without affecting the flourishing of trade in the 1400–1200 BCE era. Egyptian relations with Nubia in the Middle Kingdom era stressed the protection of a Nile-based trading system, anchored by a string of fortified trading posts (Adams 1977: 166). This approach moved into direct colonial enterprises focused on mineral extraction during the New Kingdom period. Adams (1977: 232) compares the role of Nubian gold in financing Egyptian foreign policy in Asia as similar to the later role of Mexican silver in subsidizing Spanish wars in 16th century (CE) Europe.

Late 2nd Millennium BCE

Kassite Mesopotamia fell to an Elamite invasion in 1155 (Zettler 1992). The Hittites and a number of Syrian trade entrepôts were destroyed by Sea Peoples attacks. Mycenaean and Minoan exports appeared to have ceased during this crisis period (Cline 1994: 37). Piracy increased. Palace-originated trading ventures collapsed. In general, the political infrastructure making long-distance trade possible disintegrated (Liverani 1987: 70; Muhly 1992: 19; Sherratt and Sherratt 1993: 363). Egyptian foreign contacts also contracted after the early 12th century to the 10th century before re-expanding (O'Connor 1983: 254, 270; Weinstein 1998: 188–189).

TABLE 5.6 Systemic Crises and Changes in Trade in the Ancient Near East

Systemic Crisis	Trade Changes
Late 4th Millennium (3500–3000)	Uruk retrenchment; Egyptian links to Near Eastern trade network severed
	Eastern orientation of Near Eastern trade network develops mid-3rd millennium (2600–2400 BCE)
Late 3rd Millennium (2200–2000 BCE)	Indus decline, Mesopotamian decline, Gulf trade decline, Egyptian links to Near Eastern trade network reduced
	Near Eastern trade network develops a greater western bias
	Egyptian trade links to Levant and eastern Mediterranean strengthened; Mesopotamian circumstances hazy; Western orientation of Near Eastern trade network stronger mid-2nd millennium (1800–1500 BCE)
Late 2nd Millennium (1200–900 BCE)	Near Eastern trade collapse
	Near Eastern/Mediterranean trade network expands with even stronger western orientation (1st millennium)

into parts of the Sumerian resource acquisition area. Restricted access to desired commodities in the north and east encouraged an opening to Oman and Indus through the Gulf beginning in the late 4th/early 3rd millennia that became increasingly important through the end of the 3rd/early 2nd millennia, prior to the probably interrelated decay in Mesopotamian wealth and collapse of the Indus civilization.

The other growth node of the ancient world, Egypt, had also begun to develop its own acquisition networks into Nubia and Palestine in the 4th millennium and also became linked to the Mesopotamian network in that same period.[16] While its maritime connection to Mesopotamia was broken toward the end of the 4th millennium, a now-unified Egypt continued to expand its contacts on land into Palestine until its Byblos maritime connection rendered the land caravans essentially noncompetitive. Eventually, however, Akkadian coercive expansion temporarily severed the functioning of the Byblos connection toward the end of the 3rd millennium. Egypt's maritime connections to the Levant were restored in the 2nd millennium and Egypt's linkages, whether direct or indirect remains unclear, with Minoan Crete expanded significantly.

In the 2nd millennium, trading activity was particularly intense in the Levantine/Assyrian area as the central intermediaries linking a southwest Asian-eastern Mediterranean trade now connected to Cyprus and Crete with stronger than ever Egyptian participation. The Gulf area trade languished. The movements of hinterland peoples, imperial warfare, and natural disasters (e.g., earthquakes in Crete) influenced trade patterns somewhat but did not stop trade volumes ascending to new heights in the last two centuries before the multiple catastrophes occurring around 1200 BCE and the consequent collapse of trade networks in the ancient world. However, these trading networks soon reemerged with new actors, a more pronounced western tilt (toward the western Mediterranean), and a wider playing field in the 1st millennium BCE.

Both of these abridged overviews of ancient world trade leave much to be desired. We may sense when volume and levels of interaction increase and decrease, and in which direction these changes occur, but there are a number of restrictions on any attempt at capturing trade fluctuations with precision. Still, we do not require a fully specified, International Monetary Fund-style matrix of trade flows stated in some constant, current currency to be able to make some judgments about hypothesized general patterns in the timing of dispersal and on the impact and implication of serial crises.

Toward the end of the 4th millennium, the first system crisis led to a major reorganization of the Uruk trading network. The colonial outposts in the east, the north, and the northwest were withdrawn and abandoned. What had become the customary access routes to highland raw materials were blocked or reduced. The western Mesopotamian–Egyptian link also was broken. An eastern link to the Gulf, Magan/Oman, and Meluhha/Indus was forged to substitute for the Zagros/Taurus problems.

Yet the crisis that did develop toward the end of the 3rd millennium led to, or at least was associated with, considerable political-military turmoil in Syria/ the Levant/Palestine that isolated Egypt temporarily from the rest of the south-west Asian trading network.[23] Mesopotamian aggression was also registered in all directions, including the Gulf and what appears to be the beginning of decay in the Mesopotamian-Gulf trade linkage. Indus decline appears to have begun around this time as well, indicating either an independent or interdependent atrophy of the eastern wing of the Near Eastern trade network. The trading network began to swing decisively back toward the west.

The Hyksos movement into Egypt caused or aggravated local political problems but did not have a strong impact on Egyptian trade patterns because the Hyksos were already integrated into the Levantine trading subsystem. Their movement into Egypt did not sever relations with the outside world and may even have intensified those relations. We know less about Mesopotamian affairs after the rise to power of the Kassites but one has the impression that any interruptions in the presumably already diminished Mesopotamian trade were fairly temporary. The Kassites remained in power in Mesopotamia for some time and appear to have attempted to re-establish many of the tradi-tional sources of trade, although at a lower volume than had been the case in earlier eras.

The third crisis, traditionally commemorated as the end of the Bronze Age toward the end of the 2nd millennium, reverted to the earlier pattern with a major breakdown in trade that could hardly be avoided given the mass move-ments of displaced people, the destruction of cities, and the general system-wide turmoil. That Mesopotamia was less affected by this turmoil only indicates that the center of the system had moved west. This third crisis was centered in and around the eastern Mediterranean littoral, which is precisely the direction of movement of the trading network in the centuries that immediately preceded the 1200 BCE breakdown.

One of the reasons the ancient world experienced trade crises was that its traditional productivity base had been transformed. One of its pillars or growth poles had declined significantly. This transformation was all the more critical because, perhaps unlike more contemporary processes, the growth poles of the ancient system, Mesopotamia and Egypt, were more geared to act as demand poles – as opposed to supply poles – for commodities at least.[25] The trading network was more biased in supplying their needs for raw materials than it was in finding markets for their export products. One of the corollaries of this facet of ancient international political economy, at least after the 4th millennium, is that the main agents of trade tended to be neither Mesopotamian nor Egyptian. Indus traders came to Mesopotamia and not the other way around. Entrepôts (Dilmun, Byblos, Ugarit, Mari) flourished as long as the supply and demand connections could be maintained. They floundered when the connections broke down. If demand was weakening in general, short-term crises in demand brought on by lack of water, crop failures, increased violence, and governmental collapse could be increasingly devastating, as demonstrated in 1200 BCE.

Then, too, there is probably a role for rigidities *à la* Mancur Olson's (1982) argument. Systems build up rigidities in elite stratification and interest groups wedded to traditional ways of doing things over time. To break down these rigidities, some type of revolutionary turmoil may be required to weaken traditional forces of conservatism. Again, the argument is not that the buildup of rigidities implies that transformational disorder is inevitable. Rather, the argument is only that the possibilities for transformation are greater to the extent that disorder weakens resistance to change.

What sort of rigidities characterized the ancient world toward the end of the 2nd millennium? Empires were one. Egypt became imperial early on but southern Mesopotamia had been able to avoid that prior to the rise of Akkad. Much of the Near East had become imperial territory long prior to 1200 BCE. The problem is that established empires, as a rule, are not known for their innovations. Established empires simply do not encourage experimentation and change. For example, the Levantine/Phoenician traders were most successful when they enjoyed some autonomy from their imperial neighbors. That autonomy was only possible in portions of the first halves of the 2nd and 1st millennia BCE. The windows for trading autonomy closed down later in the second halves of those millennia as trading states were subordinated and absorbed by adjacent land empires.

In addition, attention is often drawn to the ancient world's tendency to conduct trade first through the temple, and then state, agencies. Only these institutions initially had the resources to fund large-scale trading efforts. Temple and palace monopolies on trade, although perhaps never quite as absolute as is sometimes portrayed, should be counted as another rigidity of the ancient world. It is also one that seems to have been swept aside, in favor of greater private participation in trading enterprises, in the 1st millennium BCE. Here, the presumption is that private traders were more likely to be experimental and innovative in responding to trading problems than bureaucratic agents. That should encourage, in turn, a greater expansion of the network than might be the case otherwise.[26]

Finally, the role of metal for making tools and weapons was clearly a dominant factor in the development of the ancient trade networks. Initially focused on obsidian, copper, and tin (to make bronze), gradually became one of the central foci of ancient world trade. 1200 BCE is called the end of the Bronze Age because iron had begun supplanting bronze toward the end of the 2nd and into the 1st millennium. That implied that an acquisition infrastructure geared to delivering copper and tin to make bronze was in the gradual process of becoming less valued –perhaps not unlike the switch in emphasis from Indonesian spices to Chinese tea in CE18th-century commerce. Agents successful in the old emphasis frequently are not in a good position to make the shift. Their skills, investments, and contracts are not easily transferred to an entirely different enterprise. Other agents, less committed to the old emphasis, are less handicapped in this respect. [27]

Equally important is the fact that iron innovations were taking place outside of Mesopotamia and Egypt. The center for economic innovations and new sources of productivity was moving west of the Euphrates. Shifting centers of new innovation is a proclivity that will characterize the emerging world economy for millennia to follow the 1200 BCE crisis. Once again, no new center of product innovation has to emerge following a period of diffusion. But, when the diffusionary process has reached its effective limits, opportunity space for a new round of innovation opens. Moreover, the westward drift of the Near Eastern trading network, by facilitating development in Crete, Cyprus, and Greece made it more likely that any new centers of innovation would have a western location or orientation. This observation applies as well to the Phoenicians, despite their eastern location, whose opportunity space opened for a while as a consequence of the relaxation of imperial constraints in the Levant and who were then able to devote considerable resources to opening the western Mediterranean to further development.

Down-the-line trading interactions, centered on the movement of stones for tools and jewelry and stretching as far east as Afghanistan, gave way in the 5th and 4th millennia to increasingly Mesopotamian-centered resource acquisition networks (Ubaid, Halafian, and Urukian). In the 4th millennium, the Sumerian-centric network, involving a mixture of traders, trade enclaves, and colonial settlements, reached Anatolia and Iran in the north, Syria in the west, and Egypt in the south. Toward the end of the 4th millennium, the Sumerians were forced to retrench, due probably to some combination of internal and external turmoil due to the expansion of Trans-Caucasian groups, the rise of Elam, and the deteriorating climate. The Mesopotamian links to the north were taken over by intermediaries; the Egyptian link was severed for a time. Sumerian cities gradually were reoriented toward trade coming through the Gulf, via Dilmun, from Oman and the Indus.

Toward the end of the 2nd millennium a variety of problems – Mesopotamian militarism, declining agrarian productivity, the decline of the Indus – prompted a gradual shift toward the eastern Mediterranean as a focal point of southwest Asian trade. Aegean traders, Egyptian wealth, and Syrian-Levantine trading cities created a new system stretching west to Greece, Italy, and beyond in drawing in European resources to feed eastern Mediterranean demands. The reorientation toward the west was accelerated by the full collapse of the Indus by the 18th century and, ironically, by the Hyksos incursions into Egypt about the same time. The Hyksos domination of parts of Egypt further cemented Egypt's trade connections with the Syrian-Levantine coast and immediate interior. Increasing in volume and geographic scope into the 1300s, the entire system collapsed around 1200. A number of eastern cities and empires were destroyed. Migrations of people into Greece and around the eastern Mediterranean littoral were set into motion, reaching as far as Egypt and its battles with intruding "Sea Peoples" and Libyans. With the exception of some activity on the part of Phoenician cities, a two-century dark age ensued in which trading interactions were severely restricted.

To reiterate, there were three evident periods of trade crisis, reorientation, or collapse in the ancient world: circa 3200–3000, 2200–2000, and 1200–1000. The maximal period of collapse came only toward the end of the 2nd millennium, with reorientation even further to the west and then back to the east much later in the 1st millennium BCE. The earlier two crises, 3200–3000 and 2200–2000, stopped short of attaining the post-1200 dark age. The emphasis was more on gradually finding replacements for imports and exports that had become difficult to sustain.

Testing the hypotheses

Six hypotheses, derived from the model outlined in Chapter 4, are tested in this section. Focusing primarily on the direct effects of climate change and the diminished availability of water, we should expect to find a clear link between climate change and, among other processes, the movements toward economic deterioration and decline, problems in sustaining regional trading networks, the erosion of support for governments as well as increased governmental problems in maintaining order, and major hinterland incursions into sedentary areas – all ultimately resulting in population reductions.

Tables 5.7 and 5.8 provide a central focus for the test of the six hypotheses.[9] The tables' second and third columns draw attention to periods of either cool and dry climates and declining river levels. Invariably, these different but certainly related indicators of environmental deterioration overlap to some extent. But they also exhibit some independence as well, especially in the Egyptian case given the origins of the Nile in East Africa. Accordingly, we interpret periods of

TABLE 5.7 Mesopotamian Climate, River Levels, and Politico-Economic Problems

Century	Cool/ Dry	Tigris–Euphrates Levels	Trading Network Crisis	Regime Changes	De-urbanization	Hinterland Unrest
1 4000–3900	CD			Uruk		
2 3900–3800	CD					
3 3800–3700	CD					
4 3700–3600	CD					
5 3600–3500	CD					
6 3500–3400	CD					
7 3400–3300	CD	Falling				
8 3300–3200	CD	Falling				
9 3200–3100	CD	Falling	Sumerian Resource Acquisition Network Retrenchment	Uruk Retraction		Trans-Caucasian and Semitic Pressures
10 3100–3000	CD		Sumerian Resource Acquisition Network Retrenchment	Jemdat Nasr		Trans-Caucasian And Semitic Pressures
11 3000–2900				Jemdat Nasr		
12 2900–2800				Early Dynastic I		
13 2800–2700						
14 2700–2600						
15 2600–2500						

(Continued)

TABLE 5.7 (*Continued*) Mesopotamian Climate, River Levels, and Politico-Economic Problems

	Century	Cool/Dry	Tigris–Euphrates Levels	Trading Network Crisis	Regime Changes	De-urbanization	Hinterland Unrest
16	2500–2400		Falling			Declining	
17	2400–2300		Falling		Akkad Takeover	Declining	
18	2300–2200		Falling			Declining	
19	2200–2100	CD	Falling	Mesopotamian Gulf trade Disrupted and Reoriented to west	Fall of Akkad		Guti Pressure
20	2100–2000	CD	Falling	Mesopotamian Gulf trade Disrupted and Reoriented to west	Ur III	Declining	Amorite Pressure
21	2000–1900	CD			Old Babylonian (Amorite takeover)	Declining	
22	1900–1800	CD				Declining	
23	1800–1700						
24	1700–1600		Falling				
25	1600–1500				Kassite Takeover	Declining	Kassite Takeover
26	1500–1400		Falling				
27	1400–1300						
28	1300–1200						
29	1200–1100	CD	Falling	Mediterranean Trade Collapse	Fall of Kassites and Rise of Multiple Dynasties		Aramaean Pressure
30	1100–1000	CD		Mediterranean Trade Collapse	Multiple Dynasties in Babylon		Aramaean Pressure

TABLE 5.8 Egyptian Climate, River Levels, and Politico-Economic Problems

	Century	Cool/Dry	Nile Level	Trading Network Crises	Regime Change	De-urbanization	Hinterland Unrest
1	4000–3900	CD	Falling		Naqada I		
2	3900–3800	CD	Falling				
3	3800–3700	CD					
4	3700–3600	CD					
5	3600–3500	CD					
6	3500–3400	CD					
7	3400–3300	CD	Falling				
8	3300–3200	CD	Falling				
9	3200–3100	CD	Falling		Unification		
10	3100–3000	CD	Falling	Eastern trade Links severed			
11	3000–2900		Falling		Early Dynastic		
12	2900–2800		Falling				
13	2800–2700						
14	2700–2600						
15	2600–2500				Old Kingdom		
16	2500–2400		Falling				
17	2400–2300		Falling				
18	2300–2200		Falling				
19	2200–2100	CD	Falling	Trade Disrupted	1st Intermediate		Libyan/ Bedouin Pressure

(Continued)

TABLE 5.8 (*Continued*) Egyptian Climate, River Levels, and Politico-Economic Problems

	Century	Cool/Dry	Nile Level	Trading Network Crises	Regime Change	De-urbanization	Hinterland Unrest
20	2100–2000	CD		Trade Disrupted	1st Intermediate		Libyan/ Bedouin Pressure
21	2000–1900	CD			Middle Kingdom		
22	1900–1800	CD					
23	1800–1700	CD					
24	1700–1600		Falling		2nd Intermediate	Declining	Hyksos Takeover
25	1600–1500		Falling		2nd Intermediate/ New Kingdom	Declining	Hyksos Takeover
26	1500–1400		Falling				
27	1400–1300		Falling				
28	1300–1200		Falling				
29	1200–1100	CD	Falling	Mediterranean Trade Collapse		Declining	Sea Peoples/ Libyan Pressure
30	1100–1000	CD		Mediterranean Trade Collapse	3rd Intermediate	Declining	Libyan Pressure

cooling and drying, falling river levels, or both, as approximating the hypotheses' emphasis on climate deterioration and water scarcity. The fourth through seventh columns of the table record pertinent information about trade crises, regime transitions, de-urbanization, and center–hinterland conflict.

Nevertheless, these are century-long correlations. We have 30 centuries of information. Yet a lot of things go on during a century. If, for example, we code a river as falling or center–hinterland conflict as escalating, we can be sure that these activities were not constant over the entire century for which they are coded. Unfortunately, our climate information is less specific than the political-economic behavior that we are attempting to model. Thus, our effort is quite crude. We are simply looking for suggestive evidence – certainly not on the order of absolute confirmation of our hypotheses.[10]

Two other features are worth signaling before we proceed further. One is that the centuries of cooling and drying and river levels are far more common than the centuries without climate deterioration and water scarcity. That distribution creates some awkwardness for cross-tabulation purposes but it also tells us that benign environments were relatively rare in the ancient Near Eastern world. Yet strong empires continued to rise and fall, just as farmers managed to survive sometimes against the environmental odds.

This feature is reinforced by the findings that will emerge in the cross-tabulations of limited correlations. A century of cooling and drying or river levels falling did not necessarily generate turmoil, political fragmentation, or trade collapse. A sequence of such centuries is another matter. Obviously, people could adapt to environments that were less than benign. For example, they changed the types of crops they planted. Some managed to obtain more water when rain no longer accomplished what was necessary. Others migrated when no other options were available. To preclude any suspense about the findings, climate deterioration was far less than deterministic in the short run. The longer run was a different story.

H1: Significant shortfalls in water availability in the ancient world were more likely in periods of cool-dry climate deterioration.

Hypothesis 1 links cooling and drying episodes to falling river levels (columns 1 and 2 in Tables 5.7 and 5.8). In Mesopotamia, the outcome reported in Table 5.9 suggests that sometimes cooling and drying affected river flow levels and sometimes it did not. In Egypt, the correlation is negative indicating the considerable independence of Nile fluctuations from atmospheric regimes. But that was one of the gifts of the Nile. Control of the Nile gave one access to water when others in the Middle East had less access to water. One problem to note here is that the Nile discharge is not always consonant with RCCs since it is based on African rainfall. The RCC events that we use to measure periods of cooling and drying are drawn from the reading of ice cores from near the North Pole. Thus, we are not surprised that the hypothesis failed, given that our analysis was limited to estimates of river flow levels. Information on precipitation might have yielded a different outcome.

TABLE 5.9 Climate Deterioration and River Levels

	Centuries Characterized by Cooling and Drying	Centuries Not Characterized by Cooling and Drying
Mesopotamia		
River Levels Falling	6 (37.5%)	5 (35.7%)
River Levels Not Falling	10 (62.5%)	9 (64.3%)
	16 (100%)	14 (100%)
Egypt		
River Levels Falling	8 (50.0%)	10 (71.4%)
River Levels Not Falling	8 (50.0%)	4 (28.6%)
	16 (100%)	14 (100%)

TABLE 5.10 Climate Deterioration, River Levels, and Center–Hinterland Conflict

	Centuries Characterized by Combined Cooling and Falling River Levels	Centuries Not Characterized by Combined Cooling and Falling River Levels
Mesopotamia		
Hinterland conflict	4 (66.7%)	3 (12.5%)
Hinterland conflict absent	2 (33.3%)	21 (87.5%)
	6 (100%)	24 (100%)
Egypt		
Hinterland conflict	2 (25.0%)	4 (18.2%)
Hinterland conflict absent	6 (75.0%)	18 (81.8%)
	8 (100%)	22 (100%)

H2: Significant center–hinterland conflict in the ancient world was associated systematically with periods of deteriorating climate and diminished water supply.

Table 5.10 tells us that the second hypothesis linking climate and water supply to center–hinterland conflict has more support in Mesopotamia than in Egypt. The table provides a comparison between periods of climate deterioration (measured in centuries) and falling water levels versus periods without substantial climate deterioration and falling water levels. In both cases, there is proportionately more hinterland conflict in the former setting than in the latter. In fact, the conflict was three times as likely when climate deteriorated and water supply diminished in Mesopotamia. In Egypt, however, the contrast is much less and it would be difficult to argue that the proportional differences are all that different. This is probably a reminder that cooling and water levels are not all that closely related in an area with African-sourced river levels.

H3: Regime transitions in the ancient world were associated systematically with periods of deteriorating climate and diminished water supply.

TABLE 5.11 Climate Deterioration, River Levels, and Regime Transitions

	Centuries Characterized by Combined Cooling and Falling River Levels	Centuries Not Characterized by Combined Cooling and Falling River Levels
Mesopotamia		
Regime transitions	4 (66.7%)	8 (33.3%)
Regime transitions absent	2 (33.3%)	16 (66.7%)
	6 (100%)	24 (100%)
Egypt		
Regime transitions	3 (37.5%)	7 (31.8%)
Regime transitions absent	5 (62.5%)	15 (68.2%)
	8 (100%)	22 (100%)

The third hypothesis focuses on regime transitions. Table 5.11 suggests that these transitions were more common in centuries of environmental deterioration. This outcome presumably speaks to the divide among analysts as to whether regimes were likely to succumb in times of poor climate or their fall was due to mounting political problems in terms of centralized control over resources. The general answer is that it was often both. Climate adversity could be the straw that broke the political camel's back. If domestic tensions were rising over poor or corrupt governance, climate problems could ensure that the incumbent regime seemed unable to cope. Diminished governmental revenues associated with drought and famine could also further weaken a struggling regime. As in the case of hypothesis 2, however, the relationship is stronger in Mesopotamia than in Egypt.

H4: Diminished water availability and climate deterioration discouraged population growth and urbanization in the ancient world.

Environmental adversity should discourage population growth and make living in towns more difficult given the need to feed urban settlements. Yet the evidence summarized in Table 5.12 does not support any relationship between the type of environmental adversity on which we are focusing and urbanization rates. We know that this relationship was probably stronger in the prehistorical era when towns and villages were small and less permanent. Pre-pottery Neolithic A and B (PPNA and PPNB) populations were much more mobile and probably had little choice given their high levels of food insecurities. For the most part, the emergence of multiple towns of some size indicates that some of the urban dependence on the rural farming problem was solved, at least for stretches of time. We also know in the Mesopotamian case that poor growing conditions encouraged the emergence of singular and large southern cities at the expense of northern towns. If people are moving from smaller towns to a large town, we can isolate this in Table 5.4 but it will not show up in our aggregated data.

TABLE 5.12 Climate Deterioration, River Levels, and De-Urbanization

	Centuries Characterized by Combined Cooling and Falling River Levels	*Centuries Not Characterized by Combined Cooling and Falling River Levels*
Mesopotamia		
De–urbanization	1 (16.7%)	6 (25.0%)
De–urbanization absent	5 (83.3%)	18 (75.0%)
	6 (100%)	24 (100%)
Egypt		
De–urbanization	1 (12.5%)	3 (13.6%)
De–urbanization absent	7 (87.5%)	19 (86.4%)
	8 (100%)	22 (100%)

TABLE 5.13 Climate Deterioration, River Levels, and Trade Collapse

	Centuries Characterized by Combined Cooling and Falling River Levels	*Centuries Not Characterized by Combined Cooling and Falling River Levels*
Mesopotamia		
Trade collapse	4 (66.7%)	2 (8.3%)
Trade collapse absent	2 (33.3%)	22 (91.7%)
	6 (100%)	24 (100%)
Egypt		
Trade collapse	3 (37.5%)	2 (9.1%)
Trade collapse absent	5 (62.5%)	20 (90.9%)
	8 (100%)	22 (100%)

H5: Periods of ancient world trade collapse were associated systematically with periods of deteriorating climate and diminished water supply.

Major trade collapses were not all that common in the ancient Near Eastern world. But like regime transitions and hinterland conflict, collapse was more likely in periods of the most severe environmental deterioration. This generalization applies to both Mesopotamia and Egypt as seen in Table 5.13, but especially to Mesopotamia which was generally more central to the earliest trade networks.

H6: The conjunction of significant political-economic crises in the ancient world was associated systematically with periods of deteriorating climate and diminished water supply.

The last hypothesis raises the "piling-on" question. What should we expect when regime transitions, trade collapses, center–hinterland conflict, and de-urbanization all take place at the same time – our operationalization of genuine

TABLE 5.14 Climate Deterioration, River Levels, and Societal Crises

	Centuries Characterized by Combined Cooling and Falling River Levels	Centuries Not Characterized by Combined Cooling and Falling River Levels
Mesopotamia		
Crisis	4 (66.7%)	3 (12.5%)
Non-crisis	2 (33.3%)	21 (87.5%)
	6 (100%)	24 (100%)
Egypt		
Crisis	2 (19.2%)	3 (13.6%)
Non-crisis	6 (80.8%)	19 (86.4%)
	8 (100%)	4 (100%)

societal crisis? Table 5.14 indicates that societal crises – defined here, as occurring when any combination of three of the four dependent variables was present – were more than twice as likely to take place in adverse climate-water conditions than in their more benign counterparts in Mesopotamia but not in Egypt. Alternatively, societal crises were less likely when fewer than three dependent variables (i.e. 0–2 variables) were present in any given century.

The different numerical outcomes for the two pillars of the ancient Near East are rather stark. Does that suggest that Egyptian society was more immune to environmental deterioration? The answer is no. The key to the different outcomes can be found in Table 5.8. River levels were declining immediately prior to the onset of cooler climates around the times of the first and third intermediate period. There is some overlap but only some. If there had been more overlap the relationship would appear stronger in Table 5.14. At the same time, neither of the two types of environmental decline that we are focusing on appear to be linked to the second intermediate period. We think this suggests less a matter of immunity to environmental change and more a matter of (a) some immunity to cooling periods to the east and (b) responses to Nile fluctuations that varied in form from time to time. For instance, the continuing Nile declines eventually create major agricultural problems. So, too, can too much Nile water – something that is more difficult to assess with our data.

A respectable proportion of conflict is about resource scarcities. People desire more of some commodity than is readily available. Competition over resources ensues, presumably with intensity levels roughly commensurate with the level of scarcity. Without doubt, water had to have been one of the more valuable commodities in ancient southwest Asia. But it was more than just a valuable commodity. It was also critical to economic growth, prosperity, governmental legitimacy, and survival, whether one was a sedentary farmer, a goat herder in the mountains, a desert nomad, or a monarch. Kings were blamed when water was short in supply. Long periods of drought, therefore, threatened governmental legitimacy. Drought, particularly if extended over a number of years, had to have influenced the resources states could mobilize for protection purposes, just as it affected what was available to eat. Conflicts between states and cities over contested water resources and cultivable land, on the other hand, should be expected to multiply as the commodity in question became even more scarce. Mountain and desert tribes had little recourse but to drift slowly or move aggressively toward whatever areas still possessed water when their own habitats became too dry to support them. If water levels in the Tigris–Euphrates and Nile rivers are our most unambiguous indicators of environmental deterioration, the indicator system predicts maximum political-economic trouble, other things being equal, as the ancient system moved toward the ends of the 4th, 3rd, and 2nd millennia (3300/3100, 2200/2000, and 1200/1100 BCE). These centuries were also periods of cooling and drying that had been going on for some time. That is precisely when we should expect the most center–hinterland conflict, the relatively rare collapse of trade, and at least selected regime transitions. Only the de-urbanization indicator is not very cooperative at the end of each millennium – an outcome that says more about our indicator than what actually happened.

Still, Mesopotamia and Egypt, while highly dependent on central rivers, were not located in identical ecological niches. Nor did the two centers of the ancient Near Eastern world share the same type of political evolution. Egypt was moving toward unification as southern Mesopotamia was moving into a decentralized period of competing city-states that the other never fully experienced for very long. The shock brought on by a declining water level could certainly be manifested in different ways. Conceivably, it could have contributed to the relative decline of Uruk's leadership in Sumer by heightening inter-city conflicts. At the same time, it could have presented an opportunity to coercively unite north and south in Egypt, particularly if conquering the north had become more attractive to the south because of environmental changes.

But information about climate change and water supply is not sufficient to account for all of the political-economic change going on in the ancient Near East. Moreover, some phases of deterioration were more serious than others even if we are forced to treat them all similarly. For instance, the tables above give the same weight to each succeeding period of cool and dry climate. Yet the second, third, or fourth interval of cool and dry in a sequence of deteriorating climate are apt to be more devastating than the first time period because the problems and stresses associated with water and food supply have cumulated. If environmental deterioration has an impact on a number of variables including the ability of the center to organize resistance and the ability of the hinterland to survive in their natural habitats, it stands to reason that an extended period of aridification would be even more damaging than shorter-term fluctuations in weather. Consequently, it should not be surprising to find that center resistance to hinterland populations virtually disintegrated after several centuries of desiccation that also propelled hinterland populations toward the cities situated along the central rivers of Mesopotamia and Egypt.

If these climate and water level factors are intermittent, should we then anticipate a return to a phase of center concentration and toward the end of the phases
of malign conditions? The answer is no because there are still other variables
to consider. One of the consequences of water problems in the Mesopotamian
area, for instance, was an increasing reliance on complex irrigation systems to
compensate for periods of less water availability. A by-product of this activity, in
addition to the low river gradients, was high soil salinity that became especially
problematic for agrarian productivity in the period after the Old Babylonian era.
Marcus (1998) offers an interesting slant on this problem by plotting the sizes of
the leading states in the Mesopotamian and Egyptian areas over time. Whether
state size is the best indicator of the strength of the center is an issue that can
be debated profitably. For instance, one might wish to give more credit to the
degree of innovation displayed at the beginning of the series as opposed to the
bulk size of the empire toward the end. In the Egyptian case, pyramid construction figures might work better.

Nonetheless, her plots highlight the abrupt end of the fluctuation in
Mesopotamian centralization in a period of high soil salinity, widespread desiccation, and extreme pressures from hinterland populations as manifested by the
Kassite takeover of Babylon.[11] The Kassites did not remain in power forever and,
for that matter, it is easy to exaggerate the extent of their power over the general
Babylon area. Yet the point remains that strong centralization tendencies did not
reemerge immediately in Mesopotamia. The political-economic system essentially was played out for a time and the interacting cycles of climate, river levels, and political-economic concentration, as well as center–hinterland conflict,
became less sustainable after key parameters, such as agricultural productivity
potentials, were altered substantially.[12]

In contrast, the Egyptian end of the ancient world did manage to continue
to function somewhat cyclically, but not without some radical rearranging of
its elements. Unification (e.g., the first Egyptian centralization) took place
in the context of falling Nile flood levels and dryness. The Old and Middle
Kingdom re-concentrations took advantage of rising river levels and periodic
soil renewal. The New Kingdom effort not only had to work against falling
Nile levels and, eventually, a return to dry conditions, it was also assaulted
by hinterland people from throughout the eastern Mediterranean area. The
Libyan threat was greater than ever before, in part because they in turn were
being displaced by migrants from other parts of the Mediterranean world.[13]
The collapse of the ancient world in the last centuries of the 2nd millennium
BCE was not simply about climate change but climate change in the Middle
East, Europe, and Central Asia appears to have been a significant factor in
facilitating extensive turmoil and movements of people throughout the general
eastern Mediterranean area. Unlike the Hittites, the Egyptians could claim
to have survived this turmoil and fended off the invasions of the Sea Peoples
(see, for instance, Nibbi, 1975 or Sandars, 1978).[14] Yet their victory over the
Libyans proved only temporary. Subsequent dynastic recentralizing efforts

in Egypt were undertaken but they were carried out by people initially of adjacent hinterland origin and later by even more distant foreigners. Where Mesopotamia ran afoul of some of the by-products of extensive irrigation, Egypt lost its once much-touted autonomy from the rest of the Middle East and the Mediterranean.

The greatest danger in constructing an explanation that gives great credit to environmental deterioration would be to suggest that people and systems behaved as if they had no choice in responding to climate and water problems and that the outcomes were predestined. Nothing of the sort happened in the ancient world system. The many differences in Mesopotamian and Egyptian behavior (see Trigger, 1993; Baines and Yoffee, 1998) during this period reinforce that generalization. Climate and water availability were only some of the key variables necessary to reconstruct how the ancient southwest Asian system emerged and then collapsed. However, the available evidence suggests that the interaction of climate and water availability with other key variables was pervasive. Governments did not immediately fall when confronted with deteriorating environments. Continuing deterioration, however, did contribute to the collapse of trading networks, the demise of political regimes, and the onslaught of other forms of societal crisis.[15]

Similarly, hinterland people did not attack the center as soon as the temperature rose or water levels declined. A division of labor between center and hinterland first had to evolve.[16] Migrations to hinterland areas from other areas occurred. The first concentrations of power in Sumer and Egypt enjoyed the advantage of limited hinterland resistance – a factor, along with water, that was less and less available to successive regimes. Nor could the successive regimes be assured of continued access to a material foundation for political-economic concentration.

Sometimes the center defeated the attacks of the hinterland. Sometimes the hinterland overwhelmed what remained of the center. Gradually, though, the power of the center in some cases to resist the assault not only of mountain and desert tribes but also environmental deterioration diminished. This did not mean that a re-concentration of innovation and resources was out of the question.[17] Rather, it meant that these processes took place somewhere else.

In sum, though, the point is not that climate or hinterland attacks were responsible for the demise of successive regimes in the ancient Old World. However, if the data on climate, river levels, trade interruptions, regime transitions, and hinterland attacks should hold up as reasonably accurate, they appear correlated in various ways.[18] Deteriorating climate and decreasing river levels were associated with significant increases in conflict between the center and the hinterland. Collapses and reorientations of trading networks were most likely in the context of a deteriorating environment. Some regime transitions also appear to be linked to changes in climate and river levels.

The conjunction of multiple types of crises also was more probable in times of environmental problems. Correlations are not the same as causation; however, these findings suggest that we should be cautious in dismissing explanations linked to either climate or hinterland clashes with nomads. They have interacted with each other, just as, presumably, they have interacted significantly with a wide range of other variables and processes over long periods of time, including Tainter's marginal productivity interpretation of fluctuations in the level of complexity. We need to continue the investigation and mapping of these interactions as best we can if we wish to explain how the ancient Near East, and other regions like it, worked.

In the next chapter, we continue this quest for understanding just how central climate change might be to MENA life. As we move away from the ancient world toward more modern times, there is a popular myth that we have increasingly mastered the constraints associated with location and climate – so much so that they are no longer genuine constraints. Unfortunately, we regard this notion as not well-founded for understanding the Middle Eastern political economy. Not only are they very much still operative as constraints, but they may also prove to be terminal factors in some parts of the MENA. But before we get to that problem, there is more to be learned about climate change and behavior in the Late Antiquity period. One basic question is whether things changed all that radically. Can the model developed for the ancient Near East world be fitted to the post-ancient Near East world?

Notes

1　The river levels in Figures 5.1 and 5.2 are inferred (Butzer, 1995: 133) from oxygen isotope fluctuations recorded at Lake Van, the source for the two rivers. The fluctuations are measured in terms of percentage changes with each interval representing approximately a 5 percent change.
2　Nonetheless, climate problems were likely to affect nomadic economies, especially if they were of sufficient geographic scope to eliminate the option of finding different pastures and watering sources within hinterland areas.
3　Rulers won credibility and strength points for beating up hinterland groups at the outset of rule. Toward the end of a long period of rule, rulers have often alienated their allies and consumed their resources, thereby becoming more vulnerable to attacks from the hills and the desert.
4　McNeill (2013: 40) also notes that the pacification of nomads in the Middle East came later than in East Asia or North America. One reason may have been that Asian and American nomads succumbed in large numbers to diseases with which they were relatively less familiar and for which their sedentary foes had already developed some immunity. In the Middle East and North Africa, pastoralists and sedentary populations lived too close to develop different types of immunities.
5　Chariots had to be invented. While no doubt related to Sumerian battle wagons, they needed to be reduced in weight, improved in terms of wheels, and harnessed to horses – all of that was unlikely before roughly 2000 BCE.
6　Even the dating of the takeover of Babylon remains subject to debate (see Gasche et al., 1998).
7　Climate and rivers can be described in regime terms as well.
8　One analytical problem is that the minimum threshold for city inclusion changes over time. The source initially requires a minimum of 10,000 people to qualify as a countable city.

9 To facilitate the examination, more favorable climate and non-falling river level cells are left blank, as are other non-occurrences.

10 We do not subject our findings to chi-square analysis as a consequence.

11 Actually, she may exaggerate the abruptness a bit by ending her Mesopotamian plot with the Old Babylonian period. Her plots appear to be done in a free hand way as well although they also possess strong face validity. They also seem to be linked to explicit estimates of state size.

12 Jacobsen and Adams (1958) discuss two major episodes of salinization for the period in which we are most interested. The first one was in 2450–1750 BCE and a second one took place between 1350 and 950 BCE. Decreasing crop yields began by 2100 BCE without ever recovering in ancient times.

13 The attack of group X on group Y often minimizes the full scope of activity that was involved. Often, group X was set in motion by an attack or pressure from group Z. This is one of the hazards of delineating the boundaries of a system too narrowly. The ancient world was in contact directly and indirectly with areas normally considered outside its boundaries. This is also a problem for looking for climate implications under the southwest Asian lamp post. Weather patterns in the Ukraine and southern Russia have implications for southwest Asia if they set off southern tribal migrations into the Balkans, Anatolia, and Iran, that, in turn, have subsequent influences on incursions into the core of the ancient world system. Gerasimenko (1997) depicts the 2500–1000 BCE era as one of strong aridification peaking around 1500 but continuing to be dry into the 1st millennium BCE. Krementski (1997) finds a serious and rapid climate shift around 2650–2350 BCE which caused the collapse of agricultural communities in the southwest Ukraine and Moldova areas. Gerasimenko found an earlier arid period at 4100–3800 BCE but both authors describe the 4th millennium and the first half of the 3rd millennium as cool and wet, thereby favoring sedentary agriculture – but also setting up population growth for later hits.

14 It is sometimes argued that Mesopotamia was sheltered from the destruction at the end of the Bronze Age because of its location but it seems just as likely that by that time it was simply less inviting. Kassite Mesopotamia also had its own Elamite problems roughly at the same time.

15 Marcus' (1998: 88) argument is that large-scale and inegalitarian political structures are fragile, unstable, and hard to maintain. It is difficult to argue with this observation as long as it does not imply that the collapses occur randomly.

16 The river systems also had to undergo extensive evolution geologically before the types of influences that are being attributed to them could take place.

17 The 1st millennium BCE Assyrian and Persian empires clearly indicate that imperial re-concentration on a large scale was not out of the question.

18 The complications associated with de-urbanizations probably deserve more analysis as well.

6

ENVIRONMENTAL FRAGILITY IN THE MENA FROM LATE ANTIQUITY TO EARLY MODERN ERAS

The focus so far has been exclusively on the ancient world. We like to think that we have been able to overcome climate problems once sufficient technology and complexity were achieved. Unfortunately, it is not clear that we are there yet. This section of the chapter will focus on events less far back in time and the three remaining RCCs outlined in Table 6.1. However, the first focus on the two Roman Empires (western and eastern) does provide a bridge of sorts between the ancient world and late antiquity.

But if greater complexity and technology are the antidotes to being vulnerable to climate change, the Middle East and North Africa (MENA) of Late Antiquity and the Early Modern eras was passed by to a large extent by developmental trends. Why this might have been the case is no doubt subject to multiple debates. But one reason, we suggest, is that climate changes of the past 1,000 years continued to be destructive but were less likely to be linked to creative responses. Two of the reasons for this disconnect in resilience theory expectations were less mobility – larger populations could no longer migrate somewhere else in the MENA – and the advent of lingering plagues which proved difficult to combat. We do not view these factors as constituting a complete explanation of how the region was increasingly handicapped in inter-regional competition.[1] Rather, they are simply factors that deserve more attention than they have received in the past.

Roman Empire(s) and Late Antiquity Ice Age

Our transition to the post-Ancient era in western Eurasia begins with the Roman Empire. At first glance, that might seem off-target for a book focusing on the Middle East. However, nearly two-fifth of the Roman Empire's population was located in the Middle East (Anatolia to North Africa) in 165 CE. What

TABLE 6.1 Post-Ancient Era RCCs

Focus and RCC	Time Frame (CE)
Roman Empire	450–700
Late Antiquity Ice Age (RCC no. 7)	
Seljuk Turks and the Crusades	950–1070
Medieval Ice Age (RCC no. 8)	
Ottoman Empire	1400–1850
Little Ice Age (RCC no. 9)	

happened in the Roman Empire mattered a great deal to what happened in the Middle East and vice versa. Climate-wise, the rise and decline of the Roman Empire was strongly influenced by changes in its operating environment. The apex of the Empire was located within what is called the Roman Climate Optimum (200 BCE–150CE). Not coincidentally, the expansion and vitality of the Empire were facilitated by a stable mix of warmth and wetness that was highly conducive to establishing and maintaining an agrarian empire that circled the Mediterranean. The several centuries that followed the Optimum period are usually viewed as encompassing the decay of the western Roman Empire. The period between 150 CE and 450 CE was unstable with some years characterized by decent weather and other years by poor to very poor weather conditions that were also generally colder and dryer. The next 250 years (450 CE–700 CE) are now characterized as cold and ushering in the Late Antiquity Little Ice Age.

Put most bluntly, the climate trend line hardly favored the survival of the Roman Empire. A deteriorating climate was certainly not the only problem faced by late Roman decision-makers but it had to be one of the more significant problems because it was linked to some of the other culprits that are usually offered as explanations for the Roman decline. Perhaps the most overt connection is the climate roots of external pressures on the Roman eastern frontiers. In its ascent, Rome was usually in a position to select in which direction it would expand but eventually it reached a point of success in which it was difficult to press further due to a combination of logistical liabilities and indigenous resistance. Thus, there were limits to Roman expansion beyond the Mediterranean shores in North Africa and the Near East just as it proved difficult to penetrate German forests and to move beyond Hadrian's Wall in Britain. Once the Roman Empire had attained its "natural" maximum size, its borders could be defended by moving troops from one source of threat to another. But this approach only worked as long as multiple frontiers were not threatened simultaneously. Once that happened, major problems ensued in coping with the military demands of an extensive frontier. When threats emanated from the Persian southeast and the German northeastern frontiers at the same time, imperial forces found it difficult to mount an adequate military defense.

One of the reasons the German frontier became increasingly difficult to handle was the climate-induced movement of the first German/Scandinavian tribes south to warmer areas followed by the western movement of Hun-led coalitions from Central Asia toward Europe. Rome seemed to be able to manage the German tribes more or less by settling them along the frontier and recruiting barbarian soldiers from their ranks until the Huns forced the German tribes to flee into Roman territory. Neither the German refugees nor the following Huns were particularly manageable. Even if they had been manageable, the Romans had few resources left to apply to their management.

One factor that may have saved the western Roman Empire from a 5th-century takeover by the Huns was rampant disease.[2] The Huns chose to leave abruptly when they might have stayed and one easy explanation is that they realized staying longer was not a healthy option. Disease (of a different sort) within the Roman Empire had reached pandemic levels much earlier in the reign of Marcus Aurelius in the latter part of the 2nd century CE. Economic growth halted for a time, resumed, and then halted again in the middle of the next century by a combination of disease (see Table 6.2) and drought. Each time the Empire risked disintegration in the west and each time some semblance of recovery was accomplished. Yet each subsequent recovery was weaker and so was imperial resilience. By the end of the 4th century, the western Roman empire maintained few defenses against external penetration by non-Roman, tribal forces (Heather, 2005; Ward-Perkins, 2005; Hadsall, 2007).

The economic decline of Iranshahr – transitional period to Medieval Ice Age

In this context, Christensen's (1993, reprinted in 2016) argument deserves further attention. Christensen's main focus was on the economic decline of Iranshahr (a term encompassing roughly Mesopotamia and Iran) which became the base for a succession of Persian empires that petered out in the two millennia between 500 BCE and 1500 CE.[3] Achaeminid, Seleucid, Parthian, and Sassanian Empires were contenders for major power centers in their times (and in the Parthian case able to halt Roman expansion) but, with the partial exceptions of the succeeding early Caliphate and Ottoman periods, they also constituted a declining trajectory for the Middle East. What then accounts for Christensen's manifestation of "Middle Eastern" decline?

Christensen rejects relatively popular emphases on either the Arab conquests or the Mongol destruction of Baghdad in the 13th century as the agents of Middle Eastern decline. Instead, his argument is essentially that imperial strength in this part of the world came to depend on increasingly intensified efforts to irrigate farmland that could not rely on rain to grow crops. The irrigation projects succeeded up to a point when they were stymied by a staggering loss of people due to the Justinian Plague that precluded keeping up the extensive irrigation schemes, especially in Mesopotamia – the Persian breadbasket, which succumbed to

TABLE 6.2 Late Antiquity Bubonic Plague Outbreaks in the Near East

Year [CE]	Constantinople	Egypt	Palestine	Syria	Mesopotamia
542	X				
543		X	X	X	X
558	X				
561–562				X	X
573–574	X	X		X	
586	X				
592			X	X	
599–600	X			X	
619	X				
626-628			X		X
638–639			X	X	X
670–671					X
672–673		X	X		X
687–689				X	X
689–690		X			
698–700	X			X	X
704–706				X	X
713				X	
714–715		X			
718–719				X	X
725–726				X	X
729				X	
732–735		X		X	X
747	X				
743–749		X		X	X

Source: Based on information reported in Harper (2017: 237, 242).

perennial silting and soil salinity problems. Hundreds of years of subsequent and intermittent efforts to reconstruct the irrigation networks never managed to replicate the agricultural fertility that had been achieved in earlier times. The 14th-century Black Death losses only ensured that the decline would not be reversed.

Mesopotamian tax assessments of agriculture production (Table 6.3) demonstrate the outcome that persisted in the area as the leading imperial resource base. The first 7th-century amount listed in the table is almost half the assessment reported for the early 6th century. The second, late 7th-century figure presumably reflects plague effects. The numbers return in the 8th and early 9th centuries to the first 7th-century level but then drop lower after the mid-9th century and are their lowest again in the 14th century's second arrival of the plague. However, it is interesting to note that 1335 is more than a decade prior to the advent of the Black Death.

Christensen makes a good case for water scarcity problems in a fragile and arid environment being the main culprit for the downward trajectory of Mesopotamia after Late Antiquity. Where we think he goes wrong is dismissing at the outset

TABLE 6.3 Mesopotamian Agricultural Tax Assessments

Period [CE]	Assessment (million dirhams)
Kavadh I (488–531)	214
'Umar b. al-Khattab (634–644)	128
Al-Hadjdjadj (695–714)	18
'Umar II (717–720)	124
788	119.88
800	120.18
819	114.457
Middle 9th century	92.766
918	30.955
969	42
1335	13

Source: Based on Christensen (1993: 88). These figures are presented in the source with a number of caveats that suggest some caution in comparing them across time. It is never made clear whether the first five figures are representative or intended to encompass all of the years specified in each row's first column. We are assuming the former because otherwise, it would seem illogical to present them in this form in the list.

an explanatory role for climate change. His argument is that there is no evidence for climate having changed in any way over the last millennium, therefore, it can be ruled out as playing any role in accounting for what happened. Even so, he refers often to drought and abandoned settlements.

An agnostic position might have been easier to take in the early 1990s than in 2020. Understandably, it helps the disease-centric argument to be able to dismiss possibly complicating factors. But the Late Antiquity Ice Age, a phenomenon that has only begun to receive attention fairly recently, could certainly have made maintaining an extensive irrigation network difficult if less water was available to circulate throughout the system. We suspect that the argument needs to be amended to one of less water due to climate change and major population losses due to epidemics that precluded keeping up and restoring the water circulation infrastructure. The poor historical documentation for this era only compounds the analytical problem in attempting to weigh the various factors at work.

Whatever else, the lingering plague does help to explain how relatively small Arab armies that were poorly armed (swords and spears) and even more poorly armored could quickly defeat and take over much of the urbanized Middle East and beyond in short order in the 7th century.[4] One place the plague did not bother was the more thinly populated Arabian Peninsula interior.[5] Rosen (2007: 309) goes so far as to suggest that in an alternative, plague-free history, Byzantium might have been well placed to easily strangle the new Islamic movement's expansion. Thanks to the plague, however, the Byzantine army was reduced to one-third of its earlier size and even that smaller size was difficult to support due to reduced revenues, population size, and territorial control. After re-expanding around much of the Mediterranean littoral, Byzantium "lost half its territory and ¾ of its resources in 12 years" (Haldon, 2013: 481).

The emergence of Islam

The historical backdrop to the emergence of Islam is quirky. Arabia was a curious place politically in the 6th century CE. Bowersock (2017) calls it a chaotic environment and it was certainly that too. Some of the chaos may have involved turbulent weather patterns. The century began with Arab tribes that had recently converted to Judaism in conflict with other Arab tribes that had recently converted to Christianity. For a change, the Jewish tribes were massacring the Christian tribes. The losing side sought external intervention. Roman Byzantium was reluctant but encouraged Ethiopian (a former conqueror of Yemen in the 3rd century CE and also relatively recently adopters of Christianity themselves) military intervention on the side of the Christian tribes. The Ethiopian intervention encouraged the Jewish tribes to solicit Persian support for their side.

A form of proxy war was played out in both Arabia and in Syria between Byzantium and Sassanid Persia. The primary Byzantine motivation was controlling Red Sea trade in order to outflank Persia's intermediary role in east–west commerce. At that time, silk was the main Chinese export to the west and the cost to the Byzantine economy was as high as it had been in the earlier Roman economy. Yet the Red Sea corner of the proxy war could only be sustained as long as Byzantium had no other options in obtaining silk. After the mid-century, silkworms were smuggled to Constantinople allowing the Byzantines to produce their own silk. While the Byzantines subsequently lost interest in the Arabian theater, the Ethiopian intervention resulted in the Ethiopian military commander taking power in his own name and attempting to expand the amount of territory controlled. Mecca and Medina managed to stay outside of Ethiopian reach but not for want of trying.

Yet other factors do seem to have played a role in facilitating the emergence of Islam. Korotayev, Klimenko, and Proussakov (1999) suggest that a combination of global cooling and definitely above average volcano and earthquake activity in the 6th and early 7th centuries may have had a major role in clearing political space for a new type of movement. Kingdoms in Arabia had disintegrated by the 7th century after major ecological problems undermined their ability to collect taxes and considerable conflict among various religious groups which had led to the Ethiopian military intervention in the 6th century. At the same time, the Arabian setting was populated with a number of rival prophets in the early 7th century seemingly encouraged by an "end of the world" mentality perhaps stimulated by the ecological problems of extreme natural catastrophes. Korotayev, Klimenko, and Proussakov (1999) basically argue that the tribes in Arabia had eliminated allegiances to terrestrial authorities but were prepared to accept the idea of submission to a non-terrestrial, monotheistic authority. If one adds to this mixture the relatively new creation of an economic organization built around pilgrimages and shrines, albeit pagan in the Arabian setting, one could argue that the subsequent Islamic expansion was able to build an innovative form of empire based on recombining local ingredients.

Publishing in the same year as Korotayev Klimenko and Proussakoc, Keys (1999) uses some of this same material to develop one of the most complex explanations of the Islamic expansion – one centered on a Sumatran volcano, Avars, Serbs, Byzantines, Yemenis, and apocalyptic faith. This ensemble may sound like an exotic cocktail but it actually resembles a complex set of lines of falling dominoes that all feed into the likelihood of an Islamic outcome.

Starting with the Sumatran volcano, an eruption occurred in 535 CE (Krakatoa in what is now Indonesia) that was of a magnitude to rank it among the most serious eruptions of the last 50,000 years. As much as 96,000 cubic miles of rock, gas, and ash were thrown into the atmosphere that created a global dust cloud that reduced sunlight for some 18 months. Less sunlight translated into a colder climate and one that led to drought, famine, and disease for several decades on an impressively global scale.

> As the engine for extraordinary interregional change in ... Afro-Eurasia..., the Far East, Mesoamerica, and South America - the disaster [the 535 CE volcano eruption) altered world history dramatically and permanently.
>
> The hundred year period after it occurred... witnessed the final end of the supercities of the ancient world; the end of ancient Persia; the transmutation of the Roman Empire into the Byzantine Empire; the end of ancient South Arabian civilization; the end of Catholicism's greatest rival, Arian Christianity; the collapse of the greatest ancient civilization in the New World, the metropolis state of Teotihuacan; the fall from power of the great Maya center of Tikal; and the fall of the enigmatic Nasca civilization of South America.
>
> *(Keys, 1999: 3)*

How did an alleged volcano eruption in a then relatively remote part of the world bring about so much change? The answer hinges on the interpretation of events in this time period. Four causal chains are advanced by Keys (1999). The first one centers on Steppe dynamics which have repeatedly focused on transitions in rule with the losers moving from Mongolia toward Europe. This time Avars were overthrown by the Turks in the mid-6th century after ruling for a century and a half. Prolonged drought and famine were an effect wrought by the volcano eruption and the context for revolt. Keys notes that Siberian climate deterioration in 535–545 was the worst that had been experienced in 1,900 years. The subsequent Avar survivor movement to the west led to Avar conquests in Europe, initially centered on Ukraine and Hungary, during the second half of the 6th century.[6]

Eastern European climate deterioration in the 530s had encouraged the commencement of waves of attacks on Byzantine Rome by Slavs invading the

Balkans from the north. Avars and Slavs combined forces toward the end of the century to take large slices of Balkan territory from the Byzantines and to extract equally large slices of gold and silver in peace treaty payments from them as well. Financially drained and losing a third of its tax base, internal Byzantine politics moved toward an army mutiny, a palace revolution in 602 CE and ensuing civil war, and the resumption of Persian attacks on the Byzantine Empire in Mesopotamia. The war with Persia encouraged renewed Slav attacks in the Balkans and weakened Byzantine control in the Levant and Egypt. In the first quarter of the 7th century, Byzantium lost 75 percent of its territory to Slavs, Avars, and Lombards in Italy and the Balkans and to the Persians in a broad arc of territory from Anatolia to Libya. Much of the lost territory was retaken by the 630s but not without exhausting the Byzantines and Persian combatants – a second causal chain in the Keys argument.

The third causal chain centers on bubonic plague. Successive bouts of heavy rainfall and drought disturbed the normal dynamics of the breeding grounds of rodents that were hosting plague-carrying fleas. If the number of rodents multiplies quickly, so too do the fleas. If the rodents die off quickly, the fleas need to find new hosts, which can lead first to larger animals, including rats and humans. Keys posits gerbils in East Africa as the starting point of the Justinian Plague onset. A gerbil to rat to human linkage devastated port towns along the coast of what is now Tanzania. Shipping to the Red Sea introduced plague first to Yemen and then to the Mediterranean coast of Egypt (Pelusium and Alexandria), on to Constantinople, and the eastern and western Roman worlds in the 540s.

The plague remained intermittently active for two centuries. The heavy losses of life contributed mightily to the afore-mentioned physical and financial exhaustion of the Byzantines in the early/mid-7th century. It also contributed to a widespread apocalyptic mood among both Christians and Jews in the Near East. Constant conflict, climate problems, and pandemics combined to paint a picture of a disintegrating world. Islam emerged in this context as the most apocalyptic religion of the three urging believers to purify themselves and their practices before the end of the world came – still another causal chain.

Yet before Islam entered the picture, climate change problems had altered politics in western Arabia. In the 6th century, Yemen had been the center of prominence in the Peninsula but it had been wracked by the destruction of the Marib Dam, which was the central component of an extensive irrigation complex. Keys contends that the Sumatran volcano eruption led to a sequence of drought and flooding that eventually overwhelmed the dam in the 540s and again in the 550s. The dam had been broken earlier but this time the buildup of silt precluded easy repair. While still functioning in a reduced fashion until the 590s, it no longer could support a large population. Yemen's prominence declined as some groups moved north to Mecca and Medina which had had its own 6th-century climate problems. In one serious famine, Mohammed's great grandfather gained unusual prominence as a tribal leader who arranged for Syrian wheat to feed Mecca.

Thus, Keys weaves a complex tale of East African-Yemeni interactions (including the possible very early introduction of plague in the mid-6th century in Yemen) that created opportunities for a Meccan-Medina ideological movement to emerge in the 7th century that fed on the sense of a doomed world and aggressively sought to expand the population of believers. Potential resistance had been hollowed out by war and disease. Islam was able to spread quickly as a result.

Most of Keys' argument is compatible with other treatments of the 6th and 7th centuries in the Near East and eastern Mediterranean. The main exception and probably the weakest element is the very strong dependence on the volcanic eruption as the main causal agent. Volcanoes can impact climate but their impacts are usually more short term in duration.[7] In this case, the events that are being described were well embedded in the late antiquity RCC event.[8] One hardly needs to rely on one or two volcanoes to cause global climate problems that last several centuries when a cold/dry regime is in effect producing similar outcomes.[9]

However, aridity may have become less of a problem by the time that the Islamic expansion had gotten underway. Buntgen et al (2016: 5) suggest that cooling combined with increased precipitation in Arabia may have expanded the supply of scrub vegetation in the area.[10] In turn, more fodder meant larger camel herds could be sustained which would later support the Arab conquests. Something very similar has been said about the 12th-century Mongol expansion having been facilitated by increased rain expanding grass pastures for horses after the drought had encouraged attacks to the east in the first place. Moreover, in both the Islamic and Mongol expansions, some nod to greed on the part of the attackers must be acknowledged as well. Ideas can provide strong motivation but it helps if the motivation is reinforced by expectations concerning shares of the perceived gains from conquest.

None of these arguments attribute the rise of Islam directly to climate change. What they do, however, is argue that the rise of Islam was facilitated by various effects of climate change and that in the absence of these changes, there might also have been corresponding changes to developments within and adjacent to Arabia in the 7th century CE.

The "Big Chill," Seljuk Turks, and the crusades – the Medieval Ice Age

Bulliet (2011) argues that Seljuk Turks moved south from Central Asia into Iran to escape what he called the "Big Chill's" killing effects on the camels they herded. The Big Chill was a cooling episode affecting Central Asia, Iran, much of Mesopotamia, Anatolia, and Russia that began in the 10th century and persisted into the early 12th century. Initially, the Turks wanted to move from the northern fringe of the Central Asian Karakum desert to its southern fringe but various problems in finding refuge from the cold kept them moving farther south and west. Eventually, they became a powerful military group seizing control of Iran, Baghdad, much of Anatolia, and even Jerusalem.

This Turkish movement into the Middle East apparently also reinforced what had become an eastern bias in the Islamic Caliphate. The eastern portion (east of Baghdad) was wealthier than the western. As a consequence, more political and military attention was focused on the east than in the west. Not coincidentally, the west (North Africa and Spain) were the first parts of the early Islamic Empire to break free of central control. A different history of Muslim interaction with western Europe might have come about if a different orientation or bias had been in effect.

Another effect of these developments was that the early Islamic invasion of Iran had encouraged increased investment in cotton growing thanks in part to an expressed preference for wearing cotton clothes as opposed to silk and in part to incentives to invest in cultivating undeveloped land. Iranian cotton production boomed for a time in the 9th century and then gradually lost its allure at the same time cotton fields were hit by the severe cold. An additional indirect effect of the climate damage to cotton growing is that widespread famine encouraged farmers to switch from cotton to grain. An even more indirect impact of these economic vicissitudes was a large-scale flight of affluent elites from a depressed Iran in the late 11th–12th centuries. In combination with the devastation wrought by the subsequent Mongol attack, Bulliet attributes the elite migration beginning in the 11th century with stripping Iran of Sunni religious elites. Iran might not have become a Shi'ite country in the 16th century otherwise.

Ellenblum (2012) applauds Bulliet's emphasis on the role of climate but thinks he is missing a wider canvas picture of the effects of a colder climate in the 10th and 11th centuries. The Ellerblum thesis is that Central and southwest Asia and Egypt suffered from water shortages, drought, and cold weather between 950 and 1072 CE, although the situation improved a bit earlier in Iran and Iraq in the 1060s. The impact was so great because Egypt was involved. Egypt could help absorb the shock effect of prolonged drought because it relied on African monsoons as opposed to the normal Mediterranean weather system. But when both Egypt and the empires to the east of Egypt were confronted with the same water shortage problem, the impact would be all the more devastating. Between 383 and 1072 CE, Egypt had 3 years of drought in the 4th century, 2 years in the 5th century, 2 in the 6th century, 1 in the 7th century, 3 in the 8th century, 1 in the 9th century, 12 in the 10th century and 22 in the 11th century. The 34 years of drought in the 10th and 11th centuries far exceed the total number of years in the preceding 6 centuries. While areas east of Mesopotamia had turned cold in the 9th century, it was only in the 10th and 11th centuries that the Egyptian problems transformed the climate changes into a regional crisis. Equally important is the notion that these types of disasters could be dealt with to some extent as long as there was some conceivable refuge for displaced people. But if the entire region was experiencing major problems, no refuge was likely to be available and more people would starve, freeze, or die in intensified conflicts aggravated by the deteriorating environment.

Ellenblum attributes the following domino effects to the environmental problems of the 10th and 11th centuries:

1. Recurring famines and pestilence in cities and rural areas.
2. An end to the Islamic Renaissance period which valued innovation and science in favor of a more rigid adherence to religious dogma and Islamic law after the Seljuk conquest of Baghdad in 1060.
3. Byzantium, reviving in the 9th and 10th centuries, was hit with Turkish attacks in Anatolia and the Balkans, Pechenegs in the Danube, and Norman revolts in Italy and never recovered.[11] Constantinople had been one of the largest cities in the world in the early 11th century but was reduced to a population less than one-tenth that number by the 13th century.[12]
4. Rural areas lost population which reduced the ability to support urban areas.
5. Violent migration of pastoralists seeking warmth, pastures, food, and fodder.
6. Some major cities were pillaged, conquered, partially destroyed, or abandoned.
7. Collapse of bureaucratic and political institutions unable to survive declining state incomes while facing increasing state expenditures on facilitating food supply and fending off nomadic incursions.
8. Inter-religious strife intensified (sometimes encouraged by rulers seeking diversions) leading to minority persecution and the migration of minorities away from the eastern Mediterranean area thereby permanently changing the distribution of religious adherents in the area.
9. Economic crises that led to currency devaluations and failures in financing military and administrative functions that, in turn, led to military rebellions that overthrew dynasties.
10. In some places, dynastic states were replaced by nomadic states.
11. Desertification expanded throughout the region.

The main criticism of Ellenblum is that he does not spend much time presenting and discussing the climatological evidence for an extensive cooling period. For that, we turn to Preiser-Kapeller (2015) who does look at a number of relevant indicators for the time in question. While generally critical of Ellenblum, he reports empirical support for the claims that Central Asia, Armenia, and parts of Anatolia were cold if not always dry (especially in Anatolia). He finds the ideas that the Turks and other nomads may have been encouraged to move west and take on Byzantium by climate changes acceptable. Byzantine drying tendencies, however, may have extended throughout the 13th century. Moreover, he argues that the Egyptian evidence provides the strongest support for the Ellenblum thesis. The southern Balkans and Greece, in contrast, do not support the Big Chill argument and that agricultural growth in Central Europe was strong. The only problem is that Ellenblum never says that European territories were included in the cooling regime.

Curiously, Ellenblum does not add or endorse another kindred argument advanced by Issar and Zohar (2007: 222–223). The European Crusades began with a 1095 call from the Pope to liberate Jerusalem from Muslim control. Issar and Zohar stress that the Crusades were functional in terms of relieving some population and elite pressures brought about by a period of warming in Europe. Non-landed nobility were encouraged to improve their fortunes in the Near East. Unemployed males might become soldiers. Religious fervor might therefore have been reinforced by European population growth trends. But the functional western gains notwithstanding, a more direct argument might be that the westward movement of the Seljuk Turks was seen as a resurgence of Muslim threat to Christian interests in the Holy Land. Pope Urban II's call for the first crusade had responded specifically to Seljuk advances against Byzantium. Thus, regardless of the macro-functionality for European demographics, the first crusade might not have been initiated without climate change having set in motion nomadic incursions in the Near East.

Other than running out of water at a critical battle, Raphael (2013) argues that once the Crusaders were actually operating in the Near East, the only effect climate had was to prolong their presence. Both European and Muslim military commanders were often reluctant to engage in warfare at the risk of damaging fragile crops and thus postponed possible combat on a number of occasions. At the same time, though, the European colonies in the Holy Land were increasingly dependent on resources, especially men and food, from Europe as a function of diminishing access to food and water in the Near East. As European imports declined, the European presence in the Levant was endangered. The ultimate Muslim victory was achieved in part due to Arab exploitation of this vulnerability.

The Black Death – transitional period to the Little Ice Age

The Little Ice Age (LIA) is conventionally considered to have started in the late 15th century and not to have manifested full effects prior to the 17th century but there is some uncertainty about that dating. The 14th century appears to have been a transitional period leading up to the severely cold LIA episode. As a transitional period, there were short bursts of cold and dry climate onsets that appear to have been quite impactful. The earlier warm period had expanded populations throughout Eurasia, thereby increasing demand for food and inadvertently increasing the vulnerability of the expanded populations to a demographic crash in the face of abrupt food supply problems. One manifestation was the "Great Famine" of 1315–1320 in Europe that had been brought about by cold and excessively rainy conditions.

In general, the 14th century was characterized by weather extremes as in floods followed by droughts and vice versa. One of the areas impacted by these climate fluctuations is thought to have been rodent nests in Central Asia. Either the rodents were forced away from their nests or they died off or both. This meant that their fleas had to find new hosts which meant that Mongols and traders and their beasts of burden were more likely to host the fleas which carried the bubonic plague. Since Mongol trade activities had increased traffic near the rodent nesting fields, this disease vector was quickly transmitted along the land and maritime Silk Roads in both eastern and western directions. Chinese, European, and Middle Eastern deaths due to the initial onset of the Black Death (beginning in 1347) are estimated to be in the range of one-third to one-half of the existing population. Benedictow (2004: 383) puts the mortality rate closer to 60 percent. Successive onsets of disease ensured that the bubonic plague, as in the case of its preceding onset in the Late Antiquity era, was a major factor in economics and politics.

In 1347, the Black Death was centered in areas around the Black Sea but jumped to Constantinople and selected Italian and French cities, all of which signaled diffusion via trade networks by ships exiting the Black Sea, either in flight from the disease or as part of routine commercial exchanges. The next year, the plague had emanated from Constantinople into the interior to the northeast Balkans and southwest Anatolia. It had also leaped to Alexandria with subsequent movement into Palestine/Syria and down the Nile River. Algerian and Tunisian cities on the coast were also infected. On the Arabian Peninsula, the Black Death showed up in Mecca and Jedda but not in Medina. By the third year, the disease had spread to Morocco in the far west and farther into the Anatolian interior as well as down the Tigris and Euphrates to Baghdad and the Persian Gulf. The last movement of the disease in the Middle East was the 1351 transmission to Yemen by a returning king who had been imprisoned in Cairo.[13]

Just how major a factor the Black Death was in the rise and decline of the Middle East is still debated. One very useful illustration of this debate is provided by Stuart Borsch's (2005) comparison of the Black Death's impact on Egypt and England. In short, he argues that Egypt never quite recovered while England actually benefited from the Black Death. Egypt lost its economic leadership potential. England became eventually the leading technological power in the world. Why? Borsch's answer is that the nature of land ownership in the two cases differed considerably. Very different agricultural responses to the ravages of the Black Death thanks to these different ownership characteristics led to agricultural failure in Egypt and outstanding success in England, thereby setting up developments to come in succeeding centuries. We think Borsch pushes his argument a bit too far by exaggerating Egypt's economic leadership potential and misses an important ingredient in the comparison. It was actually an interaction effect between land ownership and rain versus irrigation watering supply that seems to have made the difference in outcomes. Nevertheless, Borsch's comparison is otherwise exemplary and quite useful to our examination of climate impacts.

Borsch's argument is summarized in part in Table 6.4. The outcomes in two states equally hit hard by bubonic plague were completely the opposite. The differences were not merely categorically different. Roughly two centuries after the first onset of the plague, Egypt's gross domestic product (GDP) was 40 percent of what it had been in 1300/1315 while England's GDP was 80 percent of what it had been. The price of wheat was down some 49 percent in England; in Egypt, it was up 56 percent. Wages for tradesmen were up 102 percent in England while late 15th-century Egyptian wages were 80 percent less than they had been in the first half of the 14th century.[14]

But is it fair to compare two seemingly different types of states? In the 14th century, the similarities were striking. Both had some degree of political-economic centralization. Their monarchs were constrained by powerful landholding elites. Their populations were roughly similar, as was their dependence on agricultural output. Both had some distinctive urbanization, some proto-industrialization, and were connected to long-distance trade routes. But in neither economy were these elements strongly evident as a proportion of GDP. Prior to 1347, the size of both economies was roughly equal in terms of total and agricultural product. Finally, both Egypt and England could be said to be islands or at least highly insular thanks to geography. England, of course, was an island surrounded by water. Egyptian isolation was configured by a combination of water and desert.

Yet there were important differences. Culturally, one was Christian and the other Muslim. Their locations were certainly different which gave them different access to neighbors and different geopolitical enemies. Egypt possessed less arable land than England but Egyptian soil was more fertile than in the island to the north. English agriculture was rain-fed while Egypt depended on the Nile and irrigation. The English elite generally inherited their social rank while Egyptian Mamluks could not pass along their rank and property to their offspring.

TABLE 6.4 Politico-Economic Outcomes in Egypt and England after the Black Death Pandemic

	Egypt	England
Wages	Down	Up
Land rents	Up	Down
Grain prices	Up	Down
Agricultural diversification	Down	Up
Unemployment	Up	Down
Agrarian/total output	Down	Up
Per capita income	Down	Up
Landholding system	Unchanged	Changed
Aristocracy status	Unchanged	Challenged
Economic recovery by 1500	No	Yes

Borsch contends that religious culture and location had little to do with the outcomes. The main difference, according to him, was that Egypt was ruled by Mamluks and England was not. The Mamluks were "slave soldiers" recruited from the Black Sea area. Property was provided as part of their remuneration and based on rank. Once a soldier died his property was redistributed. Areal concentration of property was denied to high-ranking Mamluks to reduce the risks of rebellion. For the most part, Mamluks were not much involved with the management of their land. They lived in urban areas, sometimes far from their property. There seems to have been little incentive to improve their landholdings. Instead, the main incentive was to squeeze as much profit as possible in order to support their standard of living.

The Black Death hit Egypt early and hard. Peasants died in large numbers. Survivors were encouraged to abandon rural farm work for urban activities. Rural areas, as a consequence, were seriously depopulated. One of the indirect effects was the breakdown of the irrigation system which required extensive labor to maintain its smooth functioning (assuming adequate Nile flow). In this case, Nile flows were not problematic but maintaining the conduits used to distribute the Nile flow was a casualty. Labor was scarce. Property owners responded to the depopulation crisis by seeking higher rents. Their collective response was unified just as their incentive to contribute to the collective good of irrigation maintenance was stymied by decentralization of the collective good just prior to the crisis and the lack of incentive for individual property owners to pick up the managerial mantle of a complex and expensive infrastructure. At the same time, the property owners controlled the state's military force and could enforce their preferences on rural peasants. The result was too much (or too little) water to redistribute in varying years. Crop losses led to reduced food supply and intermittent famine. In more remote areas where irrigation canals were allowed to decay, weeds grew that encouraged Bedouin tribes to move into formerly sedentary areas in order to feed their flocks of sheep and goats. The Egyptian military might force them back but could not keep them away forever.

Moreover, it is clear that the Black Death was not a once and done problem for Egypt. Dols (1979b) summarizes the recurrence incidence of Egypt and two of its neighbors (Table 6.5). The bubonic plague never quite went away in either country but its recurring impact was probably greater in Egypt than in England. At one time, it was thought that areas once hit by plague retained vestiges of the disease that could be re-stimulated from time to time. The latest evidence, however, suggests that most recurring bouts were initiated in Central Asia and re-communicated via trade routes (see Schmid et al., 2015).

In England, landowners were hereditary elites who often lived on their rural estates. Long-term interests in improving their holdings were variably present and acted upon. Rural holdings also tended to be geographically concentrated. English rural peasants were decimated just like in Egypt. Initially, the landowners responded similarly to their Egyptian counterparts. Profits might be maintained if surviving peasants could be forced to stay on the land and work

TABLE 6.5 Plague Recurrence in Egypt, Syria, and Palestine

	Number of Years of Plague	*Number of Major Epidemics*
1347–1515 CE		
Egypt	70	20
Syria/Palestine	40	18
1517–1894 CE		
Egypt	133	34
Syria/Palestine	79	20

Source: Based on the discussion in Dols (1979). "Major" epidemics are described as those outbreaks that penetrated beyond coastal cities exposed to trade and traders.

harder. But attempts to enforce this solution broke down in the face of peasant uprisings and abandonment of rural areas for the cities. Landowners could not maintain a unified front nor did they have sufficient military power to enforce their preferences. Individual deals were made to obtain rural labor that led to improved income for peasants. Owners were also encouraged to improve the efficiency of their agrarian operations. Owner incentives to diversify crops also increased once popular demands for different foodstuffs could be backed up by gains in income. Some of the diversification impulse translated into more rural proto-industrialization.

Borsch focuses on the very different types of power structures evident in the two cases to explain what happened. In Egypt, the landowners lacked individual incentives to do what was necessary to respond to the depopulation crisis effectively. In England, individual incentives led to bargains that worked out better for all concerned. But here Borsch seems to miss an important ingredient. What if Egypt's agriculture had been rain-fed instead of based on irrigation? We cannot say that a different outcome would have emerged but the Mamluk disincentives applied mainly to problems in maintaining a collective good, namely the irrigation canals. Rain might have helped at least some estates to improve their output, which could have stimulated other property owners to copy the successful behavior. Less economic deterioration could have resulted. Thus, we would argue that the difference between the two cases is one of an interaction between water systems and property ownership elites, and not just property ownership characteristics.

The upshot was that the English peasantry was pulled into the national economy while the Egyptian peasantry was pushed away and made more impoverished. English peasants became healthier and wealthier; Egyptian peasants not so much. The English economy was restructured in a way that facilitated greater production and affluence in the future. The Egyptian economy merely declined.

Borsch (2005) ends his story by suggesting that one important effect of the Black Death was to promote the chances of one economic leader in the European theater and demote the prospects of another in the Middle East. It is certainly true that early English agrarian productivity facilitated the rise of the industrial power that Britain became in the 19th century. To the extent that agrarian productivity was enabled by the ravages of the Black Death, it is fair to say that the bubonic plague facilitated the rise of the English/British economy. It is also clear that Egypt was hit hard by the plague. The real question was whether Egypt might have done as well as England if it had responded better to the initial crisis? The main problem here is that the ascent of the British economy to technological primacy was based primarily on developing fossil fuel-run machinery that seems most unlikely to have taken place in Egypt. Egypt lacked the fossil fuel and the weather to stimulate coal extraction for heating purposes. It also lacked involvement with the 16th- and 17th-century scientific tinkering that produced the steam engine. In this respect, location probably did make some difference. Of course, a more affluent agrarian Egypt, either as part of the Ottoman Empire or outside it, might have been able to connect with advances being made in early modern northwest Europe more readily than it did, it still would have lacked the incentives to develop coal-driven steam engines unless perhaps they needed them to maintain the irrigation dikes. Even so, they still would have lacked easy access to fossil fuels and been at a disadvantage in that respect.

The Ottoman Empire – the Little Ice Age (LIA)

One of the minor mysteries of early modern European international relations is what happened to the Ottoman Empire. At the beginning and through much of the 16th century, it ranked among the most important great powers, threatening the Hapsburgs in Central Europe and throughout the Mediterranean. Although defeated at Lepanto in 1571 with great loss of shipping, the Empire had little problem quickly assembling another fleet. Yet, a few decades later, it had retreated into decline and by the end of the next century was no longer regarded as a great power or a strong centralized empire. The inability of the imperial center to control its peripheries became a central problem in the 19th and early 20th centuries since other great powers were ultimately inclined to divide the empire among themselves. While the rise and fall of the Ottoman Empire had serious implications for the Middle East – most of which was controlled by the Turks at one time or another after the 15th century – the mystery was described as minor at the beginning of this paragraph only because the question was rarely pursued. It was taken for granted that somehow the once feared empire had become the Sick Old Man of Europe. Blame it if you must on a series of erratic Sultans and poor decision-making. More recent analyses, however, have drawn strong connections between Ottoman decline and the LIA, which is said to have hit the Ottoman Empire more severely than any other part of Europe or elsewhere (White, 2011: 7).[15]

Part of the problem was that the Ottoman Empire did what great powers have always done. Ottoman decision-makers mobilized imperial resources to fight wars in Europe and the Middle East. But resources are easier to mobilize if climate cooperates. When it does not and there is no lessening of imperial demands, various problems begin to emerge in terms of food supply, rebellion, banditry, and population shifts – all of which combine to undermine the ability of an empire to proceed as it had prior to the shift in climate.

Another part of the problem was the radical shift in climate. In the early part of the 16th century, much of the Ottoman Empire experienced a normal Mediterranean climate that features wet winters and dry summers. One index of the benign climate, albeit aided by military conquest and the abatement of the Black Death, was the near doubling in size of the imperial population in the first two-thirds of that century. The LIA brought colder temperatures, drought, and famine of a type that had not been experienced for more than a half-millennium. Beginning in the second half of the 16th century, back to back years of drought destroyed the agrarian economy, reducing the supply of grain and sheep – the two main sources of food. Ordinarily, drought in one part of the empire could be compensated by drawing in resources from other parts of the empire. But such a strategy could not work if the drought problems were experienced throughout the empire roughly at the same time. Nonetheless, it was not a descent into consistently colder temperatures and drier conditions. Rather, the proportional frequency and severity of the bad years increased. At the same time, though, Egypt, which had long served as a dependable food reserve for Mediterranean empires because its Nile water was based on African monsoons also failed in the mid-17th century due to low river levels.

With less food available in the rural countryside, migration to the cities was one response – even if the cities could no longer rely on their hinterland sources of food supply. The earlier population expansion had encouraged movement into the less fertile area which became even less productive due to cold and dry conditions. Now, these were the first areas to lose population. Bad climate also aggravated the normal tensions between sedentary and nomadic groups in some of these same areas with nomads moving into the more marginal agricultural areas – in some cases, permanently. A relatively depopulated hinterland meant more crowded cities with a greater probability of urban disease and plague outbreaks. Impoverished and landless males created fewer families and the attraction of joining large bandit gangs increased. Fewer productive peasant farmers meant less taxes for the government. Between 1576 and 1642 some parts of Anatolia lost 75 percent of their taxpaying pool and nearly half of all of the villages ceased to function (Parker, 2013: 188).[16] Currency devaluation, as a consequence, was difficult to avoid. Continuing war pressures meant no slackening of governmental demands for resources needed for prosecuting the wars. One outcome was the Celali Rebellion that persisted for 14 years (1596–1610) and threatened to bring down the Empire. The Ottomans had to juggle wars abroad with wars and banditry at home.

There are a number of moving parts in this story ranging from dying sheep to great power warfare. White (2011) who has done the most in-depth work on the 16th–17th-century Ottoman climate problem argues that the core issue was one of demographic decline. As sketched in Figure 6.1, climate change leads to

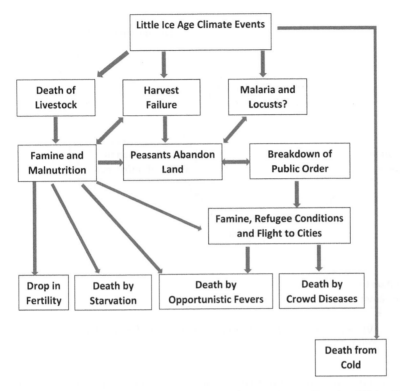

FIGURE 6.1 White's Ottoman mortality crisis diagram. Source: Based on White (201: 294).

harvest failures. Livestock die, peasants flee, famine, malnutrition, disease, and disorder become increasingly familiar. Population decline sets in and takes hundreds of years to turn around.

White's second model laid out in Figure 6.2, emphasizes how a vicious cycle emerged with declining food supply, expanded pastoralism and nomadism, and declining rural security interacted to suppress population growth. The Ottoman Empire did not disappear but, in most years, it lacked the resources to do much more than hang on to whatever control it could muster. McEvedy and Jones (1978: 137) information for Ottoman imperial populations is quite revealing on this score: 1500 = c. 8.5 million, 1600 = 28 million, 1700 = 24 million, 1800 = 24 million, 1900 = 25 million. Impressive growth occurred between 1500 and 1600. After that, not so much in what was a weakening empire. We do not have to exaggerate the weakening but it is fairly clear that central control of the empire was one casualty. Consequently, many of its far-flung peripheries developed relative autonomy from the dictates of the Empire until they were picked off by European powers during and after World War I. We are still living with some of the costs of that outcome in Middle Eastern state boundaries that have little explicit rationales.

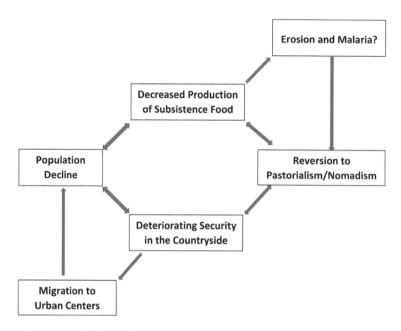

FIGURE 6.2 White's vicious cycle of Ottoman contraction. Source: Based on White (2011: 296).

There is no reason to argue that climate change alone did the Ottoman Empire in. The Ottoman Empire confronted rivals on all sides and often had to fight in every direction. The Ottomans played an impressive naval role in the Mediterranean but were far less competitive when European trade activity moved into the Indian Ocean with, initially, Portuguese armed sailing vessels. Early on, one of the advantages of the Ottomans was the ability to muster large numbers of cavalry that European armies had difficulties resisting. But these cavalry's availability was linked to agricultural activities. When it was time to harvest crops, the cavalry had to return to the farms thereby limiting the Ottoman army's ability to remain in the field. The timing of the Ottoman Empire also suffered from the subsequently diminished role in east–west Eurasian trade after various changes in the 1490s elevated the political-economic role of the Atlantic and the Indian Oceans over the Mediterranean.

There is also a healthy debate on whether the Ottoman Empire was sufficiently competitive in gunpowder weaponry. If not, as Parker (1988, 2013) argues, the decline of the Ottoman Empire owes a great deal to a Turkish deficit in military technology. But other scholars (Chase, 2003; Agoston, 2005; Johnson, 2013) disagree. Chase makes the impressive argument that the development of gunpowder weaponry depended on the distance from the Mongol threat. The greater the distance, the more likely was the time and space for the development of gunpowder weapons. The Ottomans were not very distant from the Mongols and were slow to develop cannons and infantry armed with muskets. On the slowness to copy European weaponry everyone agrees. But once the Ottomans began to copy the Hungarians that they encountered in combat, the other side of the argument contends that they did so quite successfully – at least initially. We might suggest that the main reason the Ottomans failed to maintain a competitive stance in modern weaponry may be traceable largely to a failed economy that was ruined by climate change in a way that exceeded the damages incurred by European economies of the era.

Just how much credit should be given to what was happening outside the Mediterranean in comparison to what was going on at the Black Sea end of the Mediterranean is of course difficult to say with any precision. It is conceivable that a strong Ottoman Empire might still have been overwhelmed by the external political-economic changes afoot in the 16th century and on. That the Ottoman Empire's vitality did not persist only made the "back-waterization" of the Mediterranean and the Middle East more probable.

Blom (2019) makes an argument along these lines. He claims that the LIA helped to transform Europe while having the opposite effects in Russia, China, and the Ottoman Empire. Moreover, the transformations that took place in Europe proved to be foundational for their subsequent domination of the rest of the world. Cold stimulated Europe to move ahead in breaking down historical constraints. In other parts of Eurasia, cold encouraged elites to defend and reinforce the status quo and their elite positions.

Blom puts forward a variety of events and processes that were climatologically influenced in 16th and 17th century. LIA analyses usually begin in the 1590s but Blom starts with the 1588 Spanish Armada. A large fleet was put together in Spain to ferry troops from the Netherlands to an invasion and occupation of England. The Armada encountered a number of adversities. Its ships were not designed to move in Dutch shallow waters which ruled out the movement of troops across the Channel. A number of the ships were commandeered merchant ships that were not built to carry heavy cannon and began to fall apart. English and Dutch naval attacks took their toll. Ultimately, it was decided that the Spanish fleet would sail north to Norway and around the British Isles back to Spain to fight again another day. An Arctic hurricane in northern waters put an end to that and half of the fleet sank in Irish and Scottish waters. The half that survived returned in poor shape.[17] Thus, Blom credits unusually cold weather with saving England from a Spanish occupation.

While this was a one-off event, a successful Spanish occupation of England probably would have meant far fewer European transformations due to climate change. New problems with growing grain led to a host of changes including better techniques for growing crops, and increased emphasis on trade in food and other commodities. A search for ways to move away from feudal subsistence propensities also led to pursuing alternatives such as fencing in the commons that forced large numbers of people out of agrarian occupations and made them available in urbanized areas for non-agrarian production. Increased attention to trade led to both mercantilism and the Columbian exchange. The former increased interstate tensions, the frequency of warfare, and the need for new ways to mobilize resources to support states engaged in intense security conflicts. Armies expanded as did their training and weaponry. The states that survived the warfare became stronger and more capable. The Columbian exchange led to the widespread adoption of the potato which responds to cold better than grains did.

Attitudinally, the agricultural-climate crisis led to increased numbers of witch hunts, greater religiosity – hence the Catholic-Protestant wars, and an enhanced willingness to experiment. Philosophers became more enlightened. Obviously, no single group pursued all of these paths simultaneously. Different parts of Europe were more inclined to do one or one of the others. Throughout, England and, especially, the Netherlands were vaulted into leading positions within Europe – a process that started with grain problems in northern Italy undermining the initial leadership of the Italian city-states and leading to their importing food from the north carried by English and Dutch shipping. The Netherlands became the hub of improved agricultural techniques and orienting agrarian production for the market as opposed to subsistence consumption. It was also the Dutch, confronted by larger Spanish armies, that first introduced the improvements in army training and weaponry. Nor were the Dutch ever shy about exploiting foreign markets.

Blom writes as a cultural/intellectual historian and the European transformation thesis is only part of his analysis. Otherwise, he might have made an even stronger argument. It is not always clear how or whether one event, such as the destruction of the Armada, must be attributed to the Little Ice Age. Important opportunities are missed. For instance, Blom says nothing about energy sources but it was during the LIA that movement into fossil fuels (peat in the Netherlands and coal in England) made considerable gains, in part responding to the increasing cold temperatures of the time. Dutch technology was facilitated by lower ocean levels thanks to expanding glaciers in the Little Ice Age. Dutch fishing was equally facilitated by fish and whales driven south toward the Netherlands by increasingly cold waters (Thompson and Zakhirova 2019).

The LIA is usually identified with the period between the late 1500s to the early 1800s. It is not coincidental that the Dutch and English were the European leaders in developing new ways of cultivating agriculture and introducing industrialization. Whether the epicenter of the societal transformations was in northwest Europe, as opposed to Europe writ large, major transformations took place that led much of the rest of the world becoming subordinated to the European region in the 19th and some of the 20th centuries. Perhaps one errs in crediting all of these changes to the Little Ice Age. After all, Dutch and English societal changes were underway prior to the temperature changes. But if one wishes to argue that societal transformations were at the very least accelerated by climate change, it would be hard to refute such a generalization.

Most clear is that these transformations did not take place in other places that also became colder. The Russian Czars, the Chinese Emperors, and the Ottoman Sultans were not overthrown until the early 20th century. All three have since become more competitive societies than they were when they were governed by relatively traditional rulers. Again, this is presumably not a coincidence. Blom's argument that the European response to climate adversity was much different than elsewhere in Eurasia cannot be denied regardless of what sort of caveats one might care to throw into the assertion. Thus, it is difficult to not see some role for climate change, or better yet, the differential responses to climate change as significant influences on comparative regional development. The region that changed the most came out far ahead of the other regions that changed only in marginal ways. MENA was thus a double victim of the LIA. The Ottomans lost their great power status when their agricultural economic foundation fell apart. The populations within the Ottoman Empire also lost out to the gains made by European choices during this time period.

Summarizing impacts in Late Antiquity through the early Modern era

In the ancient world, onsets of cold and dry climate reduced the supply of food and water thereby encouraging migrations, decimating cities and population sizes, and undermining the continued functioning of political regimes due in part to increased attacks from hinterland nomads. In the next two millennia, pretty much the same type of outcome occurred subject to two differences. The first major difference was the heightened role of plague, which devastated the much larger populations of the later era. While bubonic plague was not a direct function of cold and dry periods, it seems to have been stimulated by the transitional fluctuations of climate in the years between optimal warm and cold/dry climate regimes. The first major onset of plague in Late Antiquity helped seal the fate of Byzantium and facilitated the emergence and rapid spread of Islam. The second onset in the 14th century (and subsequently) rendered the likelihood of transformation toward something resembling "modernity" less likely in the Near East while it may have facilitated European changes.

The second difference was that the "last straw" effect of deteriorated and deteriorating environments in knocking over imperial control in the ancient world became less noticeable in later years. Regimes could survive and linger on for centuries, albeit controlling less and less territory. Byzantium, once threatening to restore the Roman Empire to the Mediterranean area in the 6th century, was not captured until 1453. The Islamic Empire fragmented but successor regimes retained control in most places except in southern Europe (Spain and Sicily). The Ottoman Empire that peaked in the 16th century managed to survive, albeit with some temporary protection from European great powers, into the early 20th century. While these imperial regimes might have survived long past their due dates, their ability to play vigorous roles in regional and world politics was eviscerated.

Otherwise, the core phenomena persist. Agricultural activity is diminished by cold and dry weather. Adequate food production falls off. Famine becomes more likely. Nomads continue to press into sedentary areas. Hinterland areas become more unstable. Urban sizes are rocked by pandemics but augmented by rural populations fleeing the uncontrolled interior areas. With the exception of the malaria question mark, White's model in Figure 6.2 might have been applicable to Mesopotamian developments several thousands of years earlier.

White (2013: 80–81) takes this further than other Middle Eastern analysts when he concludes:

> From the mega-drought of 2200 BCE, to the crises of the late Bronze Age and late antiquity, to plagues and regional depopulation in the Middle Ages, the Middle East has long suffered more severe environmental set-backs and recovered more slowly from each than perhaps any other part of the world. This helps to explain the long-term transition from the center of ancient civilization to a relatively thinly populated and poorly developed region by the dawn of the industrial era.

White goes on to specify four "key recurring patterns" in this evolution. First, climate shifts in and near arid zones, beginning with the Neolithic Revolution, have interacted with drought and soil degradation to make the region highly vulnerable to agricultural failure. Second, nomadic movements and conflict with sedentary areas have repeatedly hindered recovery processes. Third, famine and insecurity encouraged flight to already overly-large urban areas – a tendency accelerating after the advent of the Islamic era. Finally, epidemics have afflicted the Middle East more than elsewhere since the 6th century CE. These four tendencies:

> vulnerability to climate [change], nomadic invasions, flight to urban areas, and a higher rate of epidemics – may account for the gradual decline in the region's share of world population over thousands of years.
>
> *(White, 2013: 81)*

White (2013) does not actually provide a time series of the region's share of world population but the suggested indicator is of some interest. We made some rough calculations based on information found in McEvedy and Jones (1978). McEvedy and Jones say that most of the world's population was located within Asia prior to the last BCE millennium and that the Near East comprised about 25 percent of the Asian total between roughly 3000 and 1000 BCE. Before 3000 BCE, the share would of course have been much higher. They provide more precise estimations after 400 BCE. If we add their information for Asian Turkey, Syria/Lebanon, Palestine/Jordan, Arabia, Iraq, Iran, Egypt, Libya, and the Maghreb as "Near Eastern," the Near Eastern share falls in the vicinity of 11.5–14.4 percent of the world total between 400 BCE and 1000 CE. Note that we made no accommodation (by expanding the state set) for the high point of the Arab Empire in the 8th century.

By 1500 CE, the MENA share had declined to about 6.1 percent and to 4.1 percent by 1950. These numbers suggest a downward escalator for MENA's population share with a very high share prior to 3000 BCE, declining proportionately by about half after 1000 BCE and half again after 1000 CE. We read these numbers as indicating that the MENA share of the world's population has been declining almost from the Neolithic outset – as opposed to beginning its decline with relatively more recent developments such as the fragmentation of the Caliphate, the European Crusades, or the Mongol attack on Baghdad, the 1490s trade pivot to the Atlantic/Indian Ocean, or the onset of the Mandate system after World War I. Some proportional decline was inevitable over 12–15,000 years yet the MENA decline has been decidedly unique and rather long term.

Thus, we fully agree with White's observations but are less hesitant than he is to attribute recurring decline in the Middle East to its unusual and heightened climate sensitivities and the related political-economic processes regularly associated with coping with and responding to intermittent calamities. Fifteen thousand years of similar shocks and reactions since the epochal retreat of the European glaciers have demonstrated a rough cyclical process of repetitive setbacks which leave the Middle East and North Africa less well off than it had been before the latest onset of dry and cold climate. There may have been enough resilience to come back but not enough to get very far ahead or even to run with the competition. A promising start with commanding leads in a number of activities and spheres encountered one environmental roadblock after another. China seems to have experienced many of the same environmental problems roughly at the same time but its setbacks have been ameliorated by a different Eurasian location and a greater landscape diversity than what is found in the increasingly arid Middle East. Moreover, epidemics proved to be more devastating in the Middle East than they were in China.

Ancient climate change problems could be resisted to some extent by migration and adapting food sources. Demographic changes created larger populations which were vulnerable to new diseases introduced by trade with other regions. The repeated climate shocks and the double disease shocks failed to eliminate life in the MENA but they did manage to preclude the possibility of anything resembling the leadership in economic and political innovation once exhibited millennia before.

The resilience theory's framework, therefore, proved to be more useful in the ancient world than in the more modern world because conditions changed. Mobility opportunities slowed down considerably (after the initial expansion of Islam) and disease effects lingered. Climate changes continued to be destructive but subsequent resilience, creativity, and renewal became less likely, especially at the macro level. Farmers still experimented with crop adaptations and irrigation schemes at the micro level. Governments, with claims to limited resources and revenues, have not been particularly innovative. Staying in power demands more attention.

Timur Kuran (2011: 1) writes about what he calls the "long divergence" or how the Middle East became underdeveloped compared to parts of the rest of the world.

> At the start of the second millennium, around the year 1000, a visitor from Italy or China would not have viewed the Middle East as an impoverished, commercially deficient, or organizationally primitive region. Although the region might have seemed enigmatic, its oddities would not have painted a picture of general economic inferiority. Now, at the start of the third millennium, it is widely considered an economic laggard, and a plethora of statistics support this consensus. More than half of its firms consider their limited access to electricity, telecommunications, or transport a major obstacle to their business, as against less than a quarter of those in Europe. In the region, life expectancy is 8.5 years shorter than in high-income countries consisting mainly of North America, western Europe, and parts of East Asia. Its per capita income equals 28 percent of the average for high-income countries. Only three-quarters of the adults in the region are literate, as compared with near-complete literacy in advanced countries... .

Kuran thinks that the problem is traceable to the stagnation of institutions and organizations in the Middle East while others were developing new forms of commercial and financial institutions. Even though he recognizes that the problem of the Middle East lagging behind other regions is multi-dimensional, he chooses to emphasize the central role of Islamic law as the main "culprit." Institutional practices that once may have made sense went too long without substantial transformation to accommodate a changing environment. Whether he is right or wrong in his choice of emphasis is not an issue to be litigated in this book. He makes a strong case for his perspective but with one major flaw. The flaw is that he dismisses climate as a factor because of its "near-fixity" (Kuran, 2011: 15). That is, if the climate was stable for a long time, it cannot possibly explain either institutional stagnation or later attempts at reform. Something that varies little cannot explain something that varies a lot. Our response, not surprisingly, is that the MENA climate has not been stable. Some aspects of its instability might well help explain long-term institutional stagnation during the last millennium.

We think a series of climate shocks have intermittently set back the MENA, both in the short and long term. The next question is what to anticipate in view of the onset of still another climate shock – this time in the form of warming and drying instead of cooling and drying. Unfortunately, the MENA remains highly vulnerable to environmental vicissitudes and especially to global warming. While it remains unclear just how bad environmental deterioration will become, should some version of the worst-case scenarios come about, the region, or at least a substantial part of it, may not survive as a livable place this time around. We turn to this question in the next three chapters.

Notes

1 Different responses to some of the same impacts in other regions, as will be demonstrated, should also be part of the explanation.
2 There is a fairly new and expanding literature on the linkages between plague and climate change that focuses principally on climate changes in Central Asia being responsible for disturbing rodents repeatedly and transporting new disease vectors to the West via trade route traffic. See Schmid et al (2015); Campbell (2016); and Yue and Lee (2018) but it is too soon to tell whether or not this argument will hold up to closer scrutiny. Most of this literature is focused on the medieval Black Death and Europe but the findings should have relevance for the earlier bubonic plague pandemic and other afflicted regions such as the Middle East. One question still open is whether the Justinian plague originated in Africa or Central Asia.
3 More technically, Christensen is referring to territory lying between the Euphrates and Amu Darya.
4 With the exception of the Byzantine-Arab battle at Yarmouk, there was only limited resistance to the Islamic armies in the Near East (Bowersock, 2017). Interestingly, a similar process has been invoked in a speculative fashion for the steppe invasion of eastern Europe around 3000 BCE (Reich, 2018: 113–114). The question is how the steppe in-migration succeeded in overcoming the sedentary European population that is suggested by genetic studies. The presence of evidence for bubonic plague at this very early date (Rasmussen, Willerslev, and Kristiansen, 2015) may mean that the sedentary population was overwhelmed by a disease introduced from the steppe for which they were completely unprepared. A substantially reduced ability to resist could imply that the steppe migrants did not have to be militarily superior to dominate, particularly since this time period predated the innovation of chariots as mobile archery platforms a millennium or more later. Something similar did take place in the New World in the 16th century CE.
5 McNeill (2013: 40) cites Procopius in generalizing the basic pattern of plague as hitting "port cities hardest, cities hard, villages less hard, and mobile populations least of all."
6 The Avars were one of the many nomadic coalitions of various clans and tribes originating in Central Asia that moved west toward Europe.
7 Green (2018) notes that there may have been three major volcano eruptions in the 530s but also balks at giving them full credit for the subsequent cooling. Interestingly, Fagan (2004:143) notes that another unknown volcano erupted in 2200 BCE that may have initiated a cold winter and several years without summer. But it "coincided" with the beginning of 278 years of drought in Mesopotamia. The implication is that volcano eruptions may have something to do with the abruptness of RCCs but they cannot claim all of the credit for the ensuing climate deterioration.
8 Conceivably, volcano eruptions might lead to an RCC but the counter-hypothesis that RCCs encourage volcano eruptions seems more attractive – if one is forced to choose between the two alternatives. At the same time, there is no consensus on precisely why RCCs occur. One reason for the lack of consensus is that it is possible that different combinations of causes, as opposed to a single general cause, can improve their probability of occurring.
9 Debate also continues on the origins of the 7th century bubonic plague. While Key promotes an out-of-Africa origin, other scholars still prefer an Asian source but they are not sure where that Asian source was. It also remains unclear whether the Justinian Plague traveled as far east as China (Green, 2018). An epidemic that was restricted to a European-Persian range might have had a different source than the Black Death.
10 Korotayev, Klimenko, and Proussakov (1999) also note that Indian monsoons had returned around 500 CE after a long relative absence.
11 Pechenegs were Turkish semi-nomads pushed west by defeats by other Turkish tribes in Central Asia in the 9th century and ending up in an area north of Byzantium.

Normans had initially infiltrated southern Italy in the 11th century as mercenaries employed in local struggles with Byzantine rule. Eventually, they used their military prowess to gain political control for themselves.

12 Nonetheless, Byzantine scholars seem to be highly resistant to explanations with strong climate effect emphases (see Cassis et al, 2018; Mordechai, 2018, and Roberts et al, 2018).

13 This mapping of the transmission of the Black Death is based on the account reported in Benedictow (2004).

14 These numbers are taken from Borsch (2005: 90).

15 Ironically, McNeill (2013: 40–41) suggests that the rise of the Ottoman Empire may have been helped by the relative immunity of pastoralist tribes in Anatolia to the Black Death in comparison to the hard hit remnant of the Byzantine Empire centered on Constantinople, and to a lesser extent, the Balkans – which became the historical core of the Ottoman Empire. As he notes, this linkage is the exact opposite of the relationship small pox played in helping sedentary troops pacify nomads in east Asia and North America.

16 Simultaneous and quite major tax revenue declines were recorded for Anatolia, the Balkans, and Syria in the 17th century and Parker's (2013: 189) data indicate that they had not recovered by 1834.

17 Not discussed by Blom, at least three additional attempts at sending Spanish Armadas north took place in the 1590s but they all failed as well. Something other than Arctic storms, presumably, was in play.

7

GLOBAL WARMING IN THE MENA TODAY AND IN THE COMING DECADES

Earlier chapters, particularly Chapters 3 and 6 have argued that the Middle East and North Africa (MENA) have experienced some nine protracted episodes of cooling and drying in the past 15,000 years. Following several thousand years of warming as the northern glaciers retreated, the advent of the Younger Dryas cooling abruptly reversed the initial warming for a thousand years (10700–9700 BCE). A 500-year episode followed in the 9th millennium (8500–8000 BCE). Two more thousand year intervals encompassed the 7th and 4th millennia (7000–6000 BCE and 4000–3000 BCE). The 4.2k cooling episode in 2200–1800 BCE and the end of the Bronze Age event (1250–800 BCE) punctuated and terminated the Bronze Age in the ancient Near East. Cooling again returned in a 450 to 700 CE episode. The Medieval Climate Anomaly following some 250 years later (950–1070 CE) and the Little Ice Age (1400–1850CE) brought the list up to date.

Throughout these episodes, temperatures were not constant, spiking now and then, but in general, ushering in periods of cooling and drying climate. When they ended, temperatures warmed and precipitation increased. We emphasize the cooling periods because they did the most damage to humans and their strategies to gain adequate access to food and water. Now, climate circumstances have been altered by human activities that have radically increased the amount of CO_2 in the atmosphere, which has led in turn to rising temperatures accompanied by drying tendencies. This era of global warming is not yet rapid climate change (RCC). So far, temperatures have been rising gradually unlike past episodes that came on abruptly. But we should not assume that the processes currently at work cannot accelerate in a nonlinear fashion that causes abrupt climate change that would qualify it as an RCC.

In some respects, it may not matter whether the rising temperatures come about gradually or abruptly. As far as we can tell, their ultimate impact will be similar in magnitude and effect to the "creative destruction" impacts of the past nine RCCs. Agrarian strategies (throughout the world) will be sorely challenged to produce sufficient food. Some cities will no longer be able to function. Water will become even scarcer than it already is. Staying alive will become a great deal more difficult. Strong migratory pressures will be exerted. But there are important differences as well. The size of the MENA population today is much greater than it was in the past and it is still expanding quickly. Thus, they will need far more resources to sustain themselves than in the past. Unlike in the past, moreover, there may also be no place for climate migrants to go, especially if there are hundreds of thousands of them. These are dire prospects indeed. The parts of the MENA that were affected by the earlier nine episodes of cooling and drying were able to renew and reorganize, subject to variable lags of time. It is conceivable, however, that if some of the worst-case scenarios for global warming are realized, the negative impacts will be so great that no amount of resilience will suffice to revive the MENA region.

The aim of this chapter is to ascertain where and how badly the current episode of climate change is likely to affect the MENA. We lack a crystal ball that tells us exactly what will happen in this century anywhere. But we do have access to a number of forecasting efforts that make various explicit assumptions about the future and proceed to model processes on an if x, y, and z, what then basis. These forecasts may prove to be wrong, yet they are just as likely to be too conservative in guesstimating what is coming down the pike as they are too extreme. They tend to be linear projections and may miss nonlinear interactions and threshold effects that will make matters even worse than already forecasted.[1]

Nonetheless, the basic global warming dilemma is that we do not know how high and how fast high temperatures will climb. The latest warning from the Intergovernmental Panel on Climate Change or IPCC (2018) is that processes are moving even faster than believed a few years before. The ascent of the average temperature increase appears to be becoming steeper. Only rather drastic action that needs to be taken within the next decade or two is likely to have much impact on heading off runaway higher temperatures.[2] The Paris Agreement was not meant to guarantee a world in which temperature increases were limited to 2 degrees centigrade. Nor did the agreement seem to have much effect on decreasing carbon dioxide emissions associated with fossil fuels. It is true that world carbon dioxide emissions appeared to be plateauing in the temporal vicinity of the Paris Agreement but data for 2017 and 2018 suggest that was a brief illusion. Instead, CO_2 emissions are again on the upswing. As a consequence, increases on the order of 3–4 degrees centigrade seem far more likely. As there is little evidence of movement toward the types of drastic actions that are deemed necessary and some leading emitters of carbon dioxide even deny that there is any problem that needs addressing, the prospects for a near-term, radical response seem to have a low probability. Thus, focusing on worst-case or near-worst-case scenarios for the MENA seems highly warranted.

Accordingly, we first summarize a list of expectations about how global warming might influence life in the MENA. Then, we give more attention to selected topics, including food insecurity, water scarcity, population growth, poverty, and migration. Ordinarily, expectations about increased conflict would be included in such surveys. However, the complications associated with the climate change-conflict relationship require a separate chapter (Chapter 8).

Why the MENA is likely to be a ground zero for global warming

As we have seen in the past, the MENA region is particularly vulnerable to the impacts of climate change and that will not change in the 21st century CE. Just how bad is indicated by the World Bank's selected predictions on a 4-degree centigrade impact on the MENA. It is expected that

- MENA temperatures could average up to 7.5 degrees centigrade warmer by 2100 than they were in 1951–1980.
- Sixty-five percent of summer months will be considerably warmer by the last three decades of the 21st century and 80 percent by 2100. A large but indeterminate percentage of the region will have minimum nighttime temperatures and maximum day time temperature increase by 6 degrees centigrade by 2081–2100, in comparison to temperatures recorded in 1981–2000.
- Water availability will decrease in most parts of the region – as much as 45 percent – throughout the 21st century. Countries located on the Mediterranean (Morocco, Algeria, Egypt, Turkey) will see 50 percent less rainfall. Saudi Arabia, Iran, and parts of Libya will also experience about the same decrease in precipitation. Oman and Yemen, in contrast, will need to endure an increased intensification of extreme tropical rainfall events.
- Mountain snowpacks that feed rivers will shrink (25–55 percent and perhaps more in the Anatolian mountains that feed the Tigris and Euphrates and 70 percent in the Nahr al Kalb area in Lebanon).
- The number of days of drought around the Mediterranean should increase more than 50 percent by 2070–2099. Morocco, Algeria, and Tunisia are likely to be particularly vulnerable to drought. All land adjacent to the Mediterranean will become more arid. Ironically, drought intensification may decline in Iran and Saudi Arabia.
- Storm intensity should increase leading to more powerful storm surges.
- One hundred million people are expected to live in the MENA coastal cities by 2030. Yet coastal cities will be damaged by rising levels of seas, with flooding losses increasing as much as tenfold from 2005 to 2050. Alexandria, Benghazi, and Algiers will be especially vulnerable to the need to abandon some coastal cities. Qatar, the U.A.E., and Tunisia are also thought to be unusually vulnerable to rising sea levels.

- Salinization, of both the land irrigation and seawater intrusion varieties, will increase.
- Lower rainfall and higher temperatures will reduce crop growing seasons by 2 to 4 weeks prior to 2050.
- Increases in temperature, less precipitation, and more extreme events should reinforce the expansion of desertification.
- Some 70 percent of the MENA cropland depends on rainfall and 60 to 90 percent of the MENA water goes to agriculture. But rain-fed cropland could be reduced by 8,500 square kilometers by 2050 and 170,000 square kilometers by 2100. Crop yields, as a consequence, could decline as much as 60 percent (without taking into consideration variations in the region or possible crop adaptations). The reductions will be much greater in the eastern part (Mashreq) of the region than in the western part (Maghreb).
- Slightly more than a third of the MENA population is employed in agriculture and 13–26 percent of regional gross domestic product (GDP) is dependent on agricultural production. An increased inability to grow food will make food less available and more expensive to purchase.
- Major livestock losses – perhaps on the order of the 85 percent losses already incurred in Syria (2005–2010) – can be anticipated.
- The MENA is already dependent on food imports (50 percent wheat and barley, 40 percent rice, and 70 percent maize). Either more food will need to be imported as the population grows and local food production declines or less food will be available, regardless of source. Morocco, Algeria, Lebanon, and Egypt are especially vulnerable to food import decreases.
- Higher air and water temperatures and extreme weather events should make some types (thermal, hydro) of electricity generation more difficult. The price of energy and the likelihood of power outages should increase. Yet the MENA population growth rates suggest the need for a threefold expansion of energy supply. Air conditioning, to say the least, will seem increasingly essential to mere survival.
- Vector- and water-borne diseases (for instance, malaria, cholera, elephantiasis, bilharzia, salmonella, typhoid, dysentery) should increase due to higher temperatures. Elderly and the young will be most vulnerable to increased mortality rates in this context.
- A combination of reactions to extreme heat and increased disease could reduce overall economic productivity by 10–40 percent between 2010 and 2030.
- The tourism industry should decline quickly with a 10–25 percent decline as early as 2025.
- Ordinarily, migration would be a likely adaptation to this onslaught of environmental deterioration but it is not clear where climate migrants will be able to go. There is not likely to be sufficient variation within the region for migrants to move locally which has been the most likely variant in the past. Southern Europe will be almost equally hard hit by global warming.

More northern destinations have all expressed reluctance to take in large numbers of Middle Eastern migrants. At the same time, Sahelian and sub-Saharan African climate migrants will be attempting to move to or through the MENA as well.

- Societal crises and conflict should be expected to increase. Those states that are hardest hit by crisis and conflict will possess the least capacity to respond to the intensification of problems that need resolution in some fashion. Demonstrations of resilience or adaptation should be less likely and less operational in the absence of state capacities to address the onslaught of problems.

It seems fair to say that these expectations based on climate modeling point to a dismal future for the MENA.[3] Extreme temperatures, even further diminished water supply, rampant food insecurity, and political-economic turmoil suggest that climate change will penetrate every sphere of life and make it far less pleasant. These outcomes will not be duplicated exactly in each and every MENA state but the impacts are expected to be widespread, comprehensive, and severe. For instance, Al-Delaimy (2020) suggests that maximum summer temperatures given a 4-degree centrigrade temperature increase could result in 80% of summer months reaching warmer temperatures than 5 standard deviations above the average by 2100. Al Blooshi et al. (2019) reinforce that prospect by suggesting that a large number of MENA states could see temperatures rise by 6 degrees centigrade by the end of the century. Such unprecedented temperature increases (especially in Baghdad, Beirut, Kuwait, Riyadh, and Tehran) will, if nothing else, increase the likelihood of heat strokes and death. When climate change of this severe magnitude impacted the MENA region in earlier episodes, there were usually some local refuges. This time there do not appear to be many places in the region to go to escape what may be coming.

Nonetheless, there are specific issue areas that demand more specific attention. In this chapter, we focus on food insecurity, water scarcity, population growth, and migration.

Food insecurity

Other parts of the world are in worse shape than the MENA (Table 7.1) when it comes to evaluating food insecurity. Developing regions have 3 to 4 times as much food insecurity as developed regions. Among the developing regions, the different parts of sub-Saharan Africa lead the list of most insecure but the MENA falls into second place (albeit more in the eastern zone than in the west of the region).

Nonetheless, there are other problems than food access per se subsumed under the food security umbrella. The MENA population is becoming more heterogeneous in terms of what types of food it consumes. Table 7.2 offers a World Health Organization categorization applicable primarily to child malnutrition. The first

TABLE 7.1 The Prevalence of Modest and Severe Food Insecurity

Region	Prevalence (%)
World	19.9
Developed Regions	8.3
Developing Regions	29.5
Sub-Saharan Africa	55.6
Eastern	58.7
Western	51.8
Northern	24.0
Southern	42.5
Middle	66.4
Middle East and North Africa	31.4
Mashriq	37.6
Maghreb	17.3
Asia	23.3
Southern	26.5
Southeast	19.8
Western	29.7
Latin America	20.3

Source: Based on Food and Agriculture Organization of the United Nations (2017: 3).

TABLE 7.2 Nutritional Challenges across the MENA

Categories and Countries	Description
1 Areas with long-term complex emergencies (Syria and Yemen)	Severe child and maternal undernutrition and widespread micronutrient deficiencies
2 Areas with significant undernutrition (Iraq, Yemen, and some population subgroups in the GCC, Iran, Palestine, and Tunisia)	Particularly high levels of acute and chronic child malnutrition, widespread micronutrient deficiencies, and emerging obesity and malnutrition of affluence in certain socioeconomic groups
3 Areas in early nutrition transition (traditional diets high in cereals and fiber as in Egypt, Jordan, Lebanon, Morocco, and Palestine)	Moderate levels of overweight and obesity, moderate levels of undernutrition in specific populations and age groups and widespread micronutrient deficiency
4 Areas in advanced nutrition transition (diets of high sugar and fat content as in Tunisia and most GCC countries	High levels of overweight and obesity and moderate levels of undernutrition with some micronutrient deficiencies in some population subgroups

Source: Based on Jobbins and Henley (2015: 16).

category no doubt applies to parts of contemporary Syria and certainly much of Yemen. The impetus of the past several decades has been to work toward eliminating child malnutrition, which currently applies to some 33 million MENA children. More global warming is likely to expand the countries ensnared in long-term complex emergencies. For that matter, the entire region could be characterized as involved in a long-term complex emergency if temperatures rise sufficiently. Regardless, the clear expectation is that trends toward diminishing child malnutrition are likely to be reversed in the coming decades.

We might anticipate more countries moving into the second category as the food supply becomes more constrained. The likely consequences include greater child mortality rates, less healthy people (both as children and adults), and individuals of shorter stature than might otherwise have been the case (referred to technically as "stunting" in children). The consumption of more fruits and vegetables is the standard recommendation for addressing these problems. But locally raised fruits and vegetables could be eliminated entirely if global warming advances far enough. These nutritional problems, moreover, have been highly concentrated in rural areas – again, precisely where the most severe impacts of higher temperatures and less water will be felt first.[4]

Two of the reasons for this outcome are the region's extensive aridity and the extensive complications of food insecurity processes that are captured in Table 7.3, which generalizes the most probable ways in which climate change is expected to affect food insecurity. Land suitable for growing crops is greatly restricted in the region. There are really only a few small islands of rainfed and irrigated cropland. There are few fertile reserves to fall back on in hard

TABLE 7.3 Potential Climate Change Impacts on Food Security

Food Security	Likely Impacts
Availability	• Reduced rainfall and increased evapotranspiration reducing yields from rainfed agriculture and pastoralism • Reduced soil fertility and increased land degradation from increased temperatures, evaporation, and drought • Increased crop and livestock pests and diseases
Access	• Loss of agricultural income due to reduced yields and higher costs of production inputs • Higher global and local food prices • Increased difficulties due to displacement driven by climate extremes and disasters
Stability	• Greater supply instability due to increased frequency and severity of extreme events • Greater agricultural income instability
Utilization	• Less food safety due to increased temperatures • Less nutrition due to reduced water quality and quantity • Increased morbidity due to extreme climate events

Source: Based on Jobbins and Henley (2015: 9).

times. Table 7.3 emphasizes the availability, stability, and utilization of food, all of which are likely to be influenced by climate change.

Surprisingly, there are no strong expectations of opening up new land for agricultural purposes in the next half-century as demonstrated in Table 7.4.[5] One reason is that land that might be cultivated but has not been so far being that it is too hard to get to or lacks access to water. But Table 7.4 also underscores that the MENA has the least amount of arable land and absolutely no expectations of expanding the amount currently in use. Whether it can maintain the 84 million hectares (ha) currently in use despite land degradation, water shortages, and sea level increases remains to be seen.[6] But the prospects are not encouraging.

As for the complexity of food insecurity (Figure 7.1), the inter-connections are multiple. Climate change has direct impacts on land degradation, water scarcity, and allegedly conflict.[7] The land degradation link leads to how much arable land is available and, consequently to food supply. The climate change water scarcity track leads to the same place – food supply. Less food supply equates to greater food insecurity. Conflict is made more likely by water scarcity and climate change. But conflict also feeds into the food supply by ruining fields and making cultivation impossible. Food becomes weaponized and thus can be used to increase food insecurity both indirectly and directly. On the other side of the figure, food access competes with food supply to see which factor can outdo the other. Impoverished people are the first to lose access to food, especially in periods of conflict. Imports become far less likely no matter how dependent local populations are on external food supply. Price hikes alone drive people away from food access. Only the rich can afford access to food in times of high scarcity. Thus, food insecurity in a region characterized by climate change, water scarcity, poverty, and food import dependency is unlikely to arrive at a resolution of its many simultaneous problems. Instead, they tend to be overwhelming in their multiplicity and direct and indirect linkages.

Much of the food in the Middle East is grown by overpumping aquifers, which means that non-river irrigation ends whenever these wells go dry. Without irrigation, food production cannot be sustained if precipitation also decreases

TABLE 7.4 Total Arable Land in Use (Million ha)

	1961/63	*2005/07*	*2030*	*2050*
World	1372	1592	1645	1661
Sub-Saharan Africa	133	240	266	291
Latin America	105	202	235	251
MENA	86	84	84	84
South Asia	191	206	210	213
East Asia	178	236	241	236

Source: Based on Alexandratos and Bruinsma (2012: 109).

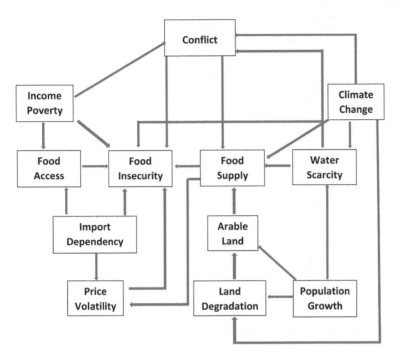

FIGURE 7.1 The dynamics of food insecurity

significantly. Overall, half of the world's population (nearly 3.8 billion) relies on overpumped aquifers for food production and many of those people happen to be in the MENA (see Table 7.5). For example, Saudi Arabia which had been self-sufficient in wheat since the 1990s announced in 2008 that it would be phasing out domestic wheat production due to the depletion of its main aquifer. Today, Saudis are dependent on imported grain to feed 30 million people.

In neighboring Yemen, prior to the current civil war, 14 of the country's 16 aquifers had been overpumped, prompting some analysts to call it "a hydrological basket case" (Brown 2011: 22). With water tables rapidly falling in a country with one of the world's fastest-growing population, the Yemenis now import more than 80 percent of their grain. Yet one cannot assume that food imports from outside the Middle East can be expected to continue at current levels let alone increase to compensate for deficiencies within the region. Unlike Saudi Arabia, which can scour the world for more land and water resources, Yemen, as the poorest of the Arab countries, is facing a bleak future. Already a failing state, with hunger and thirst mounting, Yemen's internal conflicts are likely to persist into the future and spill over onto its neighbors.

Of course, there is a more optimistic interpretation. Tony Allan (Allan, 2001; Allan, 2011) argues that the Middle East ran out of water in the 1970s.[8] However, he emphasizes that this seemingly pessimistic view is predicated on including

TABLE 7.5 Countries Overpumping Aquifers, 2015

Country	Population (in millions, 2017)
Afghanistan	35
China	1,409
India	1,339
Iran	81
Iraq	38
Israel	8
Jordan	10
Lebanon	6
Mexico	129
Morocco	35
Pakistan	197
Saudi Arabia	33
South Korea	51
Spain	46
Syria	18
Tunisia	11
USA	324
Yemen	28
Total	3,798

Source: Modified from Earth Policy Institute, with population numbers taken from United Nations (2017: 17–21).

water needed for food production as part of the definition of water sufficiency. If one adopts the position that national economies should acknowledge their endowment advantages and disadvantages and base their trade on comparative advantages, states with water deficits should import products from states with water surpluses. The imports of water-intensive products, which certainly includes food, makes up for the deficits in water via something called "virtual" water. The solution then is to expand trade in virtual water products which if pursued to the maximum extent possible would liberate water deficit countries from counting the 80 to 90 percent of their water consumption allocated to agricultural production. Water deficits are thus eliminated via trade in virtual water products.

Unquestionably, the MENA states have engaged, and increasingly so, in acquiring virtual water by importing food. Food availability has improved (Porkka et al., 2013). Yet there seems to be some limitations on the grounds for optimism in the future. Woertz (2013: 2), a virtual water optimist, inadvertently raises three problems

> The virtual water trade has added the equivalent of a second river Nile to the region's water balance. It is a necessary precondition for food security

in the Middle East and will increase in importance in coming years.... As long as other countries produce enough exportable surplus and the Middle East has the money to pay for it, there should be no problem -provided food accessibility of poor people is guaranteed by sufficient entitlements....

It remains unclear whether in fact global food production will or can be expanded sufficiently to accommodate a greatly expanded demand for food imports. In the 2008 recession, some food exporters reduced the amount of surplus food they had been placing on the open market. Why should we not expect that to happen again?[9] Or, imagine a large number of countries doubling or tripling their need for food imports more or less at the same time thanks to the ravages of global warming.[10] How would this demand be met assuming food exporter willingness? Food waste can be reduced. Food production can be expanded through technological innovation and land allocated to food production purposes might be expanded conceivably. Diet in the main food exporters (like the United States) might be reformed to permit larger food surpluses. But will all of these efforts occur magically in time to preserve the functioning of an international food market under the stress of the increased demand that will be associated with global warming? Increased prices for food seem probable as well. Can we assume that imported food will remain affordable?

If the MENA farming is allowed or forced to wither on the vine, where will all of the people involved in agriculture in one way or another find employment? In some states, the agrarian sector employs 40 or more percent of the population. Can we rely on MENA government subsidies and transfers to provide a safety net for so many? Where will the money for the subsidies come from? Gulf states may be able to afford to buy much of their food abroad but that capability is contingent on continuing oil and gas revenues. At some point in the 21st century, those fossil fuel revenues are highly likely to be considerably diminished due either to declining supply or demand or both.[11] Outside of the Gulf and other oil producers, most MENA states will have little to trade for virtual water products in the future.[12] That leaves increased food aid as the remaining possibility but the prospects for that recourse are less than auspicious. Assuming willingness of sources of food aid (which really cannot be assumed), the magnitude of what could be needed is a scary proposition. Thus, the concept of virtual water is useful and food imports have relieved some of the water scarcity problems. But it is unclear that the food trade will be able to keep up with a world of increasing demand for food imports and increasing scarcity. If the relief valve of international exchange and global food markets become overwhelmed in the future, increasing food insecurity seems likely.

Water scarcity

MENA freshwater resources are not only among the lowest in the world; they are also expected to fall over 50 percent by 2050. As we have noted earlier, much of the region's vulnerability to climate change concerns agriculture, which is most directly affected by warming temperatures and declining water tables. Ninety percent of the region already lies within arid areas and 45 percent of the total agricultural area suffers from soil erosion. At the same time, agriculture in the region uses approximately 85 percent of the total available freshwater. Table 7.6 provides a summary of various indicators used to measure the region's vulnerability to climate change particularly the availability of freshwater per capita. The data in Table 7.6 (column 3) measure water scarcity for each country by dividing each country's yearly available renewable water supply by its total population. When this number drops lower than 1,700 cubic meters per capita per year, a country is said to be "water-stressed"; anything below 1,000 is said to be suffering from chronic or absolute water scarcity, which is another way of saying "below the water poverty line"

TABLE 7.6 Summary of Water Indices in the MENA

Country	Population (in thousands, 2017)[a]	Total renewable water availability per capita, (in m³/year, 2014)[b]	Total water withdrawal per capita, (in m³/year)[b]	Freshwater withdrawal as percent of available freshwater resources (2015)[c]	Water Stress Index
Bahrain	1,493	77.70	506 (2003)	132.16%	absolute scarcity
Egypt	97,553	589.40 (2017)	999 (2010)	118.94%	chronic scarcity
Iran	81,163	1,688.00	1,356 (2004)	81.42%	water stressed
Iraq	38,275	2,348.00	2,632 (2000)	54.15%	relative sufficiency
Jordan	9,702	96.85	165 (2005)	100.07%	absolute scarcity
Kuwait	4,137	4.83	375 (2002)	2,075.00%	absolute scarcity
Lebanon	6,082	740.40	366 (2005)	57.26%	chronic scarcity
Oman	4,491	302.00	526 (2003)	90.07%	absolute scarcity
Qatar	2,639	21.98	546 (2005)	374.14%	absolute scarcity
Saudi Arabia	32,938	72.86	963 (2006)	907.08%	absolute scarcity
Syria	18,270	919.50	921 (2003)	125.96%	chronic scarcity
Turkey	78,666	2,621.00	563 (2003)	42.90%	relative sufficiency
UAE	9,400	15.96	889 (2005)	1,866.67%	absolute scarcity
Yemen	28,250	74.34	187 (2000)	168.28%	absolute scarcity

Note: The following values provide general measurement of water stress levels as indicated in column 2 above.

< 500 [absolute water scarcity]
500–1,000 [chronic water scarcity]
1,000–1,700 [water stressed]
1,700–5,000 [relative sufficiency with occasional or local water stress]
> 5,000 [abundant water resources nationally, stress possible locally]

Source: [a] United Nations (2017: 17–21); [b] Food and Agriculture Organization of the United Nations (2019); [c] United Nations (n.d.), Water Stress Indicator Database.

(Ohlsson 2000). To put things in perspective, total renewable water resources for the world average 7,453 m^3 per person per year. Per capita water scarcity will continue to increase due to population growth and climate change. Only three countries in the MENA (Turkey, Iraq, and to some extent Iran) appear to be relatively sufficient in their freshwater supply, at least for now. Others are vulnerable to economic, social, and environmental problems caused by unmet water needs.

Most of the MENA is already under severe water stress, defined by the California Water Sustainability organization as "the relationship between total water use and water availability." In other words, water scarcity refers to the lack of sufficient water resources to meet demand during a certain period. Of the 15 countries examined in the Middle East, 12 experience either absolute or chronic water shortages affecting over 200 million people. Water scarcity can be *absolute*, which according to the Food and Agriculture Organization of the United Nations (FAO), is "an insufficiency of supply to satisfy total demand after all feasible options to enhance supply and manage demand have been implemented. This situation leads to widespread restrictions on water use." Water scarcity can also be *chronic*, which refers to "the level at which all freshwater resources available for use are being used. Beyond this level, water supply for use can only be made available through the use of non-conventional water resources such as agricultural drainage water, treated wastewater or desalinated water, or by managing demand" (FAO).

The data in Table 7.6 (column 5) show just how dire water scarcity problems are. Except for Iraq, Lebanon, Turkey, Iran, and Oman, other countries in the MENA use anywhere from 100 percent of their total available freshwater sources to 2,000 percent in the case of more affluent Gulf petrostates. Such water use levels in the region are clearly unsustainable and at this rate, the region could run out of freshwater in as little as 25 years.

Some argue that assessing water scarcity on an annual basis (as presented in Table 7.6) should be taken with the proverbial grain of salt because it tends to hide the monthly fluctuations in water availability. One criticism is that annual water data could underestimate the extent of water scarcity. Unfortunately for the MENA, the difference between annual versus monthly water data does not seem to matter all that much. Mekonnen and Hoekstra's spatial analysis of global water scarcity on a monthly basis suggests that the MENA suffers from "moderate to severe water scarcity during more than half of the year" with water scarcity in the Arabian Desert as being "worse than that in other deserts because of the higher population density and irrigation intensity" (2016: 2). The Arabian Peninsula also happens to be an area that relies heavily on groundwater, drawing on fossil (i.e. nonrenewable) aquifers to meet current demands (Sowers, Vengosh, Weinthal 2011: 605), which has led to the overpumping of most of these fossil aquifers. Generally, water scarcity occurs only during part of the year when there is a mismatch between water availability and demand, but the MENA countries are water stressed pretty much all year round.

As depressing as the past and current water stress indices are for much of the Middle East, they are likely to become worse by the second half of this century (see Table 7.7). According to the World Resources Institute, the entire region, with the exception of Egypt, is expected to suffer from water shortages in the next three decades. Egypt, despite being the most populous Arab state in the Middle East is comparably well endowed with renewable water resources, since it receives 96 percent of its water from the Nile. However, Egypt's remarkable population growth is expected to put severe stress on the Nile as Egypt's current population of 91 million expands to 150 million by 2050 and 200 million by the end of this century (UN World Population Prospects). Nor can Egypt expect to utilize the Nile water without challenges from several other states upstream – one of which, Ethiopia, is currently in the process of restricting the Nile's flow to Egypt.

In another study, Luo, Young, and Reig (2015) project future country-level water stress indices for 2020, 2030, and 2040 under three scenarios: business-as-usual, optimistic, and pessimistic. Table 7.7 summarizes the water stress scores for the business-as-usual scenario for each of the Middle East states as there is no compelling evidence for optimism as of yet.[13] On a scale of 0–5, any value above 4 in Table 7.7 indicates that a given country is extremely water stressed which means that these countries use up over 80 percent of their total annual available

TABLE 7.7 Water Stress Forecast for MENA, 2020, 2030, 2040

Country	2020	2030	2040
Bahrain	5.00	5.00	5.00
Egypt	1.37	1.48	1.53
Iran	4.81	4.86	4.91
Iraq	4.28	4.49	4.66
Jordan	4.80	4.84	4.86
Kuwait	5.00	5.00	5.00
Lebanon	4.75	4.89	4.97
Oman	4.98	4.94	4.97
Palestine	4.83	4.95	5.00
Qatar	5.00	5.00	5.00
Saudi Arabia	5.00	5.00	4.99
Syria	4.27	4.37	4.44
Turkey	3.85	4.10	4.27
United Arab Emirates	5.00	5.00	5.00
Yemen	4.95	4.88	4.74

Source: Luo, Young, and Reig (2015).

Score Value

[0–1] Low (<10%)
[1–2] Low to medium (10–20%)
[2–3] Medium to high (20–40%)
[3–4] High (40–80%)
[4–5] Extremely high (>80%)

freshwater to meet their municipal, industrial, and agricultural needs. With the exception of Egypt, all of the values in Table 7.7 exceed the "extremely high" threshold by 2020 and only become worse by 2040.

Increasingly severe periods of prolonged water scarcity will lead to low crop yields and crop failure forcing farmers into already crowded cities plagued with high unemployment rates. Crop failure contributes to food price increases as well as famine and widespread starvation, a situation currently being observed in places like Syria and Yemen, which are experiencing some of the worst droughts in their recent history.[14]

As we have noted in the previous section, some observers think that the answer to water and food scarcity is obvious. If states lack sufficient water to grow food, they can import what they need from the international market for food produced by states that have fewer water problems. This type of exchange between states with water deficits and surpluses makes a great deal of sense to economists who reflexively promote relying on comparative advantages. While this "virtual water" trade has helped so far, the question remains whether the food trade is likely to persist into the future. We remain skeptical.

Population growth

Aside from the country specifics on the water supply side, most states in the MENA share one common development on the demand side: significant population growth (Table 7.8). As mentioned above, Egypt is expected to see the strongest growth with an increase of 110 million people by 2100. Iraq's population is projected by the UN Population Division to increase nearly fourfold to

TABLE 7.8 Population Growth Forecasts for MENA in the 21st Century, in Thousands

State	Population 2017	Population 2050	Population 2100
Bahrain	1,493	2,327	2,246
Egypt	97,553	153,433	198,748
Iran	81,163	93,553	72,462
Iraq	38,275	81,490	155,556
Jordan	9,702	14,188	17,319
Kuwait	4,137	5,644	6,231
Lebanon	6,082	5,412	4,350
Oman	4,636	6,757	6,572
Qatar	2,639	3,773	3,971
Saudi Arabia	32,938	45,056	44,029
Syria	18,270	34,021	38,167
Tunisia	11,532	13,884	13,321
Turkey	80,745	95,627	85,776
UAE	9,400	13,164	14,776
Yemen	28,250	48,304	53,536

Source: United Nations (2017: 17–21).

163 million by the end of the century. Even Jordan, which is already strained for resources, is expected to double its population in the same time frame. Only Iran and Lebanon should see some decline in their population growth but the rest of the region will experience both relative and absolute population growth, which will increase the demand for food and water on the national and regional scale. "Interlinked with this development are the processes of urbanization," say Schilling et al (2012: 16), "which could lead to highly localized peaks in water and food demand."

Even in the absence of climate change, it is quite conceivable that the MENA would be facing a major crisis given the combination of water scarcity and population growth. If water supply is already a problem, imagine what the MENA will look like with roughly twice the population it currently possesses and no increase in the water supply. Add climate change to the mix with its likely diminishment of what water sources currently exist and the intensity of the future crisis will be all the more difficult to resolve.

Poverty

Increased poverty is another anticipated effect. MENA is a heterogeneous region when it comes to income distribution. The Gulf states are very wealthy; the rest of MENA is the opposite. Yet the oil affluence tends to give outsiders an image of a generally wealthy MENA (Abu-Ismail, 2015). On the contrary, however, 40.6 percent of the Arab world's population was found recently to be impoverished. One in four children lives in acute poverty (Abu-Ismail et al., 2017).[15] Yemen and Sudan are among the least affluent while Egypt, Algeria, Tunisia, and Jordan tend to have fewer pockets of poverty. Iraq and Morocco fall in between these two groups. Yet the threat of global warming has two very specific implications. The hardest-hit groups will be the rural poor and more people will be pushed into extreme poverty. Again, according to Abu-Ismail et al. (2017) if nearly 41 percent of the population is already poor or severely poor, another nearly 37 percent are vulnerable to falling into either the poor (25%) or acutely poor (11.8%) categories. Thus, with a little push from climate change, two-thirds of the MENA could become impoverished.

Nor is climate change gender-neutral. Rural women constitute a disproportionate share of the agricultural workforce and as such are more likely to be prime victims of rising temperatures.

> Women, who comprise the majority of the global agricultural workforce – including between 45 and 80 percent in developing countries – must adapt to increased instances of drought and desertification. When there is water scarcity in rural settings, men tend to leave their communities to search for employment outside of cultivating crops. Women become the heads of households and assume responsibilities traditionally assigned to men, but often do not have the same authority, decision-making power, or access

to community services, education, or financial resource. They may also be undercut by discriminatory laws and customs that prevent them from being able to acquire, own or retain land or other assets, such as livestock.

(Alam, Bhatia, and Mawby, 2015: 27)

It should go without saying that in a culture characterized rather strongly by female inequality, the last thing that is needed is stronger environmental pressures to subordinate women further, to keep them away from education opportunities, and to make their lives more miserable by increased workloads.[16] Yet extreme climate change is likely to perpetuate and perhaps even intensify the marginalized role of rural women in particular. Of course, it is also fair to say that all marginalized people are likely to bear a disproportionate share of the climate change burden because poor people will have few, if any, resources to combat the deprivations associated with higher temperatures and less water.

Migration

Migration impulses can seem as complicated as the food insecurity diagram in Figure 7.1. Figure 7.2 offers a much-simplified version. Climate change can exert impacts in a variety of ways. As a consequence, the push drivers are multiple and can overlap: environmental: direct and indirect exposure to impacts, loss of livelihood, resource depletion; social: marginalization, education; economic: poverty, low prices for producers, loss of livelihood; political: discrimination, instability, weak governance, corruption, coercion; and demographic: population growth, disease. The push (and pull) drivers interact with personal characteristics (gender, marital status, wealth, ethnicity, religion, family size, age) and obstacles and facilitators that include migration costs, distance, social networks, diasporic linkages, and political/legal considerations.[17] Individuals and families make their decisions by juggling and weighting these various factors in different ways.

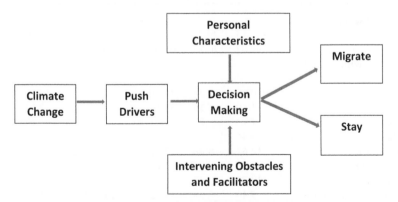

FIGURE 7.2 Migration decision-making. Source: A simplification of climate-induced, migration decision-making based on World Bank (2014: 142) which in turn acknowledges a debt to Foresight (2011).

One constraint on gauging the likely migration response to global warming in the 21st century is that we have yet to see how bad its effects will be. We are currently experiencing at least one-degree centigrade increase in temperatures. Responses to one degree may not be very similar to what will happen when the temperature increase climbs to 4 or 5 degrees centigrade. The difference could end up being similar to the response to a day of light rain versus how people react to 40 continuous days of unending downpours and flooding. Nonetheless, we do have some data on early MENA reactions.

A survey of selected communities in Algeria, Egypt, Morocco, Syria, and Yemen in 2011 (Adoha and Wodon, 2014) offers some useful comparative information given variation within MENA in terms of the extent of global warming impacts and the local abilities to absorb the environmental shocks. People surveyed were certainly aware of ongoing climate change effects. Table 7.9 lists percentage responses to whether respondents had been affected by extreme events and, if so, what kind of extreme event had had the most effect. In four of the five states, almost all respondents had experienced extreme weather events. Only in Egypt did almost a third of the respondents claim no direct link to deteriorating environmental shocks.

Exactly which type of event was most salient in peoples' memories varied considerably. In Algeria, excessive heat nosed out drought and pest infestation. In Egypt, excessive heat and rain were the two leading types of extreme events. Moroccans, on the other hand, were more impressed by flooding while Syrians were traumatized by drought. Yemenis also put drought well ahead of excessive heat. Yet Table 7.10 shows somewhat less variance in terms of

TABLE 7.9 Familiarity with Extreme Environmental Events and Rank Order of Impact (% Responding)

	Algeria	Egypt	Morocco	Syria	Yemen	All
Affected by Disaster						
No	.13	29.25			.62	5.99
Yes	99.87	70.75	100.00	100.00	99.38	94.01
Adverse Event with Largest Impact						
Drought	10.92	2.38	14.30	90.00	27.89	30.90
Flood	1.60	.13	34.56		1.38	7.54
Storms	1.72	.25			.38	.47
Mudslides	8.46				.25	1.74
Excessive heat	12.67	8.88	13.21	.75	5.28	8.16
Excessive rain	4.56	5.63	4.19		2.64	3.40
Pest infestation	10.73	.25		.13	.25	2.27
Crop and livestock diseases	7.40	5.38	.53	.13	1.26	2.94
No adverse impact	41.93	77.13	33.21		60.68	42.57

Source: Based on Adoha and Wodon (2014: 99).

TABLE 7.10 Environmentally Related Events with Most Impact (% Responding)

Events with Most Impact on Respondents	Algeria	Egypt	Morocco	Syria	Yemen	All
Lost Income	58.11	8.25	44.90	19.50	52.11	36.59
Lost crops	58.48	28.63	38.00	87.00	60.95	54.62
Lost livestock/cattle	31.21	3.75	26.92	17.00	38.18	23.43
Less fish caught	0.00	.88	14.77	1.50	25.75	8.60

Source: Based on Adoha and Wodon (2014: 100).

climate-change-induced losses. Crop losses led, followed by income losses and livestock losses. Only the role of fishing losses singles out Yemen from the other four states.

Thus, we can start with the premise that climate change deniers may not be all that common in MENA. Yet overt links to climate change do not characterize why people in these five states say that they migrated.[18] Still, the indirect linkages are not that obscure. Table 7.11 lists the first and second reasons for households migrating. The two leading explanations are variations on improved employment prospects somewhere else. Only around 6 percent cite fleeing extreme climate events (flooding and drought). The structure of the second reason is fairly similar. Better employment somewhere else factors lead, followed by incentives to save money, with escaping environmental problems in a distant fourth place. But if we combine the two reasons, it doubles the number of respondents referring to environmental problems.

The interpretation problem is that climate change is undoubtedly responsible to some extent for the dismal employment prospects at the origin. For the most part, in-country migrants come from rural areas in which agriculture is the main source of employment. Moreover, people tend to migrate to places with either better climate for agriculture or to places in which climate is not a direct obstacle to making a living (cities). Some portion of the "better employment at destination" is thus also climate related. Just how much of this movement is related to climate change is therefore hard to specify with any precision. It is clearly an important factor, however.[19]

A clever, alternative research design (Missirian and Schlenker, 2017) studied requests for European Union (EU) asylum between 2000 and 2014. By determining the climate of the requestors, the authors found that asylum requests were U-shaped – that is primarily from fairly hot or cold places. Asylum requests from places with an average temperature of around 20 degrees centigrade (approximately 68 degrees F) were relatively rare. Utilizing these data for simulation purposes, the authors found that the forecasted relationships between the number of asylum requests and increasingly greater warming were linear. If the temperature increased 3 degrees centigrade, the number of requests was likely to increase 300%. A 5 degrees centigrade increase would

TABLE 7.11 Reasons for Migrating within Country (% Responding)

	Algeria	Egypt	Morocco	Syria	Yemen	All
First Reason						
Better employment at destination	12.78	41.76	38.58	41.47	24.06	35.04
Lack of employment at origin	22.81	20.00	14.23	27.00	14.97	20.68
To accumulate savings	4.33	11.18	0.00	5.18	8.56	5.09
Transferred (job)	0.00	2.35	.41	.43	4.81	1.27
Schooling	0.00	1.76	1.67	1.08	4.81	1.74
Better infrastructure	3.75	0.00	5.52	1.08	4.28	2.80
Join family	0.00	2.94	2.23	7.34	5.35	4.38
Marriage	56.33	11.18	33.31	.43	18.72	18.92
Divorce/separation/death of spouse	0.00	.59	0.00	0.00	.53	.16
Delivery	0.00	.59	0.00	0.00	0.00	.08
Family problems	0.00	2.94	.57	.65	5.35	1.55
Accompanying patient	0.00	0.00	.72	0.00	0.00	.18
Escape flood	0.00	0.00	1.91	.22	0.00	.55
Escape drought	0.00	0.00	.86	14.90	0.00	5.60
Other	0.00	4.12	0.00	.22	8.56	1.87
Missing	0.00	.59	0.00	0.00	0.00	.08
Second Reason						
Better employment at destination	10.50	21.89	23.93	16.41	8.02	16.63
Lack of employment at origin	3.22	15.98	37.89	41.47	5.35	26.96
To accumulate savings	22.24	38.46	2.56	14.47	22.99	17.84
Transferred (job)	0.00	2.37	5.30	1.51	4.28	2.62
Schooling	0.00	0.00	1.78	.43	1.60	.76
Better infrastructure	11.90	1.18	6.29	1.3	3.21	3.80
Join family	4.18	2.96	7.54	.65	12.30	4.56
Marriage	9.33	1.78	8.36	.65	5.88	4.15
Divorce/separation/death of spouse	0.00	1.18	0.00	.22	1.60	.50
Delivery	0.00	2.96	0.00	0 .00	1.60	.67
Family problems	.62	.59	.70	1.30	7.49	1.97
Accompanying patient	0.00	.59	0.00	.22	1.07	.34
Escape flood	0.00	.59	.92	0.00	0.00	.26
Escape drought	0.00	0.00	3.38	3.17	0.00	5.78
Poor quality of land/depleted soils	0.00	0.00	1.34	.22	1.07	.51
Violence, conflict, or threat of violence	0.00	0.00	0.00	0.00	1.07	.17
Other	38.01	1.78	0.00	.86	22.46	8.61
Missing	0.00	7.69	0.00	7.13	0.00	3.87

Source: Based on Grant, Burger, and Wodon (2014: 181).

lead to a 500% increase. The usual disclaimer holds here as well. The time studied encompasses an early global warming era. When or if the temperature really climbs some 5 degrees centigrade, the urge to seek EU asylum may be much greater than that captured by linear regression outcomes based on earlier data.

Another way to look at the migration problem is to contemplate a number of generalizations that are believed to be supported by empirical evidence (although not restricted by any means to MENA).[20]

1. Rapid onset events (extreme storms and floods) usually result in short-term displacement followed by a return to the impacted areas. Nonetheless, some long-term displacement can also be anticipated.
2. Slow onset events (e.g., droughts) tend to generate slow or lagged changes in migration patterns.
3. Drought-related changes in migration patterns are more likely to emerge in areas in which land degradation is severe or drought is less common.
4. Longer-term trends in crop yields and/or water availability are likely to encourage out-migration.
5. The impact of successive climate shocks may erode household assets and therefore adaptive capacity in ways that can eventually influence decisions to migrate.

If these generalizations are appropriate and we know that natural disasters have displaced around 24–26 million people a year globally (2008–2016), we can anticipate that the migration numbers will escalate as the severity of the environmental deterioration proceeds but subject to some lag. Drought and famine will be the major drivers. People will become increasingly accustomed to this type of deprivation being normal and the build-up or onset will seem slow. Ongoing and intense conflict, of course, can accelerate these tendencies. Yet, ultimately, if the drought/famine persists long enough, adaptation and resilience are apt to be overwhelmed. Other things being equal, we should anticipate the number of climate migrants to become quite large – perhaps even resembling the type of displacement and movement that ended the Bronze Age in the 1100s BCE in the Eastern Mediterranean. One difference, though, is that there are many more people in the 21st century CE at risk than there were in the 12th century BCE.

Another difference is that the number of people currently involved in forced migration (66 million) exceeds any numbers seen before. Attempts to cope with these numbers strain the humanitarian assistance community, place a great deal of stress on neighboring less developed states (such as Jordan, Lebanon, and Turkey) that actually take in most of these migrants (94%), and confront the probability that these migrants will be away from their homes for longer and longer times.[21] Most of the current refugees are driven by conflict. Climate change will increase the numbers radically, overwhelm humanitarian assistance efforts, eliminate nearby places of refuge which will be suffering the same problems, and reduce considerably the possibilities of ever returning home. If the humanitarian system is under great duress now, it could be entirely unmanageable 20 years from now.

Concluding thoughts

Some amount of global warming in the coming decades seems quite probable. Just how bad it will get remains to be seen but we seem assured to experience at least more than the 2 degree centigrade that has been a target to avoid for so long.[22] However warmer it becomes, the effects around the world will be uneven.[23] As it happens, the MENA region lies astride an area quite likely to suffer from higher temperatures and declining precipitation. A good number of the states in the region are also among the least prepared to cope with extreme climate change. Resilience is a fine concept but it is much easier to adapt to changes if there are some resources with which to work. Many of the states in the MENA lack affluence, economic development, and access to water. Large populations in some cases and still high birth rates do not help. Situated in already relatively warm and arid places, higher temperatures will only make survival more difficult.

As parts of the MENA become uninhabitable and other parts become even more immiserated, the odds of fragile autocratic states managing the newly aggravated problems seem low. No doubt, there will be some successful adaptation to the changes in climate within the region. Yet they seem likely to be the exceptions to the rule. In the past, the MENA region has experienced repeated problems with intermittent climate deterioration. Yet things seem likely to get even worse as the 21st-century climate regime deteriorates. While some parts of the global North will be inconvenienced with food production being forced to move north, the MENA states will not be able to entertain that option. As food, water, and even the prospects for eking out livelihoods become scarcer, the problems of the region will become more acute. This time, though, there may be a half billion people at risk with no or few places to take refuge. One question is whether the region can be expected to survive in the face of so much environmental adversity. Another question is what can be done to prepare for a crisis of this magnitude? We expect that the combination of extensive aridity, high temperatures, an expanding population, and diminishing water supply will be devastating for the MENA, with some areas being hit much harder than others.

Moreover, the MENA will not be alone. If it was a matter of "only" one region that will be suffering from global warming, the problem might be readily addressed at least in theory by the rest of the world. But sub-Saharan Africa is already the weakest link in terms of food insecurity (Table 7.7). Africa's current problems will also be compounded by further warming and its climate refugees will flee north through the MENA, if they can. The hot summer days and nights in the future that will make life so difficult in the Gulf are also anticipated to affect 800 million people in South Asia (Eltahir, 2016; Im, Pal and Eltahir, 2017; Mani et al, 2018). The high temperatures and sea-level rise that are expected to affect the southern littoral of the Mediterranean will also do damage to the northern littoral. Misery may love company but in this case, miserable populations that are adjacent to the MENA will make policy problems all the more insurmountable. We return to this issue in Chapter 10.

It is one thing to put together elaborate flow diagrams of what influences what (as in Figure 7.1 showing a number of linkages centered on food insecurity). Food insecurity is already something of a problem given poverty and the dependence of MENA on imported food but in a world in which much of the Global South could lose substantial agricultural productivity, MENA's food (and water) supply will be highly uncertain. One would think increased conflict could hardly be evaded in situations involving increased scarcity of food and water. But our own evidence is mixed. The Bronze Age had plenty of conflicts and some of it was clearly related to climate-induced shortages. Even so, the conflicts that occurred were specific to various locales and not everywhere, suggesting at the least that something more than climate change alone was involved. The periods that preceded the Bronze Age (Younger Dryas to the 6.2 kya event) have not yielded much evidence of conflict. The periods that followed it (Late Antiquity and Early Modern events) encompassed a fair amount of conflict – some of which seems related to climate change while, at other times, the conflict seems as if it might have gone on regardless of the climate.

In the next chapter, we take a closer look at recent and ongoing conflicts in MENA. What should we make of the fighting in Darfur, the Syrian civil war, and Yemen? Are these climate-induced conflicts or might they have taken place in fairly benign weather regimes? We will attempt to answer these questions in Chapter 8. If widespread beliefs about their close relationship are confirmed, we have more information about what to expect from a deteriorating climate. If, however, those expectations of a close relationship are not supported, we still have more information about what to expect from deteriorating climate – at least to the extent that global warming has degraded climate so far.

Notes

1 Fischer et al (2018) suggest that warming forecasts may be missing substantial polar changes and water rising effects that lead to expected outcomes that only capture half of what is likely to occur.

2 Rockstrom et al (2017) argue that to have a chance at limiting global warming to 1.5 degree centigrade (a 50% chance) or 2 degrees centigrade (a 66% chance), it would be necessary to roughly halve CO_2 emissions every decade beginning in 2020. If we generate about 40 metric gigatons of CO_2 per year now, that amount would need to be reduced to 23 gigatons by 2030, 14 by 2040, and 5 by 2050. It would also mean doubling the share of non-carbonized energy every 5–7 years. The removal of CO_2 by technical means, not in operation at present, to 0.5 gigatons per year by 2030, 2.5 gigatons per year by 2040, and 5 gigatons per year by 2050 will be necessary. Non-anthropogenic emissions must also decrease gradually to zero by 2050. Coal should be done by 2030 and petroleum by 2040. If we do all of this and take some other steps, we might get global CO_2 back to 380 ppm by 2100, or roughly a little less than it is now. That is, assuming that it proves possible to remove existing CO_2 from the atmosphere and drastically reduce future emissions.

3 See, as well, Waha et al (2017).

4 At some point, poorly fed and under-hydrated rural people tend to become urban dwellers. Thus, there is no reason to assume that the problem will continue to be highly concentrated in MENA rural areas.

 5 The expectation, barring major technological change in food production, is that less food will be produced globally as temperatures rise. See, e.g., Zhao et al (2017) Tigchelaar et al (2018); and Scheelbeek et al (2018).
 6 Constantinidou et al (2016) argue that some parts of MENA will lose food growing capabilities while other parts may gain additional productivity.
 7 We examine this linkage closely in Chapter 8 and find much less positive evidence than we expected.
 8 See as well Woertz (2017) and D'Odroico et al (2019).
 9 See INRA (2015) and Tull (2020) on this question.
10 The reference to tripling demand is not arbitrary. Fader et al (2013) estimate that 16% of the world's population currently uses food trade to satisfy demand but the number could escalate to 51%.
11 If they are not diminished, global warming will only become a greater problem than it already is.
12 Rabinowitz (2020) suggests that oil producers could switch to harnessing solar energy to offset their future revenue problems and contribute to slowing global warming at the same time.
13 Nor are there any compelling grounds for anticipating "business-as-usual" since that assumes that temperatures will remain close to what they currently are. But the "business-as-usual" forecast is sufficiently bad to illustrate the point that more water stress should be anticipated.
14 Some recent studies have suggested that severe and prolonged drought can lead to unrest and political instability (for drought-induced instability in Syria and Yemen, see e.g., Beck 2014; Kelley et al 2015; and Halverson 2016). We explore this question further in Chapter 8.
15 The Abu-Ismail et al (2017) interpretation avoids income calculations and looks at a slate of indicators: years of schooling, school attendance, child mortality, child nutrition, early pregnancy, electricity availability, unimproved sanitation, access to safe water, floor/roof inadequacies, dung/wood/use of charcoal cooking fuel, over-crowding in sleeping rooms, and insufficient assets for accessing information, mobility, or engaging in livelihoods. The indicators are weighted differentially, aggregated and poverty/acute poverty thresholds are established when one-third of the maximum possible deprivations are exceeded. Some indicators are used for poverty and others for acute poverty; hence, the categorizations are partially independent. See Bizri (2012: 204); Ghafar and Masri (2016); and El-Khoury (2017) for discussions of alternative interpretations of poverty measurement in the MENA but all of which point to fairly widespread poverty in the region.
16 Explicit empirical support for the attitudinal foundations for women inequality in the MENA is found in El Feki, Heilman, and Barker (2017).
17 The distinction between push and pull factors often seems semantic. For instance, seeking better jobs (a push) is not much different from believing there are better jobs in the city (a pull).
18 The migrations were in-country, the most common form of migration, and not to other countries.
19 Climate change usually emerges as a significant factor in economic growth studies. See, for instance, Nelson et al (2009); Alagidede, Adu, and Frimpong (2014); Tayebi (2014); Alboghdady (2016); and Rezai, Taylor, and Foley (2017). On the other hand, Gornell et al (2010) argue that we cannot yet estimate precisely the effects of climate change.
20 These generalizations are advanced in Rigaud et al (2018: 21).
21 Apparently, the calculations are not easy to perform. However, one estimate states that the average duration of forced migrants in holding camps was 9 years in 1993, 17 years in 2013, and 21.2 years in 2018 (Yayboke and Milner, 2018: 11).
22 The implications of projected global mean temperature changes tend to under-estimate regional level changes. This is because global mean temperature implies

increases in warm and cold temperature extremes in both land and water. However, warming over land is typically stronger than over the oceans which means extreme temperatures in many regions can increase well beyond 2 degrees centigrade. In fact, the 2 degree centigrade global mean temperature target actually implies at least 3 degrees centigrade warming particularly in places like the Mediterranean region and changes in regional extremes are expected to be greater than those in global mean temperature by a factor of 1.5 degree centigrade (Seneviratne et al 2016). Analyses looking at long-term temperature data of this region suggest that the frequency of heat waves has been increasing since the 1970s (Tanarhte et al 2012).

23 But not uneven in a random fashion. Diffenbaugh and Burke (2019) estimate, for instance, that global inequality has already increased appreciably because warm Southern states have lost economic growth possibilities while cool Northern states have actually gained from global warming.

8

CLIMATE WARS – CANARIES IN THE COAL MINE?

In the ancient Near East conflict related to climate change is not prominent in the archaeological record for the first 10 millennia of the 15 that we examine in this book. It becomes fairly notable toward the end of the 3rd and 2nd millennium BCE. In the 3rd millennium, the conflict is intense but some of it is conceivably due to processes other than climate change. Some of it, as in the Gutian movement into Akkad, seems to be closely connected to cool and drying conditions that drove hill people toward the urban areas controlling the main rivers in Mesopotamia. The end of the Bronze Age at the end of the 2nd millennium BCE, on the other hand, seems closely linked to widespread and intense drought that drove refugees west in large numbers and leading to the destruction of a number of cities along the way. In this case, climate change seems closely connected to violence.

Moving forward in time, though, there is plenty of conflict centered on the decline of the western and eastern Roman empires, the early expansion of Islam, the Crusades, and the Ottoman Empire's warfare with the Hapsburgs and the Persians. To the extent that climate change encouraged imperial decline and challenges for territorial control or the movement of Turkish groups toward the direction of Jerusalem, thereby prompting a Papal decree legitimizing the First Crusade, we can probably find some indirect linkages. Yet the aridity of Palestine may have discouraged conflict between Christians and Muslims during the Crusades. The climate-induced undermining of the Ottoman Empire could also be said to have discouraged European-Ottoman conflict somewhat after the end of the 17th century CE.

In short, the longer-term history of climate change and conflict in the Near East is mixed. The historical record does not yield an unambiguous prediction about whether we should anticipate a positive relationship between climate change and conflict in the 21st century CE. However, we do have several prominent contemporary cases in the Middle East and North Africa (MENA) that are often said to be climate change induced. We seem to know enough about these cases to be able to pin down to what extent global warming has contributed to their origins. In this chapter, we examine three alleged climate-conflict crises (Sudan, Syria, and Yemen) to assess just what role climate played in their onset. We ask whether they are direct products of climate deterioration or, are they embedded in a host of interacting variables in which climate change is but one of many drivers? It is not critical to our general argument that climate change increases conflict but it is an important side question. If climate deterioration increases the probability of conflict over increasing scarcities, future immiseration will be all the more likely. Yet, even if we find no direct linkages between climate change and conflict causation, there is little reason to expect conflict to subside in the face of increasing climate deterioration. The MENA is usually judged to be the most conflictual region in contemporary international politics and we do not expect that characteristic to vanish any time soon.

Within the climate-conflict debates, the wars in Darfur, Syria, and Yemen are said to have been caused, in part, by climate change. They are often depicted as the first victims of climate wars. In this chapter, we interrogate the claimed links between climate change and conflict to ascertain whether they are in fact "canaries in the coal mine" as early warning signs of an ongoing problem or perhaps, instead, harbingers of a future marked by severe climate change and conflicts. We are aware that a case can be made regarding canaries and harbingers as being synonyms. What we have in mind is that the canary metaphor operates in the short term. A dead canary indicates that the atmosphere within a mine is fouled sufficiently to kill birds. Miners could be next and almost immediately. A harbinger is more long term in our perspective. We anticipate that the worse effects of climate change might be expected around mid-century. Are what we are seeing now early manifestations of things we will see in even more severe ways down the road?

If the canary metaphor is most apt, we should anticipate more of the same, albeit on a wider scale. As climate change proceeds, conflict propensities should expand. If the harbinger alternative seems more probable, some contemporary conflicts may simply provide illustrations of what conflict in extremely arid situations look like. Whether climate change causes conflict or not, extensive conflict only makes the effects of climate change more pervasive and difficult to manage because both increase food and water scarcities. Combined, conflict and climate change will only make matters worse for the people still alive to endure them.

In sum, the goal of this chapter is twofold: (1) to look at the evidence in Sudan, Syria, and Yemen to assess if they are indeed canaries in the coal mine or harbingers of future climate wars and (2) to ascertain whether we should anticipate more conflict in the region as the environment continues to deteriorate? We then compare the processes and outcomes involved in all three cases. But before addressing these questions we must first review some of the key climate-conflict arguments.

Climate-conflict arguments

> The scientific community is in broad agreement on the causes of climate change. However, the debate on the security implications of climate change has tended to focus on creating a security 'hook' to invest the climate change negotiations with a greater sense of urgency. This has not really translated into a more detailed understanding of how climate change might interact with existing vulnerabilities to exacerbate tensions and trigger conflict.
>
> *(Brown and Crawford 2009: 9)*

As the quote from Brown and Crawford suggests, we do not yet have sufficient understanding of how climate interacts with local conditions to produce conflict. While the field is nascent, it is growing rapidly. Just over the past decade since Brown and Crawford's assessment of the field, a number of interesting studies have come out that have examined the climate-conflict links a bit more systemically rather than anecdotally. What we know from some of these recent studies is that the evidence for the claimed links is at best mixed (Sakaguchi, Varughese, and Auld 2017). The empirical basis particularly in the area of within-country historical analyses is rather limited (Buhaug 2015). Thus, we are left with some studies advocating a positive link, some a negative link, and another group falling somewhere in between. Such academic quarrels are not unusual. They may even be par for the course. Yet it is a bit embarrassing that, collectively, we cannot seem to make up our minds. It also does not assist the grounds for mitigation/adaptation of policy-making if we make wrong assumptions about causal drivers or if we miss obvious linkages.

Among the proponents of the climate-conflict thesis, essentially there are two types of arguments: those who advocate a more *direct* link and those who advocate a more *indirect* link. The summary of some of these arguments is presented in Table 8.1, although the list is by no means exhaustive. As one can see from Table 8.1, there is a great deal of overlap among all these explanations. Basically, those who advocate a more direct link note that climate change generates some form of violence at either micro or macro levels. The micro level is associated with psychological reasons for individuals' proclivity for aggressive behavior. High temperatures are said to have "a psychological effect on humans" which in turn can create conditions that "alter social interactions to make violence more likely" (Sakaguchi, Varughese, and Auld 2017: 631). At the macro level, changes in the climatological process affect the likelihood of conflicts occurring (see e.g., Hsiang, Meng, and Cane 2011).

Perhaps the biggest proponents of the direct link between global warming and increased conflict are Miguel, Burke, Hsiang, and colleagues, who have published a series of articles since 2004 concluding that climate change is positively associated with violent conflict (Miguel, Satyanath, and Sergenti 2004; Burke et al. 2009; Hsiang, Meng, and Cane 2011; Hsiang, Burke, and Miguel 2013; Hsiang and Burke

TABLE 8.1 Climate-Conflict Arguments

No.	Causal link	Climate change leads to…	Examples
1	Direct (micro)	warmer temperature – > individual aggression/ violence	Sakaguchi, Varughese, and Auld (2017)
2	Direct (macro)	warmer temperature/extreme rainfall – > conflict	Miguel, Satyanath, and Sergenti 2004; Burke et al 2009; Hsiang, Meng, and Cane 2011; Hsiang, Burke, and Miguel 2013; Hsiang and Burke 2014; Burke, Hsiang, and Miguel 2015
3	Neutral or reverse		Buhaug 2010; Buhaug and Theisen 2012; Theisen 2012; Theisen, Gleditsch, and Buhaug 2013; Buhaug et al 2014; Buhaug 2015
3	Indirect	[vulnerability] – > resource overuse – > competition	Sterzel et al. (2014)
4	Indirect	water scarcity – > conflict	Werz and Hoffman (2013)
5	Indirect	warmer temperature/reduced rainfall – > drought – > resource scarcity – > resource wars	Barnett and Adger (2007); Scheffran, Marmer, and Sow (2012)
6	Indirect	drought – > misery – > rural to urban migration – > social unrest	Kelley et al (2015)
7	Indirect	Food scarcity – > food price hike – > social unrest – > food riots	Lagi, Bertrand, Bar-yam (2011); Sternberg (2012)
8	Indirect	reduced economic growth/ agricultural/industrial output – > political instability	Dell, Jones, and Olken (2008, 2012)
9	Threat multiplier	political instability, poverty, unemployment – > conflict proneness	CNA Military Advisory Board (2007)
10	Negative link		Bernauer, Bohmelt, and Koubi (2012); Hartmann (2010); Selby et al (2017); Selby and Hoffman (2014a and 2014b)

2014; Burke, Hsiang, and Miguel 2015). They have generally found that higher temperatures and extreme precipitation tend to correlate with a greater incidence of conflict. Because most of the inhabited world is expected to warm 2–4 degrees centigrade by 2050, Hsiang, Burke, and Miguel (2013) claim that "amplified rates of human conflict could represent a large and critical impact of anthropogenic climate change." They further argue that the magnitude of climate's influence is substantial, noting specifically that "for each one standard deviation (1σ) change in climate toward warmer temperatures or more extreme rainfall, median estimates indicate that the frequency of interpersonal violence rises 4% and the frequency of intergroup conflict rises 14%" (Hsiang, Miguel, and Burke 2013: 1). These findings are consistent with an earlier work by the Burke team that found a strong link between higher temperatures and civil war in sub-Saharan Africa "with warmer years leading to significant increases in the likelihood of war" (2009: 20670). They also predict "a roughly 54% increase in armed conflict incidence by 2030, or an additional 393,000 battle deaths if future wars are as deadly as recent wars" (Burke et al. 2009: 20670). If these predictions come to pass, weak or fragile states will be particularly prone to climate-induced instability.

Buhaug, Theisen, Gleditsch, and others have consistently advanced alternative views that challenge Hsiang, Miguel, and Burke's arguments on the claimed positive links (Buhaug 2010; Buhaug, Hegre, and Strand 2010; Buhaug and Theisen 2012; Theisen 2012; Theisen, Gleditsch, and Buhaug 2013; Buhaug et al. 2014; Buhaug 2015). Basically, this group tends to be more agnostic about the climate-conflict links. Climate change and conflict may be related, according to some in this group, but the causal arrow may not need to lead from the former to the latter. In fact, it is quite possible that armed conflicts (because they tend to be so destructive) are likely to increase societies' vulnerability to climate change. In other words, "many of the negative impacts of conflict on economic activity, education, health, and food security are major drivers of environmental vulnerability" – all of which lower social resilience to climate change (Buhaug 2015: 273).

Within the climate-conflict literature, there seems to be greater support for the indirect link between climate change and the risk of violent conflict via intermediate factors. Such arguments suggest that climate change accelerates various processes on the ground, which leads to more conflict down the road. For instance, sustained drought and extreme heat waves lead to water and food scarcity which could lead to conflict. This is commonly referred to in the literature as a "threat multiplier" which has been a popular phrase since 2007 when the US military first introduced the concept (CNA Military Advisory Board 2007: 44). Low-income countries are particularly vulnerable to climate-induced conflicts via intermediate factors such as reduced agricultural output or increased pressure on natural resources, which in turn could have adverse impacts on human security.

One notable study from this group looks at the long-term variation in temperature and precipitation over the past 50 years and finds that higher temperatures, particularly in poor countries, substantially reduce economic growth, agricultural output, industrial output, aggregate investment, while increasing political instability (Dell, Jones, and Olken 2008). They claim that "a 1oC rise in temperature in a given year reduces economic growth by 1.1 percentage points on average" (2008: 27) and this finding seems to be true mainly in poor countries. Dell, Jones, and Olken 2008) conclude that

> [w]hile higher temperatures reduce agricultural output in poor countries, we also find that they lead to contractions in industrial output and aggregate investment and to increased political instability. These results underscore the breadth of mechanisms underlying the climate-economy relationship. The results also suggest that future climate change may substantially widen income gaps between rich and poor countries, with many poor countries driven toward greater poverty, other things equal.

Barnett and Adger (2007) argue that climate change may undermine human security by reducing access to natural resources that are critical to sustaining livelihoods and by undermining state capacity to act in ways that promote security and peace. This is particularly destabilizing in low-income countries that spend a relatively high percentage of their income on basic survival needs such as food and water.

By contrast, high-income countries such as the Gulf monarchies that spend a relatively low percentage of income on food and water are likely to weather climate-induced stress with a lower risk of social instability. Wealthier economies presumably have the resources with which to manage climate-induced food price hikes, scarcity issues, and access to basic goods problems. For instance, a wave of democratization that engulfed the MENA during the 2010–2011 period, known to outside observers as the "Arab Spring," spared certain petro-states such as Saudi Arabia and the UAE, from social unrest and political instability (Sternberg 2013). It did not hurt that the same countries had access to purchased or leased agricultural land overseas in order to meet their domestic demand (ILC et al, 2014; Spiess, 2012). What this suggests is that a government that has the resources with which to respond to climate-induced events and is willing to deploy those resources to manage scarcity and price hike problems is less likely to experience social instability that will end in conflict (Gemenne et al, 2014; Adger et al, 2014).

Lagi, Bertrand, and Bar-yam (2011: 4) go even further in claiming that a persistent hike in global food prices above a certain threshold is likely to lead to "increasing global unrest." Countries that are unable to cope with local factors such as drought that affect the production and availability of food or countries that have a relatively high dependence on imported food are vulnerable to conflict proneness. "This is particularly true for marginal populations, i.e. the poor, whose alternatives are limited and who live near the boundaries of survival even in good times" (Lagi, Bertrand, and Bar-yam, 2011: 2). While we agree that food prices may be crucial for social stability and often result in riots when they fluctuate, they may not necessarily lead to a civil war, thus we need to be cautious in our extrapolation of these links.

Another noted Middle East scholar observes that "Climate change will bring about gradual, albeit negative effects. The effects will be hard to distinguish from processes already underway and to which Arab societies have become accustomed" (Waterbury 2013: 23). Waterbury further notes that "it is abrupt change that occasionally ignites the Arab 'street'." The Middle East is particularly vulnerable to such abrupt changes because it happens to be the world's youngest region. The youth bulge is both "a cause and a consequence of rapid population growth" (Brown and Crawford, 2009: 10) and conflict. Population growth, according to Brown and Crawford (2009: 10) is "fundamentally changing the facts on the ground, increasing the demand for water, food and shelter and adding thousands of new job seekers every year." In Waterbury's view,

> Unanticipated population movements might be the single most important security threat, especially if such movements are across borders. These may be caused by drought, famine, or sea level rise. By definition, to deal with them would require coordination with other states in the neighborhood, but heretofore the record for such coordination has been meager.
>
> *(2013: 22)*

Opponents of the climate-conflict thesis tend to question the direct link between warming and violence. While it is impossible to prove that climate has little to do with the onset of conflict, the available evidence, they argue, does not support the claimed positive links. Interrogating the climate-conflict linkage, Bernauer, Bohmelt, and Koubi (2012) find no statistically significant correlations between climate change and conflict. Other empirical analyses have reached similar conclusions that climate variability does not necessarily increase the incidence of violence or armed conflict whether domestic or inter-state (Gleditsch, 2012; Koubi et al., 2012). In examining the link between climate change and the Syrian civil war, Selby et al. (2017: 241) observe that the existing evidence is extremely weak and "do[es] not stand up to close scrutiny." They also note that "threat multiplier" discourse is "neither cautious nor rigorous, instead typically combining dystopian speculation and exaggerated accounts of the likely impacts of climate change" (Selby et al. 2017: 241).

Perhaps the most interesting argument among the climate-conflict thesis deniers is the one suggesting that industrial societies are less vulnerable to climate-induced stresses and subsequent violence. This is mainly because industrial agriculture has given humans the ability to generate surplus food that can be stored so that in the event of any disruptions to availability (whether climate induced or otherwise) they can be used to mitigate, making conflict less likely, not more so.[1] This thinking is consistent with Gartzke (2012) and Slettebak (2012) both of whom advanced similar types of arguments in a special issue of *Journal of Peace Research*. Yet this argument does not preclude conflict from increasing in less industrialized societies (i.e., most of the MENA economies) due to food deficit and price hikes. Others suggest that the problem may be one of accurately capturing the length of the time processes involved (Sakaguchi, Varughese, and Auld 2017: 637–638). Most of the studies have relied on fairly short periods of climate adversity to test the climate-conflict relationship. Longer-term studies are more likely to capture the connection if one exists.[2]

On the other hand, much of what we know about contemporary conflict etiology is based on analyses undertaken with data from a relatively stable and benign climate regime era. It is quite conceivable that the known relationships will simply not hold up in this century as the climate turns more malign. Thus, there is some real concern that the coming environmental deterioration could increase the risk of deadly violence in the MENA even if climate change has not caused conflict in the immediate past. Climate change may not have been the principal driver of conflict in the last half century, but if global warming continues unabated, conflict might likely increase as well. We know as well from Chapter 3 that conflict does seem to have been one of the byproducts of ancient climate change, especially toward the end of the 3rd and 2nd millennium BCE. Thus, we must keep in mind that studies that look at current conflicts in the MENA engage in past or current trends without much regard for any long-term historical climate patterns or projections. If we can no longer assume that a relatively benign environment will be around indefinitely, we need to be cautious about extrapolating historical trends from a relatively stable environment into a highly uncertain future.

We now turn our attention to the three conflicts in the MENA to assess just what role, if any, climate played in their outbreak. Our main questions concern whether these three civil wars are isolated cases, representatives of the new normal, or forecast of things to come? Moreover, whatever linkages may be found, are they direct or indirect, weak or strong, or nonexistent altogether?

Examining the three conflicts in the MENA

In this section, we examine the 2003–2005 war in Darfur, the ongoing Syrian civil war which began in 2011, and the current civil war in Yemen which erupted in 2015. In the view of many Western observers, these conflicts were caused, in part, by climate change. To assess these climate-conflict cases, we adopt Selby et al.'s (2017: 2) useful summary of the existing literature in terms of a three-step causal approach which stipulates that environmental conflicts essentially begin with (1) natural resource scarcities (due to long-term degradation, increases in demand, or short-term environmental shocks) which lead to growing environmental scarcities; (2) these environmental scarcities trigger or exacerbate socio-economic problems (such as resource competition and out-migration); and (3) these socio-economic problems, in turn, contribute to political instability and even violent conflict. The causal link is illustrated in Figure 8.1.

While this three-step linkage is our core concern, we think it would be very difficult to explain what these internal wars are about if we focus exclusively on climate as the principal driver. As a consequence, our objective is to provide

FIGURE 8.1   Three-step causal process for climate conflicts

enough information to make some sense of these conflicts without going into great detail. In the process of crafting brief overviews of the most important causal pathways, we end up asking whether the climate change drivers measure up to alternative processes leading to conflict.

Darfur civil war (2003–2005)

We begin our analysis with what is widely referred to as the "first climate war." With titles such as "Is Climate Change the Culprit for Darfur?" (De Waal, 2007), "Darfur Conflict Heralds Era of Wars Triggered by Climate Change" (Borger, 2007), "Darfur: The First Modern Climate-Change Conflict" (Mazo, 2009), and more recently, "Climate Change Could Render Sudan 'Uninhabitable'" (Britton, 2016), the climate-conflict thesis for the outbreak of the war in Darfur seems like a foregone conclusion. Much of the hype about the claimed links can be traced back to a statement put out by the UN Secretary General Ban Ki Moon who went so far as to claim that "the Darfur conflict began as an ecological crisis, arising in part from climate change" (Ki Moon, 2007). A few years earlier as the conflict was unfolding, even the US Department of Defense began to view climate change as having "a direct, negative effect on US security" and described Darfur as "the first climate war" (Waterbury, 2013: 21).

Sudan is one of those countries that never existed before British colonialism constructed a heterogeneous state from several different imperial holdings. In the late 19th century, Britain had moved into the Khartoum area as part of a series of efforts by the Turks, Egyptians, Britain, and France to control the Nile River. The British initially suppressed the Madhi Revolt just before the dawn of the 20th century. In 1916–1917, it conquered the independent Sultanate of Fur (more or less shown in Figure 8.2 as north, west, and southern Darfur or "home of the Furs") and added this western territory to what is often described as riverine (or Nile-centric) northern Sudan about a 100 years ago.[3] While the current state of Sudan is thoroughly Muslim, the state now known as South Sudan (southern Sudan in Figure 8.2) was Christian and more overtly "African" in terms of its population. As a consequence, several Sudanese civil wars focused on attempts of southern Sudan to secede from the larger Sudanese entity. South Sudan eventually gained its independence in 2011.

In the north, there is also a distinction between "Arab" and "African" tribes but it is not delineated by differences in physical appearance or necessarily

FIGURE 8.2   Map of Sudan

residential location within Sudan. The Darfur province(s) was populated by both types of tribes with the main source of differentiation being one of political economy focus. In the west, African tribes tended to be farmers. Arab tribes tended to herd camels (north Darfur) or cattle (south Darfur) which is not to say that African farmers never owned cattle or that Arab tribes never farmed.

Part of the explanation for civil war in the 21st century is the marginalization of Darfur in Sudanese politics and economics. In general, the western Darfur region was left underdeveloped in both the British era and the more contemporary post-independence era. It is not easy to move from Khartoum to Darfur or vice versa and little has been done to break down the remoteness of the mountainous western area.[4] For the most part, Darfur has been peripheral to central Sudanese decision-making. The exception to this rule emerged in maneuvering within ruling circles after the 1989 military coup, which brought to power a coalition centered on an alliance between the Sudanese Army and an Islamic movement. To simplify greatly, the military part of the coalition was centered on Omar al-Bashir and the Islamic movement was led by Hassan al Turabi. Coup coalitions can agree on the need to remove a chief executive target; they do not always remain in agreement after the target has been eliminated. In the late 1990s, the different wings of the coalition began to perceive opposing factions as threats to their political survival. It did not help that the army factions drew personnel and political support primarily from the riverine area and the Islamic movement drew some of its support from Darfur because some Darfurian individuals had viewed participation in the movement as an opportunity to overcome western political marginalization. In this fashion, central political combat in Sudan became more ethnopoliticized around the turn of the last century than it had been before if we put aside the long conflict between north and south Sudan. Al-Bashir's successful attacks on the al Turabi wing meant also purging Darfurian supporters of al Turabi from participation in central Sudanese political decision-making. In turn, this meant further alienation of leading Darfurian political actors from the central government. Roessler (2016: 183) argues that it

> not only heightened Darfurian political consciousness and intensified their political grievances… but it also contributed to an increase in the politicization of ethnicity within Darfur. As non-Arab Islamists exited the regime and… Arabs gained greater influence as brokers, Darfurians perceived that the central government was backing the latter against the former.

Another complication in the story is that while we are focusing on the 2003–2005 conflict, civil war in Darfur had been going on for some time prior to 2003. The earlier fighting was carried out by militias organized by African and Arab tribes to advance/defend their respective positions in the west. Part of this antagonism was climate driven. Droughts prior to the 1970s (see Kevane and Gray, 2008; Brown, 2010) interrupted what had been accommodative practices between farmers and pastoralists. Arab pastoralists had been allowed to move their herds into African croplands after harvest and before the next crop was sown. Prolonged drought encouraged farmers to enclose their land from camel and cattle encroachment and to restrict access to water sources that they controlled. It is not clear whether this abrupt change in practice led to physical conflict immediately but it certainly poisoned the ability of the two categorical groups to coexist peaceably.

Increased militancy by both groups was facilitated by civil wars ongoing elsewhere. Arab tribes were recruited to fight in Libyan armies waging war in nearby Chad.[5] At times, the Libyan effort was sanctioned by Khartoum, which meant that Libyan forces could use Darfur as a staging area for incursions into Chad. Arab Darfurians were thereby given arms, training, combat experience, and exposed to arguments that Arabs were naturally superior to Africans.

African militias were armed and encouraged by Chadian forces resisting Qaddafi's machinations in the Sahel and by southern Sudanese armies (the SPLA or Sudan People's Liberation Army) that attempted to ally with African tribes in Darfur and Kurdufan as part of their own secessionist attempts in the south. Khartoum, for its part, relied heavily on the African militias to help defeat the SPLA and African tribes that were viewed as giving the southern rebels cover and support. Thus, the regional diffusion of conflict played a significant role in militarizing political-economic antagonisms in Darfur several decades prior to 2003 (at least going back to 1985).

The most recent Darfur civil war broke out in 2003 with the worst of the violence lasting until 2005. It is worth recalling here the prevalent view of Western policymakers and international observers at the time about the conditions that led to the conflict:

> It is no accident that the violence in Darfur erupted during the drought. Until then, Arab nomadic herders had lived amicably with settled farmers… But once the rains stopped, farmers fenced their land for fear it would be ruined by the passing herds. For the first time in memory, there was no longer enough food and water for all. Fighting broke out. By 2003, it evolved into the full-fledged tragedy we witness today.
>
> *(Ki Moon 2007)*

This reading of the crisis by the UN Secretary General suggests that drought caused (or at the very least, significantly contributed to) the conflict by challenging traditional subsistence livelihoods, "leading to migration from areas of environmental stress, and to heightened conflict amongst nomadic pastoralists, and between them and sedentary farmers" (a useful summary of the conflict is found in Selby and Hoffman, 2014b: 365). Ki Moon's assessment seems to echo Stephan Faris' article which appeared in *The Atlantic* a couple of months earlier:

> The fighting in Darfur is usually described as racially motivated, pitting mounted Arabs against black rebels and civilians. But the fault lines have their origins in another distinction, between settled farmers and nomadic herders fighting over failing lands. The aggression of the warlord Musa Hilal can be traced to the fears of his father, and to how climate change shattered a way of life.
>
> *(Faris, 2007)*

"To truly understand the crisis in Darfur," further notes Faris (2007), one must look back to the 1980s, to the period "before the violence between African and Arab began to simmer." It is during this period that Sudan's western region of Darfur began experiencing environmental degradation, including "dramatic declines in rainfall" which negatively affected the region's vegetation. Our main questions then are: (1) Did Sudan's troubled western region of Darfur experience unusual drought during the intense period of the war in 2003–2005 or even during the prior years leading up to the conflict? (2) If so, how severe was it? and (3) Did it produce the kind of environmental scarcities that made conflict inevitable?

Data on Darfur's rainfall patterns are not too difficult to find. In fact, in a very useful analysis of Sudan's water scarcity induced conflict, Selby and Hoffman (2014b: 364–365) do just that and thereby provide a compelling response to the first question:

> [Du]ring [the 2003-2005] period, Darfur experienced above average rainfall. Moreover, rainfall levels were not gradually "declining" prior to 2003; to the contrary, rainfall levels during the 1990s and early 2000s were generally above the thirty year average, with no major droughts after 1990. This rainfall evidence is corroborated by evidence of a greening of Darfur (and the Sahel beyond) since the early 1990s which is visible through satellite imagery. Darfur was not witness to "desertification" prior to the "war" – and the war did not erupt "during the drought."

Selby and Hoffman's assessment is informed by various sources that have looked into the drought-conflict argument in great depth (see e.g., Kevane and Gray, 2008, UNEP, 2007) so we will not duplicate their efforts here. What they seem to suggest is that climate-induced drought could not have been the cause of the Darfur war's outbreak because Sudan did not experience water scarcity either during the intense years of the conflict or even in the years leading up to the war. In fact, "the worst violence took place in areas of relatively good rainfall... areas with "the highest rainfall levels" and "amongst the richest agricultural lands in Sudan" (Selby and Hoffman 2014b: 365). These facts seem to contradict the causal process laid out in Figure 8.1.

If not drought, then what induced Darfur's environmental scarcity, which in turn presumably triggered the socio-economic problems that many argue made Darfur's descent into war inevitable? The more probable culprits of most of Sudan's land degradation are likely to have been produced primarily by two factors: an explosive growth in livestock numbers (from 28.6 million in 1961 to 134.6 million in 2004); and deforestation which amounted to 8,835,000 hectares (or 11.6 percent of Sudan's total forest cover) between 1990 and 2005, according to the United Nations' post-conflict environmental assessment report (UNEP, 2007). Deforestation in Darfur seems to be particularly significant with a third of the forest cover lost between 1973 and 2006 (UNEP, 2007: 10–11). Nationally, Sudan is losing forests at a rate of 0.84–1.87 percent every year. Much of this is driven largely by energy needs and agriculture. At this rate, UNEP estimates, that "under areas of extreme pressure," parts of Sudan could cease to have forests altogether.

Nonetheless, it is a bit awkward to focus on the 2003–2005 period when the conflict had been brewing and overtly manifested much prior to 2003. Earlier drought in the west had contributed to inter-group antagonisms, which were manifested in the conflicts of the 1980s, 1990s, and in 2003. It does seem correct, however, to conclude that climate change per se did not lead to civil war in 2003. It was a contributing element in a complex of central-peripheral antagonisms playing out over a century in which ruling coalition fragmentation in the 1990s and regional conflict diffusion played the strongest roles in a civil war breaking out in the first decade of the 21st century largely because Fur political actors thought that only civil war could resolve their problem of ethnic and provincial discrimination by the central government.

To be sure, Darfur did experience many socio-economic problems prior to the war, including land disputes between pastoralists and farmers and subsequent migration and displacement of large numbers of people. But we cannot assume that conflict is an automatic product of climate-induced environmental stresses such as scarcity and migration. Scarcities were certainly involved but the civil war emerged from Fur requests for protection from Arab militia attacks in Darfur. The central government promised to help improve security in the region but only if Fur rebels were disarmed. Fur dissidents assumed that the central government was more interested in arming Arab militias than in negotiating a de-escalation of clashes between African and Arab groups. Civil war broke out when the central government attacked Fur rebels in Darfur at the end of 2002. Subsequent attempts to negotiate a ceasefire ran afoul of the al-Bashir/al Turabi falling out. Hardliners did not wish to do anything in the way of concessions to the rebels that might facilitate the re-emergence of pro-Darfur-perceived, al Turabi as a central political actor (Roessler, 2016: 195).

The lethality and viciousness of the Darfur conflict was then more likely the result of government repression of an insurgency that relied to a considerable extent on local Arab militias:

> the Darfur war was essentially a brutal counter-insurgency operation, launched and financed by the government in Khartoum, and conducted by a combination of Sudanese army, intelligence and air force units, and paramilitary janjawid brigades (who far from being independent of the state, often wore army uniforms, often operated in the company of regular army units, and would regularly undertake their attacks immediately after Sudanese Air force bombing raids.
>
> *(Selby and Hoffman, 2014b: 366).*

The causal link identified in Figure 8.1, thus, does not appear to receive much support in the Darfur civil war in the sense that it was not brought about by climate-induced environmental scarcities that triggered socio-economic problems that ultimately resulted in violent conflict. In all probability, the war itself might have broken out in some form with or without any environmental changes, as some pointed out (Butler, 2007: 1038).

The Syrian civil war (2011–?)

Syria is an autocratic state that is highly arid (only about a quarter of the land is arable).[6] Politically, it is dominated by a small minority (Alawi) that adheres to an unorthodox form of Muslim practice that has been pronounced to be akin to Shi'a tenets. Once characterized by frequent military coups somewhat in synch with its multiple defeats by Israel, Hafez al-Asaad took over in a 1970 coup that strongly reinforced the ascent of the Alawi and also largely managed to end the multiple coup sequence in part by relying on Alawi control of strategic resources. As with other autocratic MENA regimes, the al-Assad regimes (father and son, Bashar) have been noteworthy by the rewards for the elite and the lack of governmental services for most of the population.

Although the long-running Syrian civil war has claimed far more media attention than its counterparts in Sudan and Yemen, its outbreak is more simple and even classical in format than the other two. As part of the 2011 Arab Spring, teenagers writing anti-Bashar al-Assad sentiments on a wall were arrested and tortured by government security agents. Protest demonstrations were met by harsh repression escalating to the use of military fire power against unarmed citizens. The demonstrators initially had stopped short of demanding the ouster of al-Assad but evolved in that direction in conjunction with the increasing severity of the repressive tactics employed. Military defections began early on

FIGURE 8.3 Map of Syria

in the crackdown on dissent thereby feeding the possibility of armed resistance. Although protestors were killed from the very beginning, it took about a month for the resistance to the Assad regime to cross the line between unarmed dissent to armed rebellion.

One of the reasons for the high-profile media attention is the internationalization of the civil war.[7] With the Syrian regime receiving aid from Iran and Hezbollah, the civil war has become a proxy conflict in the Iranian-Saudi Arabian rivalry. The Russians have intervened substantially to preserve the al-Assad regime. The United States also became involved as a spinoff in its fight with ISIS initially in Iraq but spreading to Syria as ISIS joined the many different rebel organizations at war with the central regime. In addition to the Iranian and Hezbollah assistance to the incumbent regime, Turkey has also intervened in north Syria in part to prevent Kurdish territorial control along its border.

The human drama of a very large contingent of refugees moving into Jordan and Turkey and long lines seeking asylum in various parts of Europe and other parts of the world is another explanation for the media attention. Dead children washing up on Mediterranean beaches make for highly poignant optics but they also seem unusually close to places where civil wars no longer occur with any frequency – at least in comparison to casualties in more remote parts of sub-Saharan Africa or Central America. But the complications of a long-running conflict with many actors and considerable human displacement and deaths is not our immediate concern. Our main question is whether climate change had something to do with the initiation of conflict.

A large share of the available climate-conflict literature focuses on Syria. Of the studies that support the Syria-climate conflict thesis, one study in particular stands out in receiving more media attention than others and becoming one of the top ten most media cited climate change studies in recent years (McSweeny, 2015). It has also become a "standard reference point for all claims and reports on the subject" (Selby et al., 2017: 233). The study we are referring to here is the one led by Colin Kelley and colleagues, published in the *Proceedings of the National Academy of Sciences* in 2015. The Kelley team claimed that a severe drought which began in the winter of 2006/2007 was a catalyst for the Syrian conflict five years later. Based on climate modeling, Kelley and colleagues posit that Syria experienced "the worst 3-year drought in the instrumental record" which in turn "exacerbated existing water and agricultural insecurity" (2015: 3241). As crops began to fail and livestock began to perish, many Syrian farmers (as many as 1.5 million by Kelley et al.'s account) migrated to the cities, which in turn sparked the political unrest in 2011.[8] This thinking was so widely accepted in policy circles that many political leaders in the West including Barack Obama, Bernie Sanders, and Prince Charles of the United Kingdom began to explain the onset of the Syrian conflict in terms of climate-induced drought. Perhaps the most notable explanation of the conflict was given by the former US Secretary of State John Kerry who blamed the crisis on the country's "worst drought on record," which pushed "as many as 1.5 million people ... from Syria's farms to its cities, intensifying the political unrest that was just beginning to royal and boil in the region" (Kerry, 2015). In an attempt to break down causal linkages, we apply the same three-step approach (as illustrated in Figure 8.1) to the Syrian case: (1) was there a climate-induced drought prior to the civil war and how long did it last? (2) did it lead to large-scale out-migration of the drought-affected groups and thereby exacerbating socio-economic problems of the host communities? and (3) did the drought-related socio-economic pressures contribute to the 2011 political unrest and subsequent violent conflict?

First, was there a climate-induced drought in Syria immediately prior to the country's descent into civil war? And if so, how severe was it and how long did it last? The latter point is particularly significant because when droughts last a year, usually they can be managed politically. If they persist into several consecutive harvest seasons, however, the impacted groups are more likely to flee their homes in search of greener pastures. According to Kelley et al.:

> the region has been in moderate to severe drought from 1998 through 2009, with 7 of 11 years receiving rainfall below the 1901-2008 normal. It is notable that three of the four most severe multiyear droughts have occurred in the last 25 years, the period during which external anthropogenic forcing has seen its largest increase.
>
> *(2015: 3243)*

Based on their data, Syria experienced a severe multiyear drought (defined as "three or more consecutive years of rainfall below the century-long normal") from 2007 to 2010 immediately prior to the civil war in 2011. During the three-year drought, the 2007/2008 winter was "the driest in the observed records" (Kelley et al., 2015: 3243). Other studies have confirmed this as indeed being the case (Selby et al., 2017: 234) but points of agreement seem to end there. Indeed, the rainfall pattern across Syria during the pre-civil war period, according to more recent evidence, was "far from uniform" (Selby et al., 2017: 234) and that "most areas of western and southern Syria received close to or above-average rainfall" during the alleged three-year drought period (Selby, 2019: 2). What these recent studies suggest is that northeast Syria did experience a severe drought during the three years prior to civil war with the 2007/2008 winter being the worst, with precipitation 35 percent below the 1961–1990 average (Kelley et al., 2015). However, the severe drought did not last for several consecutive years as widely suggested nor did it affect the whole of Syria (Selby et al., 2017). It is therefore difficult to conclude that there is clear and reliable evidence that climate-induced drought caused the Syrian civil war.

Second, due to the severe drought during the pre-civil war period agricultural production in Syria's northeast region of Jazira collapsed forcing large-scale out-migration predominately from al-Hazakah to the urban centers of Aleppo in the northwest and Damascus and Dara'a in the south (see Figure 8.4). This out-migration (so the story goes) generated socio-economic pressures on host

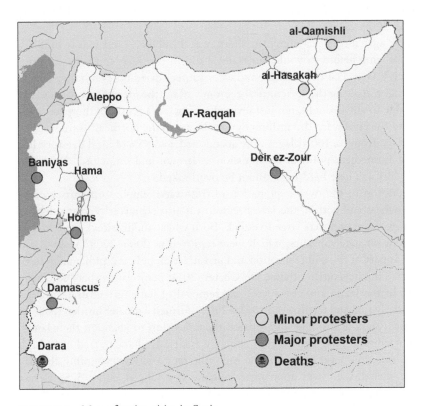

FIGURE 8.4 Map of major cities in Syria

communities that subsequently led to Syria's descent into civil war. How true are these claims?

At least three claims have considerable support: (1) that Syria's northeast region of Jazira was hit hardest by the drought (Selby, 2019); (2) that pre-civil war Jazira experienced an agrarian crisis (Femia and Werrell 2012, Gleick, 2014, Kelley et al, 2015); and (3) that a large number of Syrians migrated from the northeastern Jazira region to other parts of the country. What is less clear, however, is the extent to which Jazira's agrarian crisis was caused by the drought as opposed to political reforms of the 1980s. The Jazira region which was hit hardest by the drought, and which presumably triggered large-scale out-migration, had become by the 1990s Syria's breadbasket region thanks largely to "the systematic exploitation of the region's groundwater resources" for irrigation, producing 50 percent of the country's wheat and cotton (Selby, 2019: 7). Irrigation, unfortunately, made this region extremely dependent on groundwater, which was over-pumped at unsustainable levels, with water withdrawals some years exceeding 300 percent of sustainable levels. And how could it not exceed when irrigated wheat farming increased fourfold over the 1985–2005 period? Exacerbating the water scarcity problems, many Syrians began to dig up their own water wells which were "unlicensed and unmonitored" (Waterbury, 2013: 30). By some account, over half of Syria's 200,000 private wells were considered "illegal" (Droubi, 2009: 22).

Syria's agricultural economy, in short, was a bubble – and it was simply a matter of time before it would burst and it did in the years immediately prior to the civil war. The depleted groundwater sources dried up even faster during the drought, making the agricultural economy of northeastern Syria particularly vulnerable to any drought, not just severe drought. As a result, in the years leading up to the civil war, "1.3 million Syrians living in the northeast were severely affected by drought, 160 villages were abandoned, and 85% of local livestock lost without immediate political repercussions" (Brown and Crawford, 2009: 26; Worth, 2010). These hardships cannot be overlooked.

In addition to the over-exploitation of freshwater supplies for thirsty crops such as wheat and cotton, the Jazira region was also characterized by a disproportionate share of Syria's large livestock. Both wheat and livestock were highly dependent on state subsidies, which were removed in 2008–2009 by the Assad regime as part of price liberalization and privatization policies. In her analysis of the Syria climate-conflict thesis, De Chatel (2014: 526) claims that "for many farmers in the Jezira and elsewhere" the removal of state agricultural subsidies (particularly fuel and fertilizer) in 2008–2009 "formed a greater burden than the successive years of drought and spurred their decision to abandon their land." Without subsidies, the Jazira region, which was already characterized by "a high poverty rate, low level of healthcare, high illiteracy and few economic alternatives to agriculture," was doomed to collapse (De Chatel, 2014: 525). Syria's water shortage problems prior to its civil war as highlighted by Beck (2014) and the government's failure to respond to the ensuing humanitarian crisis seems to have formed one of the "triggers of the uprising, feeding a discontent that had long been simmering in rural areas" (De Chatel, 2014: 2). For less stable governments, severe water shortages are likely to be detrimental to political stability (Halverson, 2016).

In short, based on these observations, politico-economic factors seem to have trumped climate factors in driving people away from their homes. The confusion stems from three assumptions. First, in the view of western observers, the pre-civil war drought is talked about as if it affected the entire country when in fact it seems to have impacted only the northeastern part. Other areas of Syria actually received above-average rainfall. Second, the pre-civil war drought is often viewed as negatively affecting Syria's agricultural economy all year round when in fact it was most severe during winter months, which means it mainly affected winter crops such as wheat and barley. Summer crops like cotton were hardly affected (Selby, 2019) which means not all farmers lost their livelihood. Third, a commonly held view suggests that the drought-stricken regions experienced crop failure for several consecutive years when in reality, the drought led to total crop disaster mainly in the winter of 2008 when rainfall was its lowest. But by 2009, basic food staples had increased to a near long-term trend – two years prior to the conflict onset. In other words, Syria had been experiencing an agrarian crisis involving water mismanagement long before the anomalous drought of 2006–2009. Assad's subsidy cancellations seem to have made survival more difficult driving many people from their homes toward other parts of the country.

It is not clear if these displaced populations played any direct role in the Syrian uprisings of 2011, which brings us to our third step in the model illustrated in Figure 8.1. The commonly held story suggests that "rapidly growing urban peripheries of Syria, marked by illegal settlements, overcrowding, poor infrastructure, unemployment, and crime, were neglected by the Assad government and became the heart of the developing unrest" (Kelley et al., 2015: 3242). The protesters were essentially climate refugees who ended up in urban areas that became "the heart of the developing unrest." There are several questions to consider as we assess the evidence. How do we best characterize the people most involved in the protests and subsequent insurgency? To what extent were they motivated by rural displacement strains? Were they numerous and did they go to the cities that became the heart of the resistance? And finally, what, if anything, did the government do for the migrants? In short, were the migrants from northeast Syria "significantly involved, whether as mobilizers, participants or targets, in the early demonstrations which spiraled into civil war?" (Selby et al., 2017: 240). It is to these questions we now turn our attention.

The problem with the Syrian civil war is that we do not know exactly how many people migrated from northeast Syria in response to the 2006–2009 drought and we know even less about their involvement in the early days of the uprising. According to the limited evidence we do have, mainly based on 32 interviews with Syrian refugees in Dara'a, migrants from the northeast,

> did not participate to any significant degree in the spring 2011 protests, were not targets of the demonstrations or subsequent repression, and left as soon as the protests started and went back to the northeast.
>
> *(summarized in Selby et al, 2017: 240 based on Frohlich, 2016)*

Curiously, "none of the political demands made by Syria's early 2011 protest movements related directly to either drought or migration" (Selby et al, 2017: 240), focusing instead mainly on "economic liberalization-related grievances." Frohlich (2016) also challenges the common assumption that "migration patterns in Syria were predominantly rural to urban." Based on her interviews with the migrants in Dara'a, drought-affected farmers from the rural areas of the country's northeastern region largely went to other rural regions because there were already established "migration corridors" to rural areas, which meant "less transaction costs than non-linear moves to new destinations." Those who did migrate from rural to urban areas were often "better educated than the day workers on the agricultural projects" (Frohlich, 2016: 44). She concludes that while conditions for collective action were present "in settings such as Dara'a prior to the Syrian uprising, it was not the (climate) migrants who initiated the protests" (2016:

44). In fact, "the migrants were too marginalized… to initiate social action" and were the first ones to flee the conflict zone when violence intensified.

Based on available evidence, we are inclined to agree with the conclusions of Selby et al (2017: 241) and Frohlich (2016) that:

1. climate change was not a significant factor in northeast Syria's pre-civil war drought;
2. drought-driven migration was probably not on the scale claimed and was more likely caused by Assad's economic liberalization reforms than drought; and
3. drought migrants did not initiate the civil war.

In short, until more data become available, we have no good reason to support the claim that climate-related drought in Syria caused the country's civil war. One can argue that at the very least the drought led to the migration of some scale, which highlighted inequities and inactivity of government at a time when other processes were building to resist the Assad regime.

The Yemeni civil war (2015–?)

The ongoing Yemeni case is complicated by multiple actors and extreme climate problems which are not usually paired. That is, while observers sometimes add the Yemeni case to the list of examples of conflict within climate handicapped environments, there does not appear to be any overt claim that climate change is directly responsible for Yemen's civil war. The current internationalized civil war in Yemen is a product of earlier civil wars and elite bargaining over power sharing and patronage (Jones, 2011; Transfeld, 2016; Perkins, 2017). Conflict tends to break out when some elites are either losing ground or perceive that they will lose ground in the near future if some coercive resistance to power plays in progress is not exhibited. Whether the resistance is judged successful or otherwise, what is most clear is that the numerous civil wars do not tend to resolve who will preside over the elite bargaining system or who will replace it with some other type of structure.

In 1962, the ruling Imam was overthrown by military tanks that encircled his palace and began firing. Nevertheless, the Imam managed to escape which contributed to the subsequent eight years of republican-royalist civil war in which Saudi Arabia intervened on the royalist side despite the Imam representing Zaidi Shi'a. Egypt intervened on the republican side with a major intrusion of as many as 70,000 soldiers. The 1970 negotiated settlement kept Imamate forces in the political game by allocating them some positions and a share of the patronage pool (Byman, 2018). Within less than a decade of the settlement, Ali Abdullah Saleh, had emerged as the central player in the elite bargaining network. Relying heavily on his own family and tribe, key positions were secured just as alliances

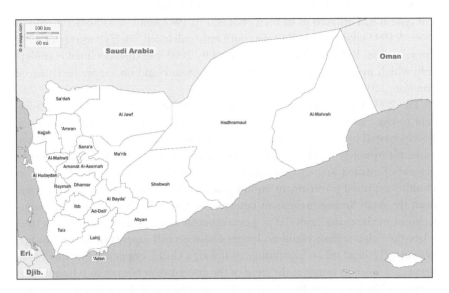

FIGURE 8.5   Map of Yemen

were made with other players representing the army, tribes, and commerce. This ruling coalition network could be manipulated to some extent by playing parts of the network against each other. For instance, Marxists in the south could be countered by encouraging Wahhabi/Salafi/Muslim Brotherhood elements in the north. Houthis could be encouraged to resist Wahhabi/Salafi/Muslim Brotherhood inroads in the north. Ties with al Qaeda (AQAP) could also be useful from time to time. The problem with "dancing on the heads of snakes," Saleh's own description for this ruling strategy (Worth, 2017), is that sometimes the snakes rebel against the dancing.

In 2004, Saada-based protests against regional discrimination, corruption, and, in particular, the introduction of conservative and anti-Shi'a lessons into the education system led to a protest movement spearheaded by the Shabab al-Mumin (Believing Youth) organization with leadership by a member of one of the leading families in the Zaidi Shi'a community, Hussein Badr al-din al-Houthi. The underlying issue might be categorized as resistance to what was perceived as reneging on the 1970 agreement and the declining influence of Zaidi Shi'a elites. Protesters were imprisoned and a Houthi insurgency was begun in which Hussein al-Houthi was killed fairly quickly. Other members of the al-Houthi family, initially the father of the martyred Believing Youth leader, have since assumed leadership of this movement.[9]

What began as a protest movement escalated into a regional insurgency and later into first a national and then internationalized civil war. Saleh's snake dancing strategy worked reasonably well in the 1990s thanks to oil revenues gained

when north and south Yemen were unified in 1990 after the south Yemeni political system had erupted in its own civil war in 1987 and lost its Soviet patron in
1989. A 1994 civil war based in the south only enhanced Saleh's control of southern resources. But the oil revenues peaked by 2003 and then declined considerably which meant that the elite patronage system could no longer be financed
(Brehony, 2015).

The 2004 Houthi insurgency was located in the northernmost region (adjacent to the Saudi border), was especially impoverished as a province, and was
poorly treated by the central government. Zaidi Shi'as, nonetheless, represented
a sizeable minority – perhaps half the population in northern Yemen and 30–40
percent of unified Yemen. The insurgency ratcheted up in scale and lethality as
Houthis captured government equipment but the turning point did not come
until the Arab Spring movement emerged in Yemen in 2011. Popular demonstrations in the main cities led to governmental repression and deaths at a time
when the Saleh regime could no longer count on full support from his national
network. To head off an impending civil war, a Gulf Cooperation Council plan
persuaded Saleh to resign as the head of the government but not to relinquish his
control of the leading political party. The interim replacement government was
supposed to devise a plan for reconciling northern and southern discontents but
instead focused on reinforcing the position of Saleh's replacement, Abd Rabbo
Mansour Hadi. That meant undermining Saleh's position, gaining control of
military units, and weakening the ability of other parts of the network to play
full roles. For instance, provincial boundaries were redrawn to submerge Saada
within a larger northern province in a new federal structure – much to the dis-
pleasure of the Houthis.

In 2014, the Houthis effectively nationalized their insurgency by taking control of the capital, Sana'a. Ironically, the Houthi escalation of the insurgency was
executed in partnership with Saleh who had switched sides. This marriage of
convenience lasted until it appeared that Saleh was switching back to the government side and was then assassinated by the Houthis in late 2017.

Prior to this turn of events, however, Saudi Arabia had begun intervening on the side of the incumbent government as early as 2009. Saudi Arabia
had permitted Saleh's regime to use Saudi territory to outflank the Houthi
rebels which led to Houthi attacks on Saudi border guards. The Saudi interest in suppressing a nearby source of instability escalated in 2015 into a much
more coercive intervention, partnering with other Gulf allies but especially the
UAE, to ensure that not only Houthis did not control the Yemeni government
but also their conservative Islamic opponents (the conservative Islamic Islah
movement which had once been a key component of the old Saleh coalition)
did not as well.

There can be little question that Yemen represents an extreme case of impoverishment (the poorest population in the MENA) and a victim of climate change,
most specifically in terms of increasing population and decreasing water supply.

Prior to the ongoing civil war, 14 of the country's 16 aquifers had been over-pumped, prompting some analysts to call it "a hydrological basket case" (Brown, 2011: 22). With water tables rapidly falling in a country with one of the world's fastest-growing population, the Yemenis now import over 80 percent of their grain.[10] But the food dependency and water scarcity have been more a function of governmental mismanagement than a source of conflict to date. Some proportion of the Yemeni water supply is privately owned and sold at high prices. Heavy reliance on food imports was only feasible as long as oil revenues were large and increasing. Once they began to drop 15 years ago, much less food could (and would) be purchased.

Without belittling Yemen's rather serious environmental problems, the easiest case to make is that Yemen's conflict problems make the effects of climate change worse more than the other way around:

> Yemen faces daunting, interconnected challenges: a failing economy, massive unemployment, runaway population growth, resource exhaustion, a rapidly falling water table, dwindling state capacity, an inability to deliver social services throughout most of the country, and interwoven corruption….
>
> *(Boucek, 2010: 2)[11]*

Boucek goes on to say that the Houthis, the threat of southern secession, and a resurgent al-Qaeda can exacerbate the many other problems of Yemen. That conclusion is difficult to escape. It is certainly conceivable that those other problems exacerbate Houthi, southern, and AQPA discontent. What remains to be demonstrated is whether the discontent would go away if all of the other problems were to vanish. We cannot run that experiment very well but it seems doubtful that Yemen's political combat can be traced causally to climate change, either directly or indirectly.

Comparing the three cases

We set out to examine three prominent MENA civil wars often linked albeit vaguely to climate change. Are they short-term canaries or long-term harbingers of things to come? As it happens, we find each of the three conflicts to be fraudulent "canaries." Direct linkages between climate change and two of the three instances seem to be nonexistent. The linkage in a third case is existent but hardly carries a great deal of explanatory weight. Nonetheless, the conflicts do make matters worse for the survivors. On this basis, we conclude that they are more like harbingers of a future plagued by the double scourges of climate change and extensive conflict. Yemen, in particular, serves as one possible prototype of what much of the MENA could become if things go as badly as the climate models suggest. All parts of the MENA are susceptible to the type of intense political in-fighting occurring in Yemen. Not all parts of the MENA resemble Yemen in terms of food and water scarcity – but they could come to resemble it. The millions of people who attempted to flee Syria constitute another type of prototype. Instead of a few million civil war refugees, the number of climate refugees in the future could multiply by a factor of 10 or possibly even 20 if global warming turns the MENA into a wasteland.

We can argue that climate-related problems certainly did not help avoid civil wars in Sudan, Syria, and Yemen but one also cannot argue that these are directly climate-related phenomena. Figure 8.6 helps pin down what we are looking for and what we found.

The Sakaguchi, Varughese, and Auld figure connects four clusters of variables. On the left-hand side are types of climate change. On the right hand are types of conflict, both internal and external. In between are "interacting factors" at the top and "mediating factors" at the bottom. It is not clear that interacting and mediating factors are wholly different concepts. They are simply intervening variables that may influence the linkages between climate change and conflict. Moreover, the figure is explicitly constructed to privilege the role of climate in accounting for conflict. There is a single driver on one side of the figure and

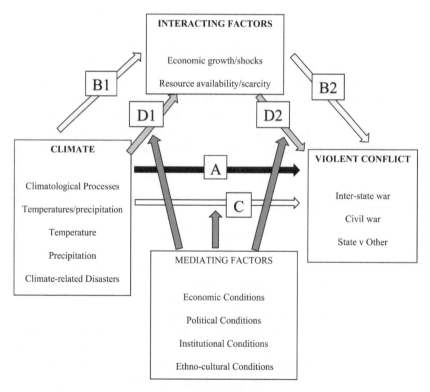

FIGURE 8.6 Climate-related interactions. Source: Based on a modified form of Sakaguchi, Varughese, and Auld (2017: 631).

everything in between is designated as intervening. What if some of the inter-acting and mediating factors are as significant or even more important drivers of conflict in their own right? Still, the figure is designed to illustrate various pathways from climate to conflict and should be evaluated as it was intended as opposed to how multiple drivers might impact conflict which Sakaguchi, Varughese, and Auld would no doubt say is a different question.

For our purposes, we are interested in utilizing the A, B, C, and D paths to illustrate what was found and not found in our three cases. Our main focus in the brief case studies was to track variation on path B (climate change to interacting factors to conflict). We found very little evidence that path B was critical in our three cases. At best, farmer-herder conflict in Darfur over sev-eral decades contributed to "African" and "Arab" polarization, hostility, and mistrust in Darfur. The problem is that this form of political-economic conflict might have been managed in a different context without the emergence of a full-fledged civil war in 2003–2005. That would suggest that path D in Figure 8.6 might best fit the Darfur case. The "mediating" factors of competition within the ruling coalition and the conflict diffusion tendencies of regional conflict (especially the long civil war in neighboring Chad and the two civil wars between northern and southern Sudan) appear to have been crucial in explaining what happened in 2003.

In the Syrian case, the evidence suggests some support for path B1. Drought did lead to rural-urban migration of some disputed size. Yet no one has made a convincing case for drought-induced migration to have tilted the scales in favor of a civil war breaking out. The B2 path seems to be missing in the Syrian case. At best, one can say that the migration did not make the Assad regime look any better than it had before the migration occurred.

In Yemen, there is no doubt that all sorts of environmental degradation and deterioration can be found. Again, this seems to be a case of support for a B1 (or D1) path without evidence for a B2 (D2) linkage. Moreover, throughout all three cases, we find nothing that underlines the significance of the direct A and C connections.[12] They may be supported somewhere else but they do not appear to be all that critical to Darfur, Syria, and Yemen.

Does that mean that climate change is destined to play only a minor con-tributory role at best in accounting for severe cases of intrastate conflict in the MENA? The answer is no. Our sample is restricted only to cases that observers have promoted as bearing considerable potential for a strong climate role. Three cases do not qualify as an adequate sample of all MENA civil wars. Yet if the strongest cases appear to actually be weak in terms of their climate-conflict link-ages, we wonder whether a larger sample would make much difference in terms of explanatory outcomes. Our major caveat, though, is that it is still early. The climate changes of the late 20th and early 21st centuries may not look so impres-sive 20–30 years down the road if the worst-case scenarios for global warming are realized. That is a comparison, however, that we can only "hold our breaths" for some time to come before we will be in a position to execute it.

Conclusions

We began our analysis in this chapter thinking that the three conflicts in MENA might represent "canaries in the coal mine" in the sense that each of the three represented in various ways climate-induced conflict. After careful examination of each case, we are now forced to conclude that none of the three cases appear to represent much in terms of early warning signs of conflict brought on by global warming. However, given what we know in 2021, we can say that these cases are more like harbingers, that is, illustrations of what might come later in the century. They are long-term harbingers because we have no reason to assume that the conflict propensity of the MENA will subside. Throughout, what we have is conflict occurring in arid areas, thereby increasing food and water insecurities, which, we think, will occur anyway with more global warming. The outcome is not more of the same but it is likely to be worse with a greater scale of human suffering and death as the likely outcome – whether or not climate change increases conflict propensity.

Yemen in particular seems like a preview of a world in which food and water are severely limited and everybody is fighting everybody else. Climate change is generally expected to cause more drought, more famine, and more mass displacement. Whatever the precise role of climate change, it seems quite conceivable that extreme warming on a scale not experienced in millennia could contribute directly or indirectly to further conflict. Another way of looking at the climate-conflict argument is that warming so far has been less acute than it will probably be in the future, which could mean that current findings underestimate the role of climate in future conflicts.

Yet if we are to have more warming, what might we anticipate? The MENA has a long history of being a conflictual region so there is little reason to assume that will change. If global warming increases scarcity – and how can it not? – conflict within the MENA as well as between the MENA and the rest of the world could be expected to at least stay the same as it is now and perhaps also to increase if scarcities become so great that the only thing people can do is fight over what remains. At present, it is mainly Muslims fighting other Muslims. But if the outside world continues to need oil, it will continue to intervene in the region to secure access to energy resources in spite of or because of local turmoil. If Syrian refugees have been causing a crisis in Europe (and increasing xenophobia and anti-immigration tendencies) and, to a lesser extent, North America today, imagine what the reaction to dealing with many more refugees might be as deteriorating environmental conditions push more people out of their homes into the neighboring countries. Increased conflict may be the least of our worries.

Of course, with the advent of global warming, there will be a host of other problems with which to deal. Coastal city populations in places like Yemen, will surely have to move inland but how much farther inland can they go without encountering stiff resistance or limited resources with which to work? Water will need to be rationed in areas where water is already scarce. But what will be done with and to the MENA population? It cannot move *en masse* nor can we build a wall around it to sequester their problems from the rest of the world. Some petrostates may be able to insulate themselves from the new hardships but most states in the region will have few resources with which to cope in an even more deteriorated environment. Can we expect much assistance from the already antagonized non-Muslim world which will be attempting to manage its own global warming-induced problems? Somehow, an increasingly liberal world characterized by widespread democratization and economic interdependence seems somewhat more remote both within and outside the MENA. What remains are mitigation and adaptation efforts that will need to be amplified at wartime speed. What are the MENA states doing to adapt to climate change, if any, and what can the developed world do to help counter some of the dangers of climate change? How much, if anything, is being done to develop policies and infrastructure for coping with a "tsunami"-level problem like severe global warming. The answers to these questions, perhaps not surprisingly, are disheartening. They are the subject of Chapter 9.

Notes

1 For instance, Lee et al (2013) and Zhang et al (2007) argue that preindustrial economies were more vulnerable to climate-related food scarcities than modern economies because the latter are better able to adapt to food shortages.
2 Bai and Kung (2011), for example, find a positive relationship between droughts and floods and nomadic incursions in China over a 2060-year period. Most contemporary studies of the climate-conflict relationship rely on shorter periods often less than 2% of the Bai-Kung length.
3 The threefold differentiation of Darfur was a post-colonial effort to dilute the ability of politicians in Darfur to be able to muster support from the full western region.
4 Road construction began in the 1990s but was still borne and never completed due to Sudanese political struggles over whose priorities would prevail (Roessler, 2016: 185–186). The non-completion of the road was one of the stated grievances underlying the 2003 war.
5 Flint and de Waal (2008) give strong attention to the regional context of the Darfur civil war.
6 The World Bank assessments of arable land portray Syria moving from about 34% in 1960 to around 25% in 2017.
7 The millions of refugees fleeing to neighboring Jordan and Turkey and the rest of the world are clearly another major reason for the media attention.
8 See, as well, Ide (2018) and Ash and Obradovich (2019).
9 One of the traditional beliefs of Zaidi Shi'ism, a sect that is often difficult to distinguish from Sunni doctrine, is that an Imam who is descended from Mohammed should rule in Yemen. Members of the al-Houthi family can claim this distinction. As a consequence, the incumbent regime early on branded the Houthi rebellion as an Imam revivalist movement even if this facet did not seem particularly salient (Freeman, 2009). The Houthis were also painted as Iranian proxies by the central government but it is not clear that was the case at the outset. As the insurgency escalated, the Houthis may have become more dependent on Iranian weapons and Iran claimed Houthi successes as its own (Hokayem and Roberts, 2016).

10 The current population (2018) is slightly less than 29 million. In 1950, it had been 4.3 million. Yemen could have a population of 60 million by 2050 if the demographic forecasts are proven correct.
11 If Boucek had written a few years later, he might have added that Yemen has had the largest cholera outbreak in history with 1 million sickened in the past three years (Erickson, 2017).
12 Since the mediating factors extend through both A and C pathways, it is not clear that both direct pathways are needed.

9

WATER, FOOD, AND ADAPTATION

With global warming (as discussed in Chapter 7), the Middle East and North Africa (MENA) states will face a host of problems, such as increases in environmental degradation, agricultural productivity losses, water shortages, and economic, societal, and political instability – all with significant potential for increasing the risk of deadly violence. These problems are already acute in the region and they are only likely to get worse. Thus, adaptation is key to the survival of the region. So what are the MENA states doing to adapt to climate change, if anything? Will their efforts make a difference, assuming we are already in the early days of the global warming crisis? This chapter looks at resilience and current adaptation efforts of the states in the MENA to assess whether they are sufficient and/or the extent to which they might be limited in the region. Climate change is not, nor will it likely to be the only cause of problems in the region. But if temperatures continue to rise while rainfall continues to decline, climate change will only aggravate existing drivers of deterioration and may make things unmanageable in some places. Sheer survival may become the primary goal for large segments of the MENA population.

Water and food again

States in the MENA draw their water from one of four sources: surface water, groundwater, treated water, or desalinated water. Historically, most states have relied mainly on the surface and/or groundwater sources to meet their needs. Those endowed with significant surface freshwater resources include Turkey, Egypt, Iran, Iraq, and Lebanon but these states have been experiencing chronic water shortages due to population growth. Those states that rely mainly on groundwater have either overpumped their nonrenewable aquifers and/or are struggling "to mobilize the kind of investments necessary to secure 'new' large-scale water sources," particularly in Jordan and Yemen (Sowers, Vengosh, & Weinthal 2011: 602). Syria has both surface and groundwater resources but the ongoing civil war has severely limited its ability to draw on these sources to meet the needs of farmers. Thus, the divide between the rich and poor states in the region remains significant with the rich economies consuming more water per capita than their poorer neighbors and meeting their water deficit through costly and energy-intensive desalination systems and by treating their wastewater. The poor economies have no choice but to continue exploiting the depleted non-renewable aquifers even though the region as a whole has long ago surpassed its carrying capacity of natural water resources.

The more affluent economies in the region such as the Gulf petrostates and Israel that can no longer rely on overpumped fossil aquifers have been able to meet their increasing water demands by investing in costly desalination plants. In a remarkable turn of events, Israel, which teetered on the brink of a decade long drought as recently as 2008, generates today more water than it needs through desalination, nationwide adoption of water-efficient appliances, and innovative water treatment systems that can recapture as much as 86 percent of consumed water and reapply it for irrigation purposes (Jacobsen 2016). Today desalination provides over 50 percent of Israel's freshwater needs. Like Israel, the rich Arab states (e.g., Bahrain, Kuwait, Qatar, Saudi Arabia, and the UAE) are able to exploit energy-intensive desalination systems to meet most of their water needs (see Table 9.1).

While desalination may solve a country's water problems, it does come with its own host of environmental problems. Among them are high energy uses from power plants that emit greenhouse gases, dumping of highly concentrated saline water and chemicals into the ocean, not to mention the use of scarce land particularly in crowded countries such as Israel (Greenberg 2014). Taking salt out of seawater is an extremely energy-intensive process. Desalination uses about 15,000 kilowatt-hours of power for every million gallons of freshwater produced (Sazak and Sukin 2015). This amount of energy is enough to power an average American household of four for a year and a half. As one of the world's largest oil producers, Saudi Arabia obviously can afford the energy to power the desalination plants. But such plants are not sustainable in the long term as oil reserves will eventually dry up and the low energy prices will not remain low forever (Sazak and Sukin 2015). Moreover, at some point in the near future, the world will transition away from carbon fuels as the cost of consuming these fuels becomes too high both for human health and the environment.[1] That said, Saudi Arabia and UAE have announced plans to develop large-scale solar-powered desalination plants, which, if fully implemented, could alleviate some of the environmental concerns associated with the current desalination plants (Freyberg 2015). Presumably, technological improvements will also further lower the cost of desalination.

Of course, one adaptation to increasing water scarcity is to capture control of what water does exist. This strategy can be especially problematic in the context of shared rivers. States upstream can dam rivers to create reservoirs for their own

TABLE 9.1 Water Resources of the MENA States

State	GDP per capita (2019)[a]	Sources of freshwater(as % of total water use)[b]			
		Groundwater	Desalinated water	Treated wastewater	Surface water
Bahrain	25,780	66%	29% (2003)	5%	--
Egypt	3,020	10%	0.3% (2010)	2%	84%
Iran	5,820	57%	0.2% (2004)	43%	--
Iraq	5,760	--	--	2.3%	98%
Israel	42,140	--	55% (2015)	13% (2004)	32%
Jordan	4,390	59%	1% (2005)	9%	31%
Kuwait	29,130	45%	46% (2005)	9%	--
Lebanon	9,610	53%	4% (2005)	13%	30%
Oman	18,080	89%	8% (2003)	3%	--
Qatar	70,290	49%	41% (2005)	10%	--
Saudi Arabia	22,510	40%	50% (2015)	--	10%
Syria	2,058 (2007)	84% (mixed with surface water)		16%	
Turkey	8,510	98% (both)	2% (2003)	--	98% (both)
UAE	39,810	70%	24% (2005)	6%	--
Yemen	918	70%	0.3% (2000)	0.2%	29%

Source: [a] IMF (2019); [b] FAO (2019).

use. States downstream receive less water, as a consequence. All of the major rivers in the MENA have some linkage to this type of behavior. Ethiopia has built a dam on the Nile, which could cut the Egyptian water supply by 25 percent while filling an Ethiopian reservoir to be utilized for generating electricity. The water shortfall would directly threaten some 60 percent of Egypt's food production (Conniff, 2017). Filling a large reservoir takes a finite number of years but the Egyptian Nile delta is already expected to decline in size and productivity by a third due to intruding saltwater linked to rising sea levels brought on by global warming (Balaraman, 2017; Conniff, 2017).

Possession of the Tigris and Euphrates, which flow from Lake Van in Turkey into Iraq and Syria have long been targets of conflict but commentators have more recently begun to discuss "the disappearance of the Fertile Crescent" in reaction to a combination of drought and reduced water supply due to damming activity in multiple states, but especially Turkey (Kitoh, Yatager, and Alpert 2008; Pearce 2009; al-Masri 2014). Even ISIS became a critical player by controlling one of the larger dams in Iraq for a while (Pearce 2014). Much the same outcome characterizes the Jordan River, which is also drying up in part due to excessive damming practices (Bromberg, 2008; Ben David, 2016).

Despite what appears to be an insurmountable problem with several contributing sources, reducing water scarcity by 2050 is thought to be possible, according to a recent study by researchers from McGill and Utrecht Universities. In a study published in *Nature Geoscience* they identify six key areas that could be utilized to effectively reduce water stress in various parts of the world. They include improving: "(1) agricultural water productivity, (2) irrigation efficiency, (3) domestic and industrial water use, (4) limiting the rate of population growth, (5) increasing water storage in reservoirs, and (6) desalinating seawater" (Wanda, Gleeson, and Esnault, 2014). But all of these require more research, more resources, more consumers prepared to try new strategies, and more governments willing and able to be proactive on water scarcity issues. So far, only a few MENA states have been prepared to develop a plan for the water scarcity issue. Ironically, one of the few successful plans has emerged from the ranks of the poorer countries – Tunisia – while entrepreneurs in some of the richer states still seek to tow icebergs to the Gulf. Desalination, similarly, appears to be restricted to some of the more developed/affluent states such as Israel and Saudi Arabia – barring a major decrease in its cost.

Water is not the only area in which adaptations might be pursued. Food is another. Unfortunately, the MENA problems with water scarcity, as we have seen, are closely paralleled by the region's problems with food scarcity. One reason for this matching is that water is critical to growing food. Some 87 percent of the MENA water consumption is currently devoted to agriculture. But the amount of available water in a region that is already highly arid has been declining for the past six decades.[2] NASA's Gravity Recovery and Climate Experiment (GRACE) mission produced data on cumulative total freshwater losses in the MENA from 2002 to 2015 . The data show an alarming rate of decrease in total water (stored in all of the snow, surface water, soil water, and groundwater in the region) of approximately 143.6 km^3 during the 13 years studied (Voss et al 2013). Groundwater depletion (particularly in Turkey, Syria, Iraq, Iran, and along the Arabian Peninsula), is the leading cause of water scarcity in the region, contributing to significant changes in total water storage in the region.

The water deficit problems will be further exacerbated in the future by increases in population growth and urbanization. So, too, is the amount of available arable land. More mouths to feed, less land to grow crops upon, and less water to allocate to agriculture suggest that the likelihood of increasing the supply of local food is not great. Storing food for bad years, a practice as ancient as the Pharaohs (Wilkinson, 2011), is certainly advisable but there are limits to food storage when it comes to feeding entire populations. Some others are rich enough to subsidize food imported into the Gulf area. But poor states such as Yemen and Jordan can contemplate none of these strategies.

Part of the food scarcity problem is that cereal and sugar constitute about 61 percent of Arab diets. Yet about half of the cereal and nearly three-fourths of the sugar must be imported (Kubursi, 2012). Prices have already increased (cereals were up 40 percent and sugar 77 percent in the 2 years after 2008). Global warming, moreover, is expected to reduce global crop production in general (Challinor et al. 2014; Porter et al. 2014). Wheat production, in particular, is expected to be reduced by 6 percent for each degree of increased warming (Asseng et al, 2014). Quality declines can also be anticipated as protein and other nutrients become less available (Marques et al, 2010; Liu, Hofstra, and Franz, 2013; Hammond et al, 2014; Campbell et al, 2016).

The MENA is already distinguished by the extent of their dependency on imported food. How will deficits in future food supply be managed? Fader et al (2013) answer that increased dependency on food imports can be dealt with in only three ways. First, a dependent population can change its diet and move toward foods that are less resource intensive to grow. Second, areas devoted to agriculture can be expanded. Third, farmers can learn to be more productive in how they produce food. If the amount of cropland is not likely to increase and only a genuine crisis will convince people to eat differently, the burden of adaptation will be placed on increased agrarian efficiency. No doubt, improvements are conceivable in this area and some gains can be anticipated. But if the world should become three times more dependent on food imports by 2050 (Fader et al, 2013), marginal improvements in agricultural efficiency will be unlikely to compensate for increased competition for food.

Even if the richer Arab states can buy cropland in Africa, Europe, and North America (a practice currently in place), that production will go to feed relatively small and affluent populations. However, Woertz (2013: 198) has criticized these efforts as both counter-intuitive and poorly executed. The counter-intuitive charge stems from the Gulf states' tendency to buy farmland in countries that have water deficiencies of their own, are food importers themselves and that have high population growth rates – not the best places to look for surplus food.

One should expect, in general, more people in the poorer MENA countries to go hungry if attempts to increase food imports are less successful on the open and increasingly competitive markets. Hungry people inhabiting areas that could become too hot to sustain life will seek to flee. But the reaction to the Syrian refugees, who are at least in part climate refugees, provides strong hints that there will be few places that provide refuge for refugees from places becoming unlivable. Even if there was an abrupt turnaround in global Northern attitudes toward taking in global Southern migrants with different languages, religions, and cultures, the sheer scale of the problem may prove to be overwhelming. Worst-case scenarios about rising sea levels are forecasting as many as 1.4 billion climate refugees by 2060 and 2 billion by 2100 (Scotti, 2017).

Such forecasts assume major losses of heavily populated coastal areas and do not take into consideration the potential loss of whole regions that become uninhabitable due to the loss of food and water access. But even if these figures are too extreme and the number of climate refugees is only a few hundred million people (Weis, 2015), the problem will still be sufficiently overwhelming to anticipate the likelihood of fairly weak responses by more fortunate places. Put more simply, there will be no place for many climate refugees to go because there will be too many to process (Lieberman, 2015; Carrington, 2016).[3] Relatively weak states, for their part, will find that they are not up to the task of managing these climate-induced problems at the source, just as they are not up to seeking proactive solutions to future crises. Internal conflict among the survivors seems unavoidable just like they were in some of the previous episodes of abrupt climate change.

Nor is the MENA alone

It seems fair to say that these expectations point to a dismal future for the MENA. Extreme temperatures, even further diminished water supply, rampant food insecurity, and political-economic turmoil suggest that climate change will penetrate every sphere of life and make it far less pleasant. These outcomes will not be duplicated exactly in each and every MENA state but the impacts are expected to be widespread, comprehensive, and severe. When climate change impacted the MENA region in earlier episodes, there were usually some local refugees. This time there do not appear to be any places to go to escape what is believed to be coming.

Some amount of global warming in the coming decades seems quite probable.[4] However warmer it becomes, the effects around the world will be uneven.[5] Moreover, MENA will not be alone.[6] If it was a matter of "only" one region that will be suffering from global warming, the problem might be readily addressed at least in theory by the rest of the world. But sub-Saharan Africa is already the weakest link in terms of food insecurity (Table 7.7). Africa's current problems will also be compounded by further warming and its climate refugees will flee north through the MENA, if they can. The hot summer days and nights in the future that will make life so difficult in the Gulf are also anticipated to affect 800 million people in South Asia (Eltahir, 2016; Im, Pal and Eltahir, 2017; Mani et al, 2018). The high temperatures and sea-level rise that are expected to affect the southern littoral of the Mediterranean will also do damage to the northern littoral. Misery may love company but in this case, miserable populations that are adjacent to MENA will make policy problems all the more insurmountable.

Just how does the MENA compare to adjacent and distant regions on this issue of future global warming problems? It cannot be alone if we know that food insecurity, water insecurity, and rising temperatures are not a MENA monopoly. What remains unclear is how extensive these problems are likely to be. If the effects of global warming are likely to be widespread, should we anticipate something on the order of a Global South wasteland developing in this century? Alternatively, how many regions will be even harder hit than the MENA? If other parts of the world will be damaged as badly or perhaps even worse, what does that imply for the likelihood of any global attention and the priorities associated with any attention?

We focus initially on the Sahel because it is immediately south of part of the MENA and on the other side of the Sahara and because it is likely to experience significantly rising temperatures in an arid and semi-arid context, not unlike parts of the MENA. Another reason for looking at the Sahel first is that, like the MENA, it encompasses more than one region. The MENA encompasses three subsystems: the Mashriq, the Gulf, and the Maghreb. The Sahel extends over large portions of West and East Africa and the African Horn or Northeast Africa. More specifically, it includes parts of Senegal, Mauritania, Mali, Burkina Faso, Algeria, Niger, Nigeria, Cameroon, the Central African Republic, Chad, Sudan, Southern Sudan, Eritrea, and Ethiopia.[7] However, we exclude Algeria and Sudan from this grouping for the current exercise because they are also members of the MENA.

The Sahel story[8]

> Africa is less responsible for the occurrence of anthropogenic climate change than any other continent but more vulnerable to its effects on account of its high population, low adaptive capacity, and multiple converging stressors.
>
> *(Thomas and Nigam, 2018: 33349)*

Much of Africa is expected to exceed the 2 degrees centigrade increase marker by 2050 and 3–7 degrees centigrade by 2100; the Sahel may experience 3–7 degrees centigrade increases as early as the late 2030s–early 2040s. Intense heatwaves in the vicinity of the Sahara are anticipated as well.

Precipitation in the Sahel, already familiar with an extensive series of droughts in the 1970s–1980s, is likely to decline as droughts reemerge. Droughts are blamed for 500,000 deaths in Ethiopia and Somalia between 1960 and 2012. A 2007 drought caused 325,000 deaths while a 2010–2011 drought in the African Horn affected 12 million-plus people and is linked to some 260,000 deaths. In general, the Sahel is still recovering from a series of droughts between 2005 and 2012. Over the rest of the 21st century, the Central African Republic, Ethiopia, Somalia, Burundi, Niger, Mali, and Chad are thought to be the most vulnerable to droughts in Africa. There is some possibility, however, of increased extreme precipitation in the eastern and western wings of the region. Oscillating wet and dry years can be expected. So, too, can the increased occurrence of extreme weather events. Later in the century, moreover, Blue Nile flow is likely to decline in response to the declining rainfall, high temperatures, and attempts to regulate the river for irrigation and hydropower purposes. The largest lake in the Sahel, Lake Chad, has shrunk by 95% since the 1960s and Mali's Lake Faguibine has been dry since the 1970s.

Agriculture in sub-Saharan Africa in general is highly reliant on rainfall. Large aquifers are scarce and tend to be located south of the Sahel (in Angola, Nigeria, and the Democratic Republic of the Congo). Water scarcity in the future seems likely to increase considerably but climate change is only one of several responsible drivers (population growth, urbanization, agricultural expansion, land use damage). However, West African river basins, some of which are within the Sahel, are expected to lose water volume in the future which will reduce possibilities for irrigation and hydroelectric projects.

Overall, less food production is anticipated, ranging from 2% to 50% decreases depending on the crop in question. Beans, maize, and banana cultivation, in particular, will be threatened in the Sahel. Shorter growing seasons by 2050 are likely. East African food production will need to move to higher elevations to survive. Chad and Niger might lose all of their rainfed agricultural production by 2100. Drought conditions will imperil livestock. Higher temperatures will reduce their feed supply, water availability, fertility, milk production, fitness, calving rates, and longevity. At the very least, heat-resistant animals will need to be increasingly relied upon as an adaptive response. But livestock die-offs can be anticipated in any event. At the same time, some weeds and even diseases (malaria for instance) may disappear in the Sahel thanks to increasing temperatures, while other diseases such as meningitis that thrive in dry conditions may become more common. Cattle ticks in the east may migrate in a southern direction. Despite an increasing demand for fish, fisheries in the Mauritanian and Senegalese coastal areas are in a decline that is expected to persist. West African losses in fish supply could range between 15% and 50%.

Population growth in Africa will be very high. Doubling since 1980, it could double or even triple again by 2050. The proportion of the African population, in general, that is undernourished could also increase by 25%–90%. The number of stunted growth children should expand as well. Burkina Faso, Mali, Mauritania, Niger, and Chad already possess some of the highest mortality rates for children under 5 years of age. Given the likelihood of population growth outpacing the food supply which is under attack from climate change, the Sahelian death rate is likely to climb, especially among the young and old who are usually the first to succumb to temperature-food insecurity crises.

An increase of 4 degrees centigrade could correspond to 1 meter of sea-level rise. Coastal flooding and population displacement, however, are most likely to be concentrated in parts of west and east Africa but should not be a major problem for the Sahel. Half or more of Chad, Mauritania, Mali, and Niger are already within the Sahara Desert zone. The Sahara is moving southwards at a rate of 1–10 kilometers per year, eliminating arable land as it expands. Population movement will increase, focusing primarily on migration from distressed rural areas to urban areas and neighboring states. Half of the African population could live in towns as early as 2030. Perhaps a million Sahelian people are currently in a displaced status. These numbers are likely to increase exponentially in the future. Long-term warming has and will continue to decrease economic growth in warm countries in general and in the Sahel in particular. The economic inequality gap with cool countries which have tended to benefit from long-term warming should be expected to expand as a consequence.

Analysts remain divided on whether warming increases the probability of domestic conflict.[9] In general, warming should aggravate resource scarcities, inter-ethnic interactions, and farming-pastoral activities. The question remains whether African civil wars would occur in the absence of long-term warming. There is also the secondary question of whether even more prolonged misery linked to protracted and intensive warming will impact behaviors in ways not yet observed. At the same time, governance and state capacity are quite limited in the Sahel. Sahelian states do poorly on the expectation that the state is supposed to monopolize armed force within its boundaries. Rebellions and terrorist groups are hardly strangers to the Sahel. State failures along Somali lines are certainly conceivable in the not-so-distant future in the context of accelerating and overwhelming economic and socio-political crises.

Thus, the future Sahel story resembles the one constructed for the MENA. Extremely high temperatures, less precipitation, and increased food insecurity show up in both sets of forecasts. Population growth is likely to be greater in the Sahel than in the MENA. On the other hand, more MENA cities are coastal and thereby threatened by rising seas and flooding than in the Sahel. We are hard-pressed to say which region has a more dismal future and if forced to call one a "bigger loser," our inclination is to call it a draw.

Yet this inability to rank order one region over another on which ones face relatively more dismal futures is based on a subjective comparison of lists of forecasted problems. Fortunately, there is an alternative route. We can construct a vulnerability index that should facilitate regional comparisons. Such an index will not mean that we can do without the lists of forecasted possibilities. They are needed for interpretation. But we should be able to create ordinal rankings of lesser and greater damage probably coming in the future.

Global warming vulnerability assessments already exist (see Table 9.2) but we find that they do not quite meet our needs.[10] Maplecroft is a political risk firm that has been generating lists for at least a decade. They do focus on the next 30 years which is appropriate and presumably indicator based but

TABLE 9.2 Most Vulnerable Lists

Maplecroft 2010	Maplecroft 2016	Germanwatch 2017	Notre Dame 2019
Bangladesh	Central African Republic	Puerto Rico	Somalia
India	Democratic Republic of the Congo	Sri Lanka	Chad
Madagascar	Haiti	Dominica	Eritrea
Nepal	Liberia	Nepal	Central African Republic
Mozambique	S. Sudan	Peru	Democratic Republic of the Congo
Philippines		Vietnam	Sudan
Haiti		Madagascar	Niger
Afghanistan		Sierra Leone	Haiti
Zimbabwe		Bangladesh	Afghanistan
Myanmar		Thailand	Guinea-Bissau

their product is proprietary and not transparent to non-clients. We have found two listings online which seem to be based on quite different criteria in 2010 and 2016. While the lists are country oriented, we also found a 2016 average regional list that appears to have considerable face validity: Europe (8.13), North America (7.81), South America (5.69), Asia (5.55), Oceania (4.93), Caribbean (4.56), Central America (4.05), and Africa (2.89). Nevertheless, some of the most interesting regions such as "Africa" encompass too much heterogeneity to be fully useful while others such as the MENA seem to have been buried in Asia and Africa.

Germanwatch (2017) is predicated on current costs from natural disasters. This orientation helps explain how Puerto Rico heads the 2017 list. Presumably, Puerto Rico will be replaced by the Bahamas in 2019 unless some other island is hit even worse in the hurricane season. Given the increasing intensity of Caribbean and African-originated hurricanes, Caribbean islands will no doubt maintain their places on the top 10 vulnerable lists. But a focus on the present (or immediate past) does not work for our interest in longer-term, future vulnerability.

The Notre Dame (2019) list is focused on 45 current indicators, 36 are oriented toward vulnerability and 9 speak to readiness or resilience. The list is rank ordered on the basis of the following formula: (readiness − vulnerability +1) × 50 = gain, with gain or net vulnerability ranging from 0 to 100 (low scores connote higher vulnerability). While the generation of the Notre Dame list is manned by a large number of internationally based analysts, the emphasis on both vulnerability and resilience in the current time frame does not necessarily set us up to address 30 years into the future. Our suspicion is that global warming will ultimately overwhelm most, if not all, Global South resilience thought to be in existence today.

If existing lists do not serve our purposes all that well, the solution is to develop our own. We prefer a simple accounting scheme as opposed to an extensive stable of indicators to cover all conceivable parameters. Global warming will do a number of different things. Coastal cities will be flooded perhaps permanently. Intense heatwaves will kill the elderly and children. Climate migrants will abandon rural areas for more urban settings. All of these processes are important but they skirt the core of the issue. We assert that the most central issues involve water scarcity, food insecurity, and the number of people needing water and food. It is possible to suffer damage from global warming despite an abundance of food and water and a negative growth rate. Rising seas, for instance, could still be a major problem. But increasing temperatures in the absence of sufficient food and water and an explosive growth rate suggests famine and death, especially in countries in which a large proportion of the population work in agriculture. Not only will food and water become more scarce but so too will the ability to earn a livelihood in agrarian employment.

In this context, we propose a simple five-indicator index that employs available indices on the percentage of the population employed in agriculture (World

Bank), water scarcity (World Resources Institution, 2019), food insecurity (The Economist Intelligence Unit, 2018). To this combination, we add a fourth, more future-oriented, population growth rate forecast for the years 2025 to 2050 utilizing 2001 United Nations population data (Coutsoukis, 2001). The fifth indicator focuses on the fragility of state and society (Fund for Peace, 2019). We add the fifth indicator because less fragile states and society are more likely than fragile states to develop collective responses to climate crises and emergencies. Whether the responses prove to be effective may not be as critical as the extent to which governmental problem solving is even attempted. In the absence of such problem solving, crises will be even nastier to the people suffering through them than might otherwise be the case.

Vulnerability to global warming = agrarian employment + food insecurity + water scarcity + future population growth + state fragility.

Keep in mind that the global warming vulnerability we are attempting to assess empirically is a specialized sort of vulnerability – not an all-comprehensive type. A more comprehensive effort could encompass such risks as coastal flooding, wildfires (as in Australia or the western United States), or intensive hurricanes in the Caribbean.[11] Instead, the vulnerability upon which we are focusing emphasizes food and water scarcity. People in states with affluence and strong technology may be able to cope with this type of vulnerability. People in places without wealth and technology will be devastated by these increased scarcities. Therefore, our index will stress problems that will be felt most acutely in the Global South. Yet even there, there is variation. As we will see, it is possible to differentiate not only between the Global North and South but also areas within the Global South that will be hit hardest.

Food insecurity encompasses affordability, availability, and quality/safety of the food supply. The emphasis is on whether populations have adequate access to sufficient and nutritious food that enables living a healthy life.[12] The WRI water scarcity index (Aqueduct) taps into the extent to which the available water supply is consumed each year. More than 80% consumption is considered very high (4–5), high (3–4) requires more than 40%, while a low rating equals less than 10% consumption. State fragility also encompasses several foci, including societal cohesion, economic, political, and social pressures.[13]

Ideally, we would add the extent to which temperature will increase. The higher the sustained temperature increase, the greater is the customary expectation of damage to ecology, livestock, and the population. In addition, the greater the change in temperature, the less the probability of escaping the harmful effects. Of course, it need not always work this way. The UN's Food and Agricultural Organization argues that warming below 3 degrees centigrade could benefit temperate areas while punishing tropical areas. But above a 3-degree centigrade increase, all regions will suffer to varying extents.

The ultimate extent of temperature increase in different parts of the world is clearly a known unknown and one complicated by the likelihood that temperatures will vary even within countries, let alone whole regions. We would also expect our vulnerability index to correspond to a linear function with temperature increase. The interactions among water scarcity, food insecurity, population growth, and state fragility, especially in agrarian political economies will all be aggravated considerably by rising temperatures. In actuality, this linear assumption is probably highly conservative since more extreme temperatures are likely to have nonlinear consequences that we can only guess at.

Nonetheless, the components of our vulnerability index require some preliminary manipulation to facilitate calculating their additive effect.[14] Since the agrarian employment numbers could run from 0 to 100 percent, the original numbers are retained. The water scarcity index is a scale running from low (0) to very high (4–5). The food insecurity index runs from 0 to 100, with higher scores indicating little insecurity. Since these scales move in opposite directions and the food insecurity scale has double the range of the water scarcity index, we double the water scarcity index numbers (after dividing by 10). The food insecurity number is subtracted from 1.0 to change what its higher scores imply.[15] That way higher or lower scores on the water and food indexes run in the same direction – higher scores are less desirable while lower scores are less dangerous. Population growth rates are not revised in form (other than to translate a 1.5 percent increase into a .015 number). The state fragility score is normalized since its highest score is 113.5 for Afghanistan. Each country's score is divided by the Afghan score so that scores range between 0 and 1.0. All scores are first calculated on a national basis (adding the five scores and dividing by 5) and then aggregated according to perceived regional membership.

The regional scores calculated for these are reported in Table 9.3.

In some respects, one can argue that there is nothing new revealed in Table 9.3. It is not a secret that the parts of the world that have been least responsible for global warming will be hit the hardest if and when the more severe scenarios play out. At the same time, though, Table 9.3 does underline that this outcome is not simply a matter of geography. If we are right, it is and will be a complex matter combining rising temperatures with food and water scarcities, agrarian political economies, population growth, and minimal governance capabilities.

Nonetheless, there are different combinations of these attributes. East Africa fares the worst of the 17 regions because it has the highest scores on 4 of the 5 categories: agrarian workers, food insecurity, population growth, and sociopolitical fragility. The overlapping Sahel region comes in second worst because it has high scores on all of the five but more water scarcity than East Africa. The third most vulnerable region has far fewer agrarian workers and less population growth than East Africa and the Sahel but shares extremely high water scarcity scores with the Middle East, the Gulf, and Central Asia.

TABLE 9.3 Regional Vulnerability Scores, Arrayed from Low to High Vulnerability[16]

Region	Agrarian Employment	Water Scarcity	FoodInsecurity	Future Population Growth	State Fragility	Regional Vulnerability
Europe	.045	.290	.229	0–79	.278	.143
FSU	.163	.330	.330	−134	.617	.207
North America	.015	.273	.159	.123	.250	.217
South Pacific	.290	.185	.175	.122	.429	.232
South America	.145	.299	.387	.210	.584	.277
C. America	.210	.253	.422	.241	.582	.288
SE Europe	.211	.418	.365	−109	.524	.289
East Asia	.194	.443	.428	.030	.504	.293
SE Asia	.358	.286	.463	.207	.684	.320
Central Asia	.298	.640	.519	.168	.639	.417
Gulf	.030	.873	.269	.275	.474	.418
West Africa	.504	.176	.543	.467	.789	.423
Middle East	.149	.794	.488	.310	.767	.442
Southern Africa	.497	.345	.581	.306	.683	.445
South Asia	.453	.574	.514	.317	.640	.462
North Africa	.226	.630	.489	.214	.695	.463
Sahel	.514	.461	.622	.531	.825	.465
East Africa	.699	.324	.713	.595	.840	.564

More generally, one can delineate three broad sectors of vulnerability. There are two natural breakpoints in the vulnerability score array. The most obvious one is the numerical gap between Southeast Asia (.320) and Central Asia (.417). After .417, there is not too much-aggregated differentiation among the next eight regions until the East African outlier at .564. The most vulnerable areas encompass a large portion of western Afro-Eurasia from Central and South Asia through the Middle East to all of Africa.

A less overt breakpoint could be said to separate South Pacific from South America. Latin America (South and Central America), Southeast Europe (and the Caucasus region), and East/Southeast Asia appear to fall into an intermediate zone of vulnerability. These regions tend to possess a mix of low scores on one or two attributes, medium scores on most attributes, and the occasional high score on one or more attributes. For instance, South America is low on the number of agrarian workers, relatively high on socio-political fragility, and medium on food and water scarcity/instability and population growth. Southeast Asia begins to resemble the most vulnerable regions in terms of agrarian political economy and socio-political fragility but scores lower on water scarcity.[17]

The least vulnerable regions in terms of food/water insecurity, not too surprisingly, are Europe, basically the Russian Federation, and North America.[18] We might regard the South Pacific region as a transitional area from low to medium vulnerability.[19] As in the case of other regions, a low aggregate score does not imply that the entirety of any of the regions so depicted will escape global warming problems. Sub-regions within these regions, such as Mediterranean Europe or the American southwest will be hard hit by climbing temperatures. But in general, the global North will not suffer the devastating outcome apparently in store for much of the global South.[20] That could be one of the reasons it has been difficult to mobilize sufficient enthusiasm for seriously addressing global warming in some parts of the global North.

Yet, as we have suggested earlier, our index does not capture everything worth considering when it comes to global warming effects. It ignores the threats to coastal cities and the warming of the oceans both of which will not be monopolized by the global South. Natural catastrophes influenced by climate change, and said to have quadrupled in the last four decades by the Institute for Economics & Peace (IEP, 2020: 72) should be taken into consideration as a major source of displacement.[21] The index also ignores the more ambiguous threats to global peace, security, and well-being that could well be the product of mass starvation, displacement, and migration pressures. A world of increased scarcities does not bode well for either the global North or South.

The general point is that we can make some headway in differentiating what to expect where. If the emphasis is placed on food and water problems, Central/South Asia to Africa are likely to be devastated if the temperature rises well above the mythical 2-degree centigrade boundary – as they now seem likely to do. Unfortunately, there will not be a whole lot that can be done about altering the number of agrarian workers, water and food scarcities, population growth, or socio-political fragility in Afro-Eurasia in the next few decades. Adaptations will and are taking place. But changing the types of crops grown or developing new (or falling back on old) irrigation techniques are unlikely to avert the kinds of problems that will come with increased temperatures in the 21st century. In addition to mobilizing for serious mitigation efforts in the economies that still generate the lion's share of carbon dioxide and other gases, we also need to be thinking about dealing with rather massive humanitarian problems unlike anything experienced before in terms of numbers of people.

In the interim, though, the relative status of the MENA is also made more clear. The MENA's future climate woes are not exceptional. The Sahel faces similar problems and, in many respects, even more problems. But we also find a string of regions from Central/South Asia to Africa in a somewhat similar and difficult boat. A considerable portion of the periphery may be re-marginalized by climate change just as it looked as if the Global South was beginning to make some headway in bridging the gap between its economic prospects and those of the Global North. This is admittedly a pessimistic conclusion but there appear to be ample grounds for anticipating the realization of some of the more pessimistic climate change scenarios. There is not much the Global South can do about it. Unfortunately, neither it nor the Global North is fully prepared to accept global warming as a critical existential threat. Recognition of that threat will probably come but by then it may be too late to head off the kinds of future problems that appear to be coming down the proverbial pike.

Adaptation

The IPCC defines *adaptation* as "the process of adjustment to actual or expected climate and its effects." In human systems, adaptation "seeks to moderate or avoid harm or exploit beneficial opportunities." In natural systems, "human intervention may facilitate adjustment to expected climate and its effects" (IPCC 2014). Given the vulnerability of MENA to recurrent droughts and temperature extremes, are the states in the region seeking to moderate or avoid harm heading their way? Over the past 15,000 years, societies have coped with climate change, with varying degrees of success. As the discussion in previous chapters has emphasized, the impact of global warming will not be equally distributed, and much will depend upon national resources and adaptive capacities. The most affluent states are more likely to weather the storm while the hardest hit are also the ones with the least amount of resources with which to adapt.

One of the challenging aspects of adaptation to climate-related risks involves making decisions in "a changing world, with continuing uncertainty about the severity and timing of climate-change impacts and with limits to the effectiveness of adaptation" (IPCC 2014). In the presence of persistent uncertainties, the first step toward adaptation to anticipated climate change, according to the IPCC, is to adopt strategies that help reduce national "vulnerability and exposure to present climate variability." Such strategies can ideally "increase resilience across a range of possible future climates while helping to improve human health, livelihoods, social and economic well-being, and environmental quality" (IPCC 2014; see also Durrell, 2018). The IPCC also warns us that many global risks of climate change will be concentrated in urban areas, which means "risks are amplified for those lacking essential infrastructure and services or living in poor-quality housing and exposed areas" (IPCC 2014). Yet apart from the Arabian Peninsula and Israel, the rest of the MENA lacks the essential infrastructure to reduce vulnerability and exposure in urban areas as they lack the resources to improve basic public services, to provide and improve housing, to build resilient infrastructure, to provide clean water, and so on. As heatwaves and droughts continue to severely limit the farmers' ability to grow food, those farmers will continue to migrate to the already crowded cities (as seen in Syria and Yemen already) in search of better economic opportunities. Conditions in the urban areas will further deteriorate as they become more crowded.

There can be no doubt that some adaptation to climate change is going on at different levels. At the individual and family level, Table 9.4 summarizes the responses of a five-state survey that applies to activities in 2009–2010. The answers vary from state sample to state sample somewhat but this variation stems from different degrees of climate problems. Egypt was the least affected by climate change duress at that point and Algeria and Yemen were perhaps the most affected. No doubt, the Syrian responses would have been different if the survey had been conducted at a later point. Coping at the household level means what it means pretty much anywhere around the globe. One falls back on selling whatever assets are available or tapping into whatever personal cash reserves, including international remittances (not mentioned in the table) to get by. The most popular response is using one's savings, followed by selling or pawning non-livestock assets and seeking loans. Clearly, there are limits to how long such coping can persist before the assets are exhausted and the loans are no longer available.

The least popular responses in Table 9.4 have to do with altering how agriculture is cultivated. Changing the ways farming is executed is risky, expensive, and requires overcoming traditional habits. Without generous state subsidies, it also may simply be beyond the reach of many poor farmers. Seeking nonfarming work and training may be easier to at least contemplate, if not to embrace.

Community or local adaptation at the next level up from families, summarized in Table 9.5, ranges from planting trees to slow desertification and land erosion to providing seeds and animals to farmers when needed. Regardless of the different types of climate change represented in the five-state survey, not much is going on at the community level, although Algerian communities may be a partial exception. Weak local governance and lack of funding are two easy and generic explanations.

Table 9.6 continues the summarization of the five-state survey response this time focused on the national governmental level. Again, the aggregate activity level is low. Providing drinking water presumably in emergency situations is the most likely response. No other activity comes close to the respondent's memories. The implication is that the MENA governments, for the most part, have yet to accept an interventionary mode when it comes to climate change. Activity occurs but mainly along traditional lines such as road building or emergency relief.

TABLE 9.4 Household Activities to Cope with Climate Change in Five States (% of Respondents)

Activities	Algeria	Egypt	Morocco	Syria	Yemen	All
Selling or pawning livestock	68.96	21.00	41.41	33.75	37.94	40.61
Selling or pawning other assets	50.65	20.25	35.26	65.50	62.19	46.79
Withdrawing children from school	60.15	5.13	31.12	54.00	31.72	36.42
Using one's savings	78.42	26.88	46.62	90.38	60.45	64.43
Asking for loan	50.48	13.75	42.04	60.25	64.43	46.21
Change in production technology	48.61	2.13	21.43	5.38	21.95	19.35
Change in crops mix, varieties	42.45	4.50	16.04	4.38	12.44	15.53
Change crops vs. livestock	15.25	2.50	8.93	3.38	15.10	8.89
More fertilizers, pesticides	42.16	4.63	31.47	5.88	23.48	21.12
Seeking nonfarm work	57.04	4.13	25.33	1.13	29.06	22.67
Training for nonfarm work	43.30	4.00	1.67	2.00	27.28	15.09

Source: Based on Adoho and Wodon (2014: 127, 133).

TABLE 9.5 Community Activities to Cope with Climate Change in Five States (% of Respondents)

Activities	Algeria	Egypt	Morocco	Syria	Yemen	All
Planting trees and soil protection	47.62	4.88	2.53	14.63	26.72	19.06
Banks against flooding	38.40	1.63	3.43	1.63	12.58	11.41
Boreholes, wells, irrigation roads	21.02	2.38	4.09	4/13	19.73	10.19
Information on how to reduce losses	14.27	8.25	1.97	2.00	13.23	7.90
Preparation for future disasters	32.40	3.13	2.18	1.50	12.36	10.15
Market access for products	21.84	7.13	4.96	.88	12.98	10.47
Seeds, animals, and farm equipment	39.88	8.13	4.22	1.50	20.10	14.58

Source: Based on Adoho and Wodon (2014: 140).

TABLE 9.6 Governmental Activities to Cope with Climate Change in Five States (% of Respondents)

Activities	Algeria	Egypt	Morocco	Syria	Yemen	All
Planting trees/soil protection	19.30	8.25	6.00	10.75	17.75	12.36
Banks against flooding	16.46	5.00	5.00	10.88	15.75	10.57
Boreholes, wells, irrigation roads	19.78	4.63	6.19	21.88	22.60	14.98
Seeds, fertilizers, livestock fodder	19.19	6.38	8.31	23.88	9.24	13.35
Storage facility for crops	17.17	4.88	2.04	21.38	6.87	10.41
Cash or food for work programs	14.69	7.38	1.13	18.13	8.49	9.93
Cash for food during floods/droughts	16.67	7.38	2.37	13.88	10.36	10.08
Provision of drinking water	27.82	7.38	29.31	30.75	28.21	24.67
Provision of skills training programs	11.12	4.38	0.70	2.88	14.36	6.65
Provision of credit during crop losses	38.21	5.75	4.67	4.38	7.87	11.98
Improved access to markets, transportation	14.9	6.63	4.80	10.75	14.73	10.33
Price support when agricultural prices are low	18.8	8.00	1.94	15.38	6.74	10.10

Source: Based on Adoho and Wodon (2014: 141).

If one scans the MENA horizon on a country-by-country basis, there are quite a few national plans to combat climate change as well as recommendations about how to proceed.[22] But most states, while recognizing some of the realities of climate change, do not give it a high priority. Most are content to organize planning groups and to collect information on present and future environmental problems. As Waterbury observes, the MENA governments are more oriented to mitigate carbon dioxide emissions than they are to take explicit steps to adapt to climate change. Of course, affluent governments search for alternative food supply sources buying farmland on other continents. Coastal states fret about the sea-level rise. All state decision-makers ponder whether they are adequately managing land and fishery use. Moving to actually regulate farming and fishing activities explicitly and forcefully is not something that has taken place yet.

Why that might be the case is not difficult to explain. MENA conflict is widespread as are economic problems. Even the Gulf states have to cope with budget adjustments in view of declining oil receipts. But the pressure of other problems distracts decision-makers and the generally limited resources governments control outside of the oil-producing places constrain doing much to address climate problems that are both familiar and yet somewhat distant in the future. That climate problems are familiar tends to make them less remarkable even if they connote an unfolding disaster if the most severe forms are experienced. Governmental decision-makers everywhere always prefer dealing with the crises of the moment than the ones brewing in the future that will occur on someone else's watch.

Moreover, the MENA governance capabilities tend to be less than strong. Figures 9.1–9.3 capture recent World Bank estimations of government effectiveness.[23] The scale employed ranges from a high of 2.5 to a low of −2.5. In the east (Mashriq – Figure 9.1), only one state is consistently evaluated on the positive end of the scale (Israel). Jordan and Turkey manage to score above 0 in most years. The rest have negative scores. In the west (Maghreb – Figure 9.2), again only one state earned any positive evaluations (Tunisia) but was unable to maintain a score greater than 0 after the Arab Spring. Moreover, there is a noticeable negative trend in most of the scores. In Figure 9.3, the Gulf encompasses several states with positive effectiveness evaluations but they are almost uniformly very small, affluent city-states. Saudi Arabia's scores fluctuate around zero while the two largest states, Iraq and Iran, possess consistently negative evaluations of their governmental effectiveness. It is only marginally conceivable that these governmental effectiveness scores will improve substantially in future years. Impressive standards of national governance have not been a prominent characteristic of the MENA politics.

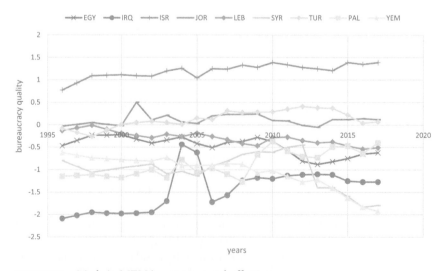

FIGURE 9.1 Mashriq MENA governmental effectiveness

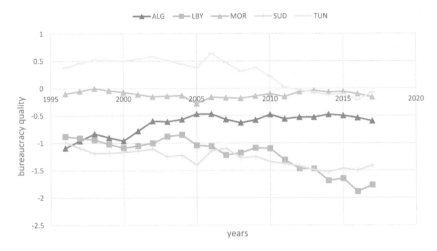

FIGURE 9.2 Maghreb MENA governmental effectiveness (world bank bureaucracy ranking)

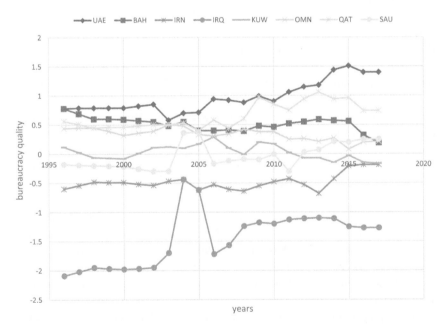

FIGURE 9.3 Gulf MENA governmental effectiveness (world bank bureaucracy ranking)

The problems governments encounter in attempting to develop adaptation plans are easy to list (see Table 9.7). Overcoming weak institutions for political–economic problem solving is not impossible but hardly probable in the near short term.

What they need to be doing is also easy to enumerate. Table 9.8 offers one checklist of things to do in response to ongoing and impending climate

TABLE 9.7 Factors Inhibiting Climate Change Adaptation

Problems operationalizing National Plans
Lack of institutional coordination
Lack of engagement and involvement of stakeholders and partners
Lack of structural monitoring and evaluation systems
Weak local and territorial institutions and structures
Difficulties in implementing regulatory and institutional reforms
Gaps in financing mechanisms and instruments
Short-term planning (2020 or 2030) versus long-term planning (2050 or 2090)
Conflicts for resource management around interstate boundaries
Diversion of political attention and will toward immediate concerns and business influenced by international and regional factors

Source: Based on Oualkacha et al. (2017: 22–23).

deterioration. Yet even if everything is actually done that can be done, there is always the nagging doubt that it will not suffice to prevent a major, region-wide catastrophe. The MENA governments are more comfortable discussing what they can contribute to reducing carbon emissions because that is something that they can address even if or in part because the MENA share of global carbon emissions is not particularly great. Short of turning off the oil spigots completely, the MENA success in reducing carbon consumption will not save the global warming day. There are also clear advantages in developing renewable energy infrastructure in areas with a great deal of sunshine and wind. Doing something about the full impact of rising temperature and less water, on the other hand, is a task that probably can only be nibbled at within the MENA. Just as effectively curbing global carbon dependence lies outside the MENA so, too, does dealing with the severe impact of global warming. None of that implies that the MENA governments should not be addressing mitigation issues but, if forced to choose between mitigation and adaptation, the latter should probably receive higher priority.[24]

Of course, we do not mean to give the impression that no one is working on these issues. For instance, the World Bank nearly doubled its funding for the MENA climate change projects for the 2016–2020 period (World Bank Group, Middle East and North Africa, 2016). Aside from energy mitigation programs, water management, and resilience in most vulnerable eco-systems (such as coastal and desert areas) projects appear to be receiving the most attention, and primarily in Egypt, Jordan, Lebanon, Morocco, Tunisia, and Yemen. Nonetheless, the programs announced to date seem to have modest goals, which is not altogether surprising since they are often pilot efforts. Thus, they represent steps in the right direction, but, at the same time, they are not exactly major steps.

If the scale of contemporary crises overwhelms the resources available for relief, it is hard to imagine the situation will improve in the coming decades. Calling for concerted regional cooperation on climate change issues, as many policy surveys do, may be understandable, if rather unlikely to be realized. Demanding

TABLE 9.8 Climate Change Adaptation Options

Adaptation Foci	
	Agricultural/Farming Level
Crops	Improve varieties that are drought, heat, salt, pest resistant-tolerant
	Alter mix and/or planting techniques
	Adjust maturation times for shorter or longer growth seasons
	Change fallow timing
	Increase soil additives (carbon, fertilizer)
Livestock	Expand drought-hardy and pest resistant breeds
Water	Expand use of irrigation where feasible
	Improve irrigation technology and management
	Utilize marginal water for watering purposes
	Reduce overwatering
	Improve rainwater capture and storage
	Expand desalination technology as feasible
Fisheries	Improve fisheries management
	Governmental Level
Finance	Reevaluate and enhance targeted subsidies
	Increase investment and microfinance
	Improve social safety nets
	Introduce climate-related insurance
Services	Increase funding and support for research
	Improve farmer extension services
	Improve monitoring of food availability (local and global)
	Improve drought, disease, and pest monitoring
	Build strategic food stocks and storage
	Improve bilateral and multilateral food supply chains
	Support education and skills development for non-agricultural employment
	Increase support for local and global migration
Institutions	Improve connections for agricultural marketing and supplies
	Strengthen local institutions and governance
	Regulate and reduce land degradation
	Support women's groups and local nongovernmental organization development

Source: Based primarily on Verner (2012: 220–222, 225).

more proactive participation in climate change programs from underfunded and beleaguered, Bretton Woods institutions operating at the global and regional levels is certainly justified but equally unrealistic. There are only so many actors inside and outside the region that are prepared to do anything at this time. When they are prepared to do more, it may prove to be too late.

Even if there was some place for the MENA climate refugees to go, the conventional humanitarian assistance in disaster-like circumstances is oriented to providing temporary shelter until refugees can return home. In a high temperature–insufficient water situation, there will be no going home. The conventional approach to humanitarian aid will need to be reconsidered and probably replaced. Another way of looking at this problem is that between 2008 and 2016 roughly 25 million people were displaced globally by environmental disasters each year. That number adds up to 196 million displaced in less than a decade. Imagine the full 196 million refugees needing assistance in 1 year and just from the MENA. The World Bank (Rigaud et al, 2018) estimates the likelihood of another 143 million climate refugees from Africa, South Asia, and Latin America. The global humanitarian effort is already stretched to the limit and greatly underfunded. Increasing the caseload eight to sixteenfold would require a genuinely massive undertaking unlike anything ever attempted before.[25]

Unfortunately, there are very few choices for relocation and, presumably, even those places will be subject to some of the same environmental pressures. Yet if there is nowhere to go and few preparations to cope with the possibility of an impending environmental catastrophe, many people in the MENA can be expected to perish. Vulnerable children and the elderly will be the first to succumb as is customary. Yet there is little reason to expect the damage to stop there. Pockets of air-conditioned affluence may prevail but the rest will be impoverished and undernourished at best.

Figure 9.4 sketches a two-variable diagram of four possible future scenarios for humanitarian assistance (Maietta et al, 2017: 15) predicated on contrasting future humanitarian needs (in 2030) and global governance considerations. The horizontal axis differentiates smaller, more localized crises from larger-scale crises. The vertical axis runs from the complete withdrawal of attempts at global governance to the other extreme of new types of humanitarian intervention. Combining the adjacent areas in each of the four cells (e.g., the upper right-hand quadrant represents the combination of global governance withdrawal and

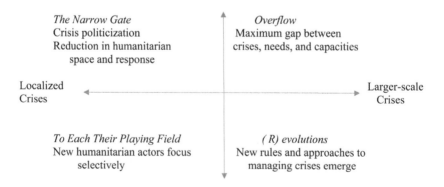

FIGURE 9.4 The future of humanitarian assistance (world bank bureaucracy ranking)

larger-scale crises while the lower left-hand cell combines smaller, localized crises with new types of humanitarian intervention).

The two scenarios in the upper portion of the figure seem only too familiar. The two in the lower half of the diagram seem rather remote and less than capable of meeting the type of humanitarian needs that extreme climate change will create. In the upper left corner of the diagram, the Narrow Gate scenario for 2030 envisions a world in which the rise of nationalism leads to increasing pressure on minorities and targeted groups by fragile governments committed to dealing with their own problems, often by making use of their military forces. Protracted crises ensue in which global governance institutions can make a little dent. If one were to add widespread and intensive climate change to this scenario, the unlikelihood of fragile states handling these crises on their own is obvious. Protracted crises would suggest that they probably would not be able to handle the crises even without adding intensive and extensive climate change to the mix.

The Overflow scenario in the upper right-hand corner of Figure 9.4 does add climate change problems. Compounded crises combining state fragility, environmental deterioration, and great humanitarian needs overwhelm the temptations for local states to try taking care of their own problems. Boundaries become meaningless as the problems spread throughout whole regions. Yet global governance has no more capability than the local governments to deal with the compounded crises. Migration becomes the only escape possibility. But migration flows to the Global North only lead to political destabilization, increased resistance to more migration, and increased uncertainty about whether migration is an escape possibility. While the scenario is set in 2030, it does not seem all that different from what we observe in 2021.

In the To Each Their Playing Field (lower left-hand corner of Figure 9.4) non-traditional actors (corporations, foundations, megacities, citizens' movements) are introduced as the principal actors in implementing humanitarian assistance. The problem here is that each group or alliance will have to be focused on very specific causes or places given more limited resource foundations. What is apt to emerge is a two-tiered structure in which some causes/places receive most of the attention while other causes/places are ignored entirely. While the structural outcome seems quite plausible, it is by no means clear how non-traditional actors will be able to mobilize the types of resources needed to address large regions attempting to cope with extreme global warming.

Perhaps in response to the above criticism, the fourth scenario, (R)evolutions, starts with the intensification of environmental crises leading to region-wide disruptions and distress. In some unstated way, non-formal actors (presumably incorporating the non-traditional actors of the Playing Field scenario) will construct a new paradigm for managing severe crises. The conventional approach to coping with disasters will be unable to meet the escalated needs associated with widespread climate change. A new structure of global governance will have to emerge.

But here is the rub. Unless a new structure of global governance emerges, we are consigned to the inadequacies of the present (see the Overflow scenario). The inadequacies of the present humanitarian intervention capabilities have been demonstrated to be insufficient to deal with the kinds of crises we have today. These crises are horrific enough but they are likely to be far worse in the future. It follows, therefore, that we are unlikely to be able to do much should the bigger crises start coming in a decade or two. Extreme crises and emergencies are often said to be necessary to develop new ways of solving problems. Perhaps we will find out.

Yet even if global warming should prove to be some kind of a hoax, the MENA is slated for severe problems in the coming century given its food insecurities and water sufficiency problems. If global warming turns out to be something less than a full-blown catastrophe, it is conceivable that some of the resulting problems will be handled by local governments. But which ones? There is an index for this called the readiness against (climate change) vulnerability. It involves combining 45 indicators that are intended to capture the extent of exposure, sensitivity, and adaptive capacity relating to problems pertaining to health, food, ecosystem, habitat, water, and infrastructure – the vulnerability domain – and economic, governance, and social readiness to act on the problems. Basically, the sum of the vulnerability scores is subtracted from the readiness scores to achieve the readiness against the vulnerability index (Chen et al, 2015).

Since the index relies strictly on current indicators, it makes no effort to anticipate lesser or greater degrees of warming in the future. Much of what we have discussed assumes, contrarily, that we are heading for something beyond the 1.5/2 degree centigrade increase on which so much attention has been paid. But if that assumption proves to be wrong, either because the climate modeling is all wrong or radical policy steps are actually taken to head off greater warming, the readiness index now prepared at Notre Dame University can make some predictions about which states fare better or worse. Table 9.9 summarizes the 181 states that have been ranked and divides them into three groups (the first 60, second 60, and last 61) ranging from most readiness to least. The MENA states are underlined in the table.

The outcome is not too surprising. The most affluent and developed MENA states might get by best if temperatures do not climb through the roof in the Gulf. Half of the Gulf states (the ones not already listed in the first group) and some of the larger MENA states (plus Jordan) fall in between the most and least ready states. Presumably, that means some of this in-between group may do almost as well as the first group in Table 9.8 while the efforts of the rest will resemble the fate of the third group. Libya, Iraq, Syria, Yemen, and Sudan – all large, poor, and quite arid states – fall into the last group. So, even with a relatively weak global warming, a number of MENA states are likely to be unsuccessful in coping.

Of course, it is most unlikely that global warming will turn out to have been a hoax and, unfortunately, all the signs suggest global warming beyond the 1.5/2

TABLE 9.9 Coping with Climate Change According to the Notre Dame Index

Most Readiness to Cope with Climate Change
Norway, New Zealand, Finland, Sweden, Switzerland, Denmark, Austria, Iceland, Australia, Singapore, United Kingdom, Canada, Luxembourg, United States, Republic of Korea, France, Netherlands, Slovenia, Ireland, Japan, Czech Republic, Poland, Spain, Estonia, Portugal, Belgium, Chile, _**Israel**_, Lithuania, Latvia, Italy, Russia, _**UAE**_, Slovakia, Greece, Hungary, Cypress, Belgium, Kazakhstan, Malta, Bulgaria, Georgia, Croatia, Dominica, Malaysia, Georgia, _**Turkey**_, St. Vincent & Grenadines, Mauritius, St. Kitts & Nevis, Brunei Darussalam, Macedonia, Barbados, _**Saudi Arabia**_, Uruguay, Montenegro, _**Qatar**_, China, and Armenia

Medium Readiness to Cope with Climate Change
Costa Rica, Mongolia, Romania, Thailand, Bahamas, St. Lucia, Ukraine, _**Oman**_, Argentina, Serbia, _**Iran**_, Azerbaijan, Albania, _**Morocco,**_ Brazil, Mexico, Colombia, South Africa, Panama, Portugal, _**Jordan**_, Kyrgyzstan, Moldova, _**Kuwait**_, Tunisia, _**Bahrain**_, Bosnia & Herzegovina, Jamaica, Fiji, Antigua & Barbuda, Seychelles, Trinidad & Tobago, Botswana, Paraguay, Dominican Republic, Vietnam, Uzbekistan, _**Egypt**_, Sri Lanka, Samoa, Ghana, Bhutan, Namibia, El Salvador, Indonesia, Cuba, _**Lebanon**_, Suriname, _**Algeria,**_ Ecuador, Guyana, Guatemala, Philippines, Tajikistan, Nicaragua, Rwanda, Venezuela, India, Sao Tome & Principe, and Gabon

Least Readiness to Cope with Climate Change
Belize, Honduras, Tonga, Maldives, _**Libya,**_ Bolivia, Lesotho, Turkmenistan, Timor-Leste, _**Iraq**_, Laos, Nepal, Senegal, Cambodia, _**Syria**_, Comoros, Cameroon, Pakistan, Zambia, Solomon Islands, Djibouti, Vanuatu, Togo, Swaziland, Nigeria, Gambia, Cote d'Ivoire, Guinea, Sierra Leone, Tanzania, Kenya, Micronesia, Equatorial Guinea, Burkina Faso, Myanmar, Mali, _**Yemen,**_ Zimbabwe, Madagascar, Liberia, Burundi, Guinea-Bissau, Haiti, Afghanistan, Niger, _**Sudan**_, DNC, Central African Republic, Eritrea, and Somalia.

degree centigrade threshold, the question then revolves around just how warm the world in general, and the MENA in particular, will become. The forecasts are ominous in this regard. Yet the efforts to deal with global warming are relatively miniscule. So is the available capacity that will be needed to cope with its victims. Places with large and growing populations, water and food scarcity, and agrarian economies – a description that applies to most of the MENA – will encounter something more than turbulent headwinds if climate conditions continue to deteriorate.

Should a worst-case climate change scenario be realized in the MENA, local government capacity, in most instances, is unlikely to be up to the task of dealing with widespread famine, impoverishment, and displaced people. As Waterbury (2013) contends, the MENA politicians have little incentive to do much to work on a problem that will take several decades to become most overt. Other than selling a disproportionate amount of fossil fuel, the Arab world contributes little to CO_2 buildup. In this respect, it might not make much difference what local governments or global governance try to do to protect or insulate the MENA populations from environmental catastrophe. The payoff for adaptation is too far down the road to appeal to the short-term perspectives of most rulers – even assuming that there were spare resources to spend on human welfare issues in a region that favors spending on military hardware.

On this pessimistic note, we turn to wrapping up our treatment of the MENA environmental crises through the ages in the next chapter. Regrettably, its tone can be no more optimistic than the flavor conveyed in this dour chapter.

Notes

1 For more on global energy transition, see Thompson and Zakhirova (2019).
2 Per capita renewable water resources have declined by 75 percent in the Middle East since 1950 (Akhtar, 2012). Kubursi (2012) reports another 40 percent decrease is anticipated by 2050.
3 The basic conceptualization of climate refugees is quite novel and has yet to penetrate international law and the development of norms about how to classify and treat people fleeing uninhabitable areas either because there is too much or too little water (Curtis, 2017).
4 The implications of projected global mean temperature changes tend to underestimate regional level changes. This is because global mean temperature implies increases in warm and cold temperature extremes in both land and water. However, warming over land is typically stronger than over the oceans which means extreme temperatures in many regions can increase well beyond 2 degrees centigrade. In fact, the 2-degree centigrade global mean temperature target actually implies at least 3 degrees centigrade warming particularly in places like the Mediterranean region and changes in regional extremes are expected to be greater than those in global mean temperature by a factor of 1.5 (Seneviratne et al 2016). Analyses looking at long-term temperature data of this region suggest that the frequency of heat waves has been increasing since the 1970s (Tanarhte et al 2012).
5 But not uneven in a random fashion. Diffenbaugh and Burke (2019) estimate, for instance, that global inequality has already increased appreciably because warm Southern states have lost economic growth possibilities while cool Northern states have actually gained from global warming.
6 Bastin et al (2019) suggest as rules of thumb that all Northern Hemisphere cities will seem to have moved 1,000 kilometers south in terms of average temperatures while cities in the tropics will shift to drier conditions accompanied by more extreme oscillations between precipitation and drought.
7 A difference, however, is that while the MENA tends to be modeled by climate analysts as a unit, climate modeling on sub-Saharan Africa tends to adopt an African unit of analysis with references (sometimes) to distinct sub-regions behaving differently from the overall model.
8 Our summary of possible climate change effects in the Sahel are based on the following studies: Shongwe et al (2011); Holtuijzen and Maxmilian (2011); Peasron and Niaufre (2013); Potts et al (2013); Niang et al, (2014); World Bank Group (2014); Serdeczny et al (2016); USAID (2017); Ahmadipour and Moradkhani (2018); Fischer et al (2018); King and Harrington (2018); Psima et al (2018); Rigard et al (2018); Sylla et al (2018); Weber et al (2018); Gebrechorkos, Bernhofer, and Hulsmann (2019); and Girvetz et al (2019).
9 See, e.g., Burke et al (2009); Buhaug (2010); Hsiang and Meng (2014); and Uexkull et al (2016).
10 Busby and von Uexkull (2018) have generated a fourth series based on three indicators: the extent of economic dependence on agriculture, a recent history of conflict,

and an index of ethnic/religious political discrimination. But this approach is clearly geared to an expectation that climate change leads to conflict (and humanitarian crises) in a path-dependent way. Agrarian economies have been strongly susceptible to civil warfare in recent decades and civil wars to have a tendency to recur, presumably because the earlier grievances were never fully resolved. The climate change-conflict relationship, however, is strongly contested. It remains unclear whether conflicts in deteriorated environments would have taken place in the absence of deteriorated environments. Moreover, we are not restricting our examination to conflict-induced humanitarian crises. We anticipate substantial humanitarian crisis with or without an increasing number of civil wars. Alternatively, the Institute for Economics & Peace (IEP, 2020) are developing an "Ecological Threat Register" that bears some similarity to what we have in mind. Their index is described as combining water stress, food insecurity, droughts, floods, sea level rise, and population growth with indicators for coping capacity. We discovered only after developing our own index and, unfortunately, at the time of this writing they had not fully revealed how they combined their scores or their outcome. Presumably, subsequent reports will do so.

11 Conceivably, a separate natural catastrophe vulnerability index might be constructed to cover these types of problems.

12 Affordability encompasses assessments of the ability to produce, the vulnerability to price shocks and the existence of support programs for when shocks to the food supply occur. Availability assesses the adequacy of the national food supply, the risks of supply disruption, and the national capacity to distribute food and develop research for expanding food supply. Quality/safety translates to the variety and nutritional quality of average diets. These three factors are adjusted for the susceptibility to risks to national resources and general resilience. Twenty-eight indicators are enlisted in total.

13 Twelve indicators, three each for the four foci, are constructed from a triangulation of quantitative data, content analysis, and qualitative expert analysis. The cohesion category encompasses the security apparatus, elite factionalization, and group grievances. Economic indicators include ones for decline, uneven development, and human flight or brain drain. The political indicators cover state legitimacy, public services, and human rights/rule of law assessments. The three social factors are described as demographic pressures, refugees and IDP's, and external intervention.

14 A case could certainly be made for their multiplicative effect but that might work against the desire for simplicity.

15 If we did not double the water scarcity score, the implicit effect would be to give double the weight to food insecurity as accorded water scarcity.

16 Multiple regional membership lists exist. Therefore, we are obligated to explicitly identify the states we assign to the different regions rank ordered in Table 9.3. Moreover, not all states have water scarcity and food insecurity scores. The averages reported in Table 9.3 represent the states for which we do have scores. Europe = Austria, Belgium, Czech Republic, Denmark, Estonia, Finland, France, Germany, Hungary, Ireland, Italy, Latvia, Lithuania, Luxembourg, Netherlands, Norway, Poland, Portugal, Romania, Slovakia, Spain, Sweden, Switzerland, and the United Kingdom. FSU (former Soviet Union) is restricted to Belarus, Moldavo, and the Russian Federation. North America encompasses Canada and the United States. South Pacific is a heterogeneous grouping of Australia, Fiji, New Zealand, and Papua New Guinea. South America includes Argentina, Bolivia, Brazil, Chile, Colombia, Paraguay, Peru, and Venezuela. Central America is Belize, Costa Rica, El Salvador, Guatemala, Honduras, Mexico, Nicaragua, and Panama. Southeastern Europe (and the Caucusus) incorporates Albania, Armenia, Azerbaijan, Bosnia and Herzgovina, Bulgaria, Croatia, Georgia, Greece, North Macedonia, Serbia, and Slovenia. East Asia spans China, Japan, North Korea, South Korea, and Mongolia. Southeast Asia focuses on Cambodia, Indonesia, Laos, Malaysia, Philippines, Thailand, and

Vietnam. Central Asia includes Kazakhstan, Kyrgyzstan, Tajikistan, Turkmenistan, and Uzbekistan. The Gulf region incorporates Bahrain, Kuwait, Oman, Qatar, Saudi Arabia, and the UAE. West Africa is based on data from Benin, Burkina Faso, Cameroon, Central African Republic, Cote D'Ivoire, Gambia, Ghana, Guinea, Guinea-Bissau, Mali, Niger, Nigeria, Senegal, and Sierra Leone. The Middle East encompasses Cyprus, Egypt, Iran, Iraq, Israel, Jordan, Lebanon, Syria, Turkey, and Yemen. Southern Africa equals Angola, Botswana, Lesotho, Madagascar, Malawi, Mozambique, Namibia, South Africa, Zambia, and Zimbabwe. South Asia is centered on Afghanistan, Bangladesh, Bhutan, India, Nepal, Pakistan, and Sri Lanka. North Africa spans Algeria, Libya, Morocco, Sudan, and Tunisia. The Sahel incorporates territory from Burkina Faso, Eritrea, Ethiopia, Mali, Niger, Nigeria, Senegal, and Sudan. East Africa (including some Central African states) is Burundi, Chad, Djibouti, Eritrea, Ethiopia, Rwanda, Somalia, Tanzania, and Uganda. In most cases, countries are assigned to only one region. The main exception involves the Sahel which overlaps with parts of the North, East, and West African regions.

17 Corresponding to Table 9.1's list of most vulnerable states, our index nominates the following top 10: Yemen (.735), Niger (.722), Somalia (.711), Afghanistan (.669), Eritrea (.691), Georgia (.634), Chad (.580), Pakistan (.566), Guinea-Bissau (.537), and Zimbabwe (.541). However, some of the states for which we lack food insecurity and water scarcity scores, such as Congo-Kinshasa, might rival these states for a position on the most vulnerable list.

18 The Russian Government announced a plan at the beginning of 2020 to take advantage of the positive effects of global warming (The Guardian, 2020).

19 We only have four states with sufficient information for index purposes in our South Pacific grouping and it is a highly heterogeneous grouping ranging from New Zealand at .057 to Papua New Guinea at .448.

20 The ten least vulnerable states based on our index are : Austria (.038), Norway (.039), Sweden (.049), Finland (.053), Switzerland (.054), New Zealand (.057), Latvia (.081), Slovakia (.083), Ireland (.100), and Canada (.112).

21 The Institute for Economics & Peace count earthquakes, tsunamis, volcanic activity, cyclones, floods, extreme temperatures, drought, and forest fires as natural catastrophes.

22 See, for instance, Ministry of Energy, Mines, Water and Environment – Morocco (n.d.); Sahoune et al (2013); Oualkacha et al (2017); Reliefweb (2017); Verner (2013), Amanou et al (2018); McCarl et al (2015); Fahim et al (2013); Gordon (2017); Traerup and Stephen (2015); Gohari et al (2013); Karimi, Karami, and Keshuarz (2018); DeNicola et al (2015); Gunel (2016); UAE Ministry of Climate Change and Environment (2017); and Elmi (2017).

23 The MENA states are broken into three subsystems here because it would be impossible to read the charts if all the MENA states were included in one figure.

24 Some policy responses do blur the lines between the two. For instance, accelerating the transition to renewable energy reduces emissions just as it lowers the cost of electricity, which could free financial resources for other adaptation purposes.

25 Estimates pertaining to the likely number of climate refugees in 2050 range from 25 million to 1 billion people, with a median of 200 million (European Political Strategy Center, 2017). Empirical evidence for the linkages between climate change and refugees seeking asylum can be found in Abel et al (2019).

10
THE END OF THE MENA?

Climate change is certainly not new to the MENA, which has been through transitional processes before in a way that few other places have been. What we know from history is that abrupt climatic changes have happened numerous times over the past 15,000 years (see Chapters 3 and 6) – all centered in areas that became critical to the contemporary functions of the MENA. Table 10.1 lists them one more time. This time, however, they are clustered in groups of three with a different type of outcome.

The 15,000 years we examined began with a warming trend as the far northern glaciers receded. Flora and fauna expanded in the ancient Near East thereby making it possible for hunter-gatherers to expand their populations and become more sedentary. The Younger Dryas cold spell put an end to the expansion of vegetation and wild food. Sedentary settlements became much more difficult to sustain. Natufian societies returned to their former hunting and gathering strategies. Contrary to a popular academic thesis, we think the transition to agriculture was probably delayed by the Younger Dryas cold spell, not accelerated by it as a matter of urgency to feed an expanded population without ready access to wild food sources. Necessity can be the mother of invention but not when it is cold and dry for an extended period of time.

Pre-pottery Neolithic A and B (PPNA and PPNB), the Levantine groups that archaeologists focus on after the earlier emphasis on the Natufian culture receded, seem to have struggled mightily with intermittent cold and dry episodes and may have been seriously set back themselves by adverse climate change. What the first three cases share in common was that these groups became more sedentary in years of better climate, seemed to prosper and expand in numbers, and then became less prosperous and healthy. Years of adverse climate did not help but they also contributed to their food insecurity problems by killing more animals than was sustainable and, no doubt, consuming extensively nearby edible

TABLE 10.1 Nine Cooling Periods, Duration, and General Outcomes

RCC	Name	Timing	Duration in Years	Outcome
1	Younger Dryas	10700–9700 BCE	1,000	Deceleration and decentralization
2	10.2 kya	8500–8000 BCE	500	Deceleration and decentralization
3	8.2 kya	7000–6000 BCE	1,000	Deceleration and decentralization
4	5.9/5.2–5.1 kya	4000–3000 BCE	1,000	Sumer retreat
5	4.2 kya	2200–1800 BCE	400	Imperial decline and conflict
6	End of Bronze Age	1250–800 BCE	450	Urban breakdown and destruction
7	Late Antiquity Ice Age	450–700 CE	250	Regional handicapping
8	Medieval Climate Anomaly	950–1070 CE	1,020	End of the Islamic Renaissance
9	Little Ice Age	1400–1850 CE	450	Ottoman Empire decline as a great power

vegetation. But prolonged climate change also seems to have overwhelmed these groups' ability to cope with their environments. As a consequence, their settlements simply faded away in the archaeological record as most were abandoned for more decentralized survival strategies. A few, like Jericho, managed to survive based on easy access to a water supply.

While ancient Egypt may have gained from Saharan immigrants forced to leave what had become a desert, Mesopotamia was less advantaged in the 4th millennium BCE rapid climate changes (RCCs). The Ubaid culture receded and was supplanted by the Uruk culture which, in turn, was set back by a loss of its trade and colonial resource acquisition network. Toward the end of the 3rd millennium BCE, Akkad, the Egyptian Old Kingdom, and Indus disintegrated more or less at the same time in another cold and dry period. The Old Kingdom's demise may be traceable to Nile fluctuations that are based on a different climate regime centered on East African rainfall. But even the demise of the Egyptian Middle Kingdom may have had at least an indirect link to climate change to the east. The next cold and dry episode contributed considerably to the End of the Bronze Age in the Eastern Mediterranean.

All of these imperial/civilizational breakdowns are susceptible to rival top-down and bottom-up causation emphases. Climate, the ultimate top-down process, clearly played some significant role in pulling down these regimes. Yet climate change was probably not sufficient as a leveler although it may have been necessary. In all of these cases, with perhaps the exception of the least known Indus, governance had been breaking down prior to the onset of climate change. With time, empires tend to fragment as coercive centers weaken or lose their formula for holding things together. Short-term deals to raise revenues turn into long-term resource drains. Dynastic succession problems become more difficult to manage. Peripheral groups develop stronger incentives to resist subordination to a weakening center.

It is not clear that we must choose between assigning causal primacy to top-down or bottom-up processes. Regimes naturally fall apart over time but they are capable of lingering on indefinitely. Climate change can be both a trigger and more if it alters the political economy so radically that it is difficult to recreate a new centralized regime for hundreds of years. This type of breakdown does not happen often. But protracted cool and dry onslaughts are not frequent either. The contours of the ancient Near Eastern Bronze Age were shaped strongly by climate change. Ultimately, the Bronze Age died at the hands of an extensive and intensive drought, among other fatal processes. After several hundred years of little population growth and urbanization, the Iron Age emerged. Assyrian and Persian empires reemerged in the Near East but so too did Greeks, Romans, and Chinese expansions. Near Eastern activities had lost a good deal of their distinctiveness. For a time, much of the region also lost its autonomy to outside forces.

The expansion of Islam and the Caliphate might be viewed as a sort of dependency reversal of a long-term western and eastern Roman Mediterranean primacy. Rome had for a time controlled much of the Mediterranean littoral and was stopped only by Parthian resistance to penetrating farther inland in the east. The Byzantines attempted unsuccessfully to reconstruct Mediterranean primacy. The Islamic expansion fared better in this respect gaining control over much of conventional MENA, parts of southern Europe, and extending into Central Asia. While the rise of Islam was abetted by climate change, it is not clear that its initial reign was well supported by a benign weather regime. More work needs to be done in this area to be able to say more but, in any event, Islamic expansion was unable to stem tendencies toward fragmented rule and attacks from Europe and the Mongols.

While civilizations were less likely to succumb in late antiquity through the early Modern era (in comparison with earlier millennia), plague joined the problem set associated with cold and dry periods. The Byzantine Empire was set back tremendously by plague while the Muslim expansion was facilitated by the Late Antiquity Ice Age. A Medieval Ice Age may have encouraged the Seljuk movement into Iran and Mesopotamia, which made the First Crusade more likely. The return of plague in the transition from the Medieval Warm era to the Little Ice Age constituted still another setback to political-economic development in the Mideast while it and the plunging temperatures are seen as contributory elements in the advance of northwestern Europe toward technological and economic centrality in the world system. The initial and short-lived success of the Ottoman Empire in the 16th century was stopped cold, no pun intended, by an

RCC, which undermined the ability to mobilize resources after the end of the 17th century CE. While the Ottoman Empire stagnated, leading edges of the rest of the world moved far ahead politically and economically.

With the advantage of hindsight, we view a long string of climate adversities contributing to the lagging behind of what was once the world's first innovative leader. The first three cooling periods in our 15,000-year schedule set back early efforts to develop villages and agriculture. Nonetheless, the setbacks did not halt the movement toward greater complexity. They only slowed the advance down. Bronze Age cooling periods interrupted urbanization, empire, and long-distance trade networks. It did not stop them from developing but, in the long term, it did make it more likely that these developments would thrive somewhere else subsequently. In western Eurasia, they moved toward the Mediterranean and southern Europe. The Bronze Age gave way to the Iron Age. By the end of the first millennium BCE, much of MENA had lost its autonomy to foreign intruders.

Attempts to reverse regional subordination, as manifested by the expansion of Islam and the Ottoman Empire, had initial success that did not persist. MENA remained largely Muslim but not Ottoman. The region as a whole, beset by European and Asian invaders, lagged behind certain other parts of the world that had made, in part, better use of plague and cooling climate in the 2nd millennium CE. By no means do we wish to attribute everything that happened to climate change alone. But climate change keeps showing up at critical times in MENA history. It deserves more credit or blame than it has hitherto received as a major explanatory factor in the many iterations of the rise and demise of the MENA. That has been true of the past and it looks likely to be true in the future.

What does all of this say about our resilience theory framework? We have relied on an organizing device, signified by a recumbent "eight" figure showing the flows from growth to conservation to creative destruction to renewal (Figure 1.1). Growth leads to problems in the conservation phase. Some type of destructive influence such as climate deterioration provides an opportunity to renew or rebuild, thereby setting up another run at growth and conservation or rigidification problems. The x and y axes in Figure 1.1 are resource potential and connectivity/organization with a third dimension centered on varying resilience. Generally, resilience should be greatest when resource potential is high and connectivity/organization is low. This combination is found in the renewal phase that follows creative destruction. Resilience, therefore, is not a constant. It fluctuates according to the context in which actors operate.

The ecological applications of this framework tend to be focused on subjects like forests hit by fire or lakes clogged by algae. Forests and lakes are fixed places characterized by successive phases of growth, destruction, and renewal. Our application is also fixed on one rather large region and therein lies a problem. The framework is geared optimistically toward eventual renewal even though forests and lakes presumably disappear from time to time and the principal questions are how is renewal accomplished and at what pace. In our applications, renewal occurs but not usually in the same place. To be sure, there are Ubaid/

Uruk/Akkad/Babylonian and Old/Middle/New Kingdom sequences but they do not appear to be the rule. Nor do they persist forever. Either societies are more fragile and less resilient than forests and lakes or something else is afoot.

We think the problem is located in the creative destruction zone. Small forest fires can clear out deadwood and debris and make large forest fires less likely. But large forest fires can be more destructive and less creative when they eliminate the foundation for forest sustainability. Similarly, a series of droughts can encourage farmers to shift crops to ones that require less moisture or it can encourage more efficient water management policies. Decades of drought and cold or hot temperatures are something different. They can be so destructive that they overwhelm any resident resilience or mitigation/adaptation responses. As long as we allow the destruction phase to vary – as opposed to viewing it as a convenient trigger for change – the application of the resilient theory can be sustained.

Part of the renewal process may be moving to places with more resource potential and less organization after older sites have "used up" their potential and choked on excessive organization. One does not think of forests and lakes doing that but it is not clear that it is inconceivable. It is certainly possible for human societies to do that as manifested in moving to new frontiers and the oft-repeated ascent of fringe marcher states throughout human history (and prehistory).

Moreover, the theory offers a compromise between top-down and bottom-up interpretations. It does not tell us which end has explanatory primacy. That is something we still need to nail down. But it does suggest that we can avoid wasting time arguing in a zero-sum fashion that it is either top-down or bottom-up factors that are most responsible for what takes place. Both sets of factors are likely to be present and interacting. Better explanations will take advantage of their interactions – as opposed to wasting time dismissing the rival emphasis altogether.

Yet being able to move to a new frontier or anywhere for that matter is not something that can be taken for granted. In each of our ancient cases, given sufficiently protracted environmental deterioration, pretty much the same thing happened: in some parts of MENA, governments collapsed, urbanization ceased, trade stopped, population growth ended for a while, and survivors migrated. But where can the current region's 550 million climate refugees go in contemporary global warming circumstances? Refugees usually flee from rural areas to cities or neighboring countries. The practice is well known in the eastern part of MENA. In an environmental "meltdown" scenario, however, cities and neighboring countries will have the same problems. There will be no nearby refuge. Farther away refuges seem to be disappearing as well as xenophobic and populist regimes seize the opportunity to restrict migration and asylum. Renewal, as a consequence, can never be taken for granted.

The largest humanitarian and refugee crisis in modern times is ongoing in MENA at the present.[1] While the numbers involved change from year to year and need to be viewed as rough estimates, something like half or more of the population of Syria (once about 22 million) has been displaced by years of internal warfare. Casualties to date approximate a million injured and perhaps nearly half a million dead. Close to six million people are refugees outside of Syria. Thousands of people have drowned or disappeared attempting to cross the Mediterranean. Still, most external refugees reside in relatively poor states adjacent to Syria (Jordan, Lebanon, and Turkey). Less than 10 percent of these refugees live in camps set up to house them. The rest make do in crowded urban settlements or temporary, ad-hoc shelters.

It is difficult to find overall estimates of the spending required to cope with refugee needs. The immediate burden falls on the governments hosting them, buttressed by aid and donations from external groups and individuals. Lebanon has received some 1.2 billion US dollars in humanitarian assistance through 2017. EU states, for instance, report that they have spent 9.4 billion euros through 2017, with a pledge of another 3.7 billion after 2017. Other states have pledged another 5.5 billion euros. UN agencies requested 8 billion dollars for Syrian relief in 2017. Yet pledges are never quite fulfilled and each year sees increased resistance to providing more humanitarian aid. In the meantime, millions of Palestinians, Iraqis, and Syrians remain in camps somewhere.[2] Despite international aid efforts, refugees go hungry, children do not go to school, and overall conditions remain unsanitary. Since many of the refugees fear returning to Syria because their homes have been destroyed, health services, education, and food are scarce within Syria, and/or because they fear government retaliation if they do return, their future remains uncertain.

Writing in 2020, whatever happens to Syrian refugees will be characterized by a largely uncoordinated and expensive humanitarian effort that has strained and even overwhelmed government aid, NGOs, and the adjacent states that house the majority of refugees. The flood of refugees from one civil war has encouraged extreme variations on populism and anti-Muslim fervor in North America and Europe. Less governmental assistance in the future seems likely, as illustrated by the sharp Trump era reduction in US acceptance of Syrian immigrants. What will happen when the numbers are multiplied by many million more climate refugees who cannot flee to adjacent states in MENA that are no better off? If the large numbers from one civil war overwhelm the relief system, the likelihood of crises of a much greater scale in the future receiving even minimal humanitarian attention in advance seems dim. On the other hand, waiting for the full-blown crises of the future to materialize before responding seems futile.

What is the MENA doing to adapt to climate change and is it enough? The region is doing some adaptation but efforts are modest and scattered (as discussed in Chapter 9). At the same time, it is not clear that residents of the MENA can do much about mitigating the effects of climate change, short of refusing to sell any more oil and gas. Thus, the world outside the MENA at some point will have to decide ultimately what to do about the impact of climate change in MENA. Some might suggest that sustained military intervention is key to stabilizing the conflictual region before anything meaningful about climate change could be implemented. But a recent experience with military intervention in

Afghanistan and Iraq suggests that this strategy has not exactly made these places more peaceful or more stable. Whether these interventions have resolved any problems remain an open-ended question as well. Foreign aid is badly needed to restore damaged infrastructure in the region but oftentimes aid ends up being just another form of resource curse at the hands of corrupt and authoritarian leaders. So the MENA will need a great deal of international assistance to help mitigate the catastrophic impact of climate change but that is difficult enough to achieve in good times and will likely become less probable if the environment continues to deteriorate. Some local adaptations will no doubt occur but it will be insufficient to preserve anything resembling the contemporary MENA.

At some point in this century, the number of climate migrants will no doubt escalate. The number of people going without adequate supplies of food and water will increase. Inter-state conflicts over access to water could increase in a region, which has a greater dependency on shared water resources than elsewhere. Famine and disease will be difficult to evade. None of these problems will be easy to contain within MENA borders. Nor will the problems be confined to MENA. Central and South Asia and sub-Saharan Africa, to name the most obvious places, will be experiencing similar problems nearby. The climate-induced crises, after all, will not be regional in scope, but global. Already, Central Asia's glaciers, which provide freshwater to millions in the region, are retreating at an alarming rate (Fountain and Solomon, 2019). Melting glaciers land dams located in Chinese territory will affect what water is available for both South Asia and Southeast Asia. South Asia, sub-Saharan Africa, and parts of Latin America will be experiencing similar environmental problems and competing for the same food, water, and humanitarian assistance. Unfortunately, the outside world does not have a good track record of responding to crises on any scale, let alone one of this possible magnitude.

The most appropriate regional organization, the Arab League, does not quite span the entire region. Nor does it have any track record in resolving difficult regional problems. The closest regional organization with some capacity, the European Union, may not be around in its current configuration two decades from now. Even if it survives almost intact, it, too, stumbled in dealing with Syrian refugees. Imagine a 10- or 20-fold increase in the number of immigrants seeking relief somewhere in Europe. Southern Europe could also be hard-pressed by similarly severe climate change problems. The United Nations already has its hands full dealing with conflict refugees. How will it cope with three or four times the current number, which is already higher than it has been in the past? Moreover, in the past, war refugees sometimes could return home. Climate refugees in this century will not have a home to which to return. Finally, in another major source of global governance, governmental decision-makers in the United States no longer exhibit as much interest in systemic leadership activities as they once did. The odds that this disinterest may change sometime in the near future remains to be seen. There may also be some major difficulties in rebuilding burned bridges for allies to overcome. When all is said and done, the United States is no longer the behemoth it was at the end of World War II. The leader of the "Free World" has given way to more of a role emphasizing coordination skills that are not always evident. China, on the other hand, will have its own global warming problems with which to deal. The mid-21st century may also be too soon to expect an ascending China to pick up the global governance mantle.

The poor prospects for systemic leadership are most unfortunate because some form of systemic leadership will probably be critical to tackling the looming problems associated with global warming. Whether the warming problems become as severe as we fear or not, systemic leadership will be absolutely critical to organizing a meaningful attack on decarbonizing the world economy. Without rapid decarbonization, the warming problems will only become worse.

If the warming problems do become as severe as we fear, two additional global policy responses may become necessary and they will also require considerable systemic leadership. If a considerable portion of the Global South runs out of food and water, a new global food regime could be constructed to at least ameliorate the problem. Some analysts (e.g., McMichael, 2009) have noted that Britain constructed a global food regime in the 19th century in part oriented to supply the metropole and in part to develop a division of labor for food supply outside of Europe. A second global food regime was built by the United States to distribute surplus food after World War II. The 21st century is likely to experience the need for a global food regime to provide virtual water products to MENA and large segments of the Global South. If the MENA can create a capability to help pay for the increased volume of food that will be demanded, so much the better. Whether or not that capability is forthcoming, there will still be a great need for organizing the distribution of food and water more equably.

A second possible, but perhaps less likely, response is even more unprecedented. No parts of the world will go untouched by global warming damage. Yet the Northern Hemisphere may not be as hard hit as the Southern Hemisphere. Some parts of the Global North will even become more inhabitable. One way to respond to areas made less inhabitable by global warming involves moving some inhabitants from those areas to places less densely occupied and more livable. Chi Xu et al. (2019) argue that for the past 6,000 years, most humans have been concentrated in areas with an average temperature of 13 degrees Centigrade. Global warming could mean that 1 to 3 billion people will find themselves no longer living in this comfort niche. Granted, it may not prove possible to move a third of the world's population and things will have to get very bad before such a radical "solution" on this scale becomes in any way palatable. But things could become that bad.

We can list all of the governments that are unlikely candidates for achieving some type of response/resolution for the global climate crisis. We probably cannot or will not do much to reverse the forces leading to climate change in time to head off the immediate consequences. Let us hope that we can construct some type of organizational response before the climate changes hit full bore.

Otherwise, we will have demonstrated that we could neither head off a major catastrophe nor deal with its continuing aftermath. That will only mean the crisis will prove to be all that much more grave than it might have been.

Writing on the current crisis involving forced migration, Yayboke and Milner (2018: v) claim that

> People in almost every region of the world are being forced from home by armed conflict and violence, persecution, political oppression, economic malfeasance, environmental, climate, and human-induced disasters, or food insecurity and famine. Common threads among these root causes of forced migration are underdevelopment and poor governance. Yet, our efforts to confront the crisis continue to be disproportionately reactive rather than proactive in addressing these and other core issues. All too often this more narrow focus leads to one-off foreign assistance programs that rely on yearly replenishments as armed conflicts and other root causes entrench themselves.

When these drivers of forced migration intensify exponentially a few decades from now, as we anticipate they might because of global warming, reactive and one-off foreign assistance programs will not prove very efficacious – assuming they will even be forthcoming. People who are less affected by global warming will wish the miseries associated with hotter and dryer climates would just go away. They will build walls and use gunboats to keep away the surviving victims seeking an escape to the global North. They may become more receptive to autocratic leaders and xenophobic programs. The odds are, though, that none of these defenses will prove all that effective. People less affected by global warming will have to learn how to help people who are more affected. Some pro-action would be all to the good even if it need be confined to planning for mega-crises that might be coming in the next several decades. Should those mega-crises occur, there should be little surprise. We have had 15,000 years of climate-induced crises in the MENA to serve as warning.[3]

We do not enjoy donning the "Chicken Little" mantle and proclaiming the sky may be falling in the not-so-far-future. John Waterbury (2017: 57), referring to water management issues, takes a more optimistic stance:

> The MENA is the region of the world likely to be the most highly and negatively impacted by climate change. The challenges of the coming century will be severe but also familiar to the policy-makers and technical experts of the region. The challenges have characterized the region throughout history. They differ now only in degree not in kind.

We certainly agree and hope to have demonstrated in this book that the MENA of the 21st century will be at least one of "the most highly and negatively impacted by climate change" and that these "challenges have characterized the region throughout history." We hope he is also right about the challenges differing only in degree and not in kind.

While it is customary to end an analysis such as this with pleas for preventive governance programs capable of fending off disaster, it is not really clear that such a message could or would do much good. Global governance efforts remain rather rudimentary. They struggle to cope now with what will be seen in the future to have been limited-scale problems. The future seems to promise humanitarian assistance disasters of unprecedented scale. The rational answer is that the world will need to rise to the occasion and either develop new institutions or expand and reinforce existing institutions. What might seem rational, however, does not always correspond to international reality. Nevertheless, while we are waiting for local, state, and international governmental capacity to catch up to the demands of the near future, we might as well seek to encourage the acceleration of the following:

- Substituting non-carbon fuels for fossil fuel and committing to developing the comparative advantage in solar and wind energy
- Developing more centralized and monitored water management capabilities
- Making and distributing potable water less expensively
- Growing or making much more food using about the same or even less land
- Cultivating plants and animals using less water
- Storing surplus food and distributing it globally as needed
- Insulating large population centers from extreme heat and/or making air conditioning much more efficient than it is now
- Resettling environmental refugees on a large scale
- Negotiating explicit agreements on the use of shared water resources

We will have a dire need of these technological improvements in MENA, and not just there. Nor would it be difficult to find applications for these technologies and policy commitments in the present.

James Hansen, one of the earliest climate specialists to warn about climate change, finds that people who think we are doomed by global warming "to be very tiring and unhelpful" (Schwarz, 2018). Smart people, he contends, should be able to figure out a way to avoid doom. We hope our assessment of the past 15,000 years of intermittent climate crises with associated costs and consequences will help persuade some smart people to come up with solutions to avoid a total collapse of a MENA civilization that once gave the rest of the world's complexity in human systems. The only problem is that the MENA's climate future looks to be highly resistant to smart solutions. On repeated occasions, going back at least 15 millennia, the region has been wounded severely by radical climate change. Each time it struggled back from periods of environmental catastrophe. It could do so again but those revivals took centuries to make transitions back to more endurable environments. This time there is no guarantee that extremely heightened and widespread aridity will permit a return to something more endurable no matter how many centuries might be available for a regional revival. Sometime in the 21st century, the MENA may simply be a ground zero of future global warming effects run amuck. Short of some type of radical geoengineering or Hail Mary technological breakthroughs in decarbonization, food production, and making potable water, the region appears to be headed for another interval of societal, political, and economic collapse that will resemble the previous Dark Ages in style. But given the number of people involved, the outcome could prove terminal for a functioning region.

Notes

1 Useful information on various aspects of the ongoing Syrian relief problem can be found in Rabil (2016); Barnard (2017); Tyyska et al (2017); Chulov (2018); UNHCR (2018; and Yassin (2018).
2 As of 2015, some 5.5 million Palestinians, representing much older crises than those pertaining to the Iraqis or the Syrians, were official residents of 58 camps (Swain and Jagerskog, 2016: 137).
3 Frank Herbert's once-popular Dune novels, which featured an arid landscape in which people had to wear suits to preserve all of their liquids in order to survive, borrowed from Middle Eastern history. The ultimate irony might be if his science fiction landscape became the MENA reality.

BIBLIOGRAPHY

Abbo, Shahal, Simcha Lev-Yadin, and Avi Gopher (2010) "Yield Stability: An Agronomic Perspective on the Origin of Near Eastern Agriculture." *Vegetation History and Archaeobotany* 19, 2: 143–150.

Abel, Guy J., Michael Brottager, Jesus Crespo Cuaresma, and Raya Muttarak (2019) "Climate, Conflict, and Forced Migration." *Global Environmental Change* 54: 239–249.

Abu-Ismail, Khalid (2015) *The Reality of Extreme Poverty in Arab Countries is Much Harsher than Implied by Data*. Washington, DC: International Food Policy Research Institute.

Abu-Ismail, Khalid, Bilal al-Kismani, Arthur van Diesen, Tarek Nabil El-Nabulsi, Lucia Ferone, Verena Gantner, Bilal Malaeb, Beatrice Mauger, Alberto Minujin, Ottavia Pesce, and Maya Ramadan (2017) *Arab Multidimensional Poverty Report*. Beirut: United Nations Economic and Social Commission for Western Asia (ESCWA). Available at https://phi.org.uk/wp-content/uploads/LAS_et_al_2017_Arab_ MP_ Report.ENG .pdf.

Adams, Matthew J. (2017) "Egypt and the Levant in the Early/Middle Bronze Age Transition," in Felix Hoflmayer, ed., *The Late Third Millennium in the Ancient Near East: Chronology, C14, and Climate Change*. Chicago, IL: Oriental Institute, University of Chicago.

Adams, Robert McC. (1981) *Heartland of Cities*. Chicago, IL: University of Chicago Press.

Adams, W. (1977) *Nubia: Corridor to Africa*. London: Allen Lane.

Adger, W.N., Pulhin, J., Barnett, J., Dabelko, G.D., Hovelsrud, G.K., Levy, M., Spring, Ú.O., Vogel, C., Adams, H., Hodbod, J., Kent, S., and Tarazona, M. (2014) "Human Security." *Climate Change 2014: Impacts, Adaptation, and Vulnerability*. Contribution of Working Group II to the Fifth Assessment Report of the Intergovernmental Panel on Climate Change. Chapter 12. Final Draft. Edited by C. Field, V. Barros, K. Mach, and M. Mastrandrea, IPCC AR5 WGII. Cambridge, UK: Cambridge University Press.

Adoha, Franck and Quentin Wodon (2014) "Perceptions of Climate Change, Weather Shocks, and Impacts on Households," in Quentin Wodon, Andrea Liverani, George Joseph, and Natalie Bougnoux, eds., *Climate Change and Migration: Evidence for the Middle East*. Washington, DC: World Bank.

Agoston, Gabor (2005) *Guns for the Sultan: Military Power and the Weapons Industry in the Ottoman Empire*. Cambridge: Cambridge University Press.

Ahmadipour, A. and H. Moradkhani (2018) "Multi-Dimensional Assessment of Drought Vulnerability in Africa, 1960–2100." *Science of the Total Environment* 644: 520–535.

Akhtar, Shamshad (2012) "Keynote Address II," in Atif Kubursi, ed., *Food and Water Security in the Arab World: Proceedings of the First Arab Development Symposium*. Washington, DC: The International Bank for Reconstruction and Development, World Bank, and the Arab Fund for Economic and Social Development.

Akkermans, Peter M.M.G., Johannes van der Plicht, Olivier P. Nieuwenhuyse, Anna Russell, and Akemi Kaneda (2015) "Cultural Transformation and 8.2 Ka Event in Upper Mesopotamia," in Susanne Kemer, Rachael Dann, and Pernille Langsgaard, eds., *Climate and Ancient Societies*. Copenhagen: Museum Tusculanum Press.

Al Blooshi, L.S., S. Alyam, N.J. Joshua, and T.S. Ksilksi, (2019) "Modeling Current and Future Climate Change in the UAE Using Varius GCMs in MarksimGCM." *The Open Atmosphere Science Journal* 13: 56–64.

Alagidede, Paul, George Adu, and Prince Boakye Frimpong (2014) "The Effect of Climate Change on Economic Growth: Evidence for Sub-Sahara Africa." *Wider Working Papers, no. 17*. Vienna: United Nations University. Available at https://ideas.repec.org/p/unu/wpaper/wp2014-017.html.

Alam, Mayesha, Rukmani Bhatia, and Briana Mawby (2015) *Women and Climate Change: Impact and Agency in Human Rights, Security, and Economic Development*. Washington, DC: Georgetown Institute for Women, Peace and Security. Available at https://giwps.georgetown.edu/resource/women-and-climate-change/.

Alboghdady, Mohamed (2016) "Economic Impacts of Climate Change and Variability on Agricultural Production in the Middle East and North Africa Region." *International Journal of Climate Change Strategies and Management* 8, 3: 463–472.

Al-Delaimy, Wael K. (2020) "Vulnerable Populations and Regions: Middle East as a Case Study," in Wael K. Al-Delaimy, Veerabhadran Ramanthan, and Marcelo Sanchez Sorondu, eds., *Health of People, Health and Climate Change, Air Pollution and Health*. Cham, Switzerland: Springer.

Alexandratos, Nicos and Jelle Bruinsma (2012) "World Agriculture Towards 2030/2050: The 2012 Revision." *ESA Working Paper, no. 12-03*. New York: Agricultural Development Economics Division, Food and Agricultural Organization of the United Nations. Available at http://www.fao.org/fileadmin/templates/esa/Global-perspectives/world-ag_2030_50_2012.rev.pdf.

Algaze, Guillermo (1986) "Kurban Hoyuk and the Late Chalcolithic Period in the Northwest Mesopotamian Periphery: A Preliminary Assessment," in Uwe Finkbeiner and Wolfgang Rollig, eds., *Gamdat Nasr: Period or Regional Style?* Weisbaden: Dr. Ludwig Reichert Verlag.

Algaze, Guillermo (1993) *The Uruk World System: The Dynamics of Expansion of Early Mesopotamian Civilization*. Chicago, IL: University of Chicago Press.

Algaze, Guillermo (1995) A Fourth Millennium BC Trade in Greater Mesopotamia: Did It Include Wine?" in P.E. McGovern, S.J. Fleming, and S.H. Katz, eds., *The Origins and Ancient History of Wine*. Amsterdam: Gordon and Breach.

Algaze, Guillermo (2000) "Initial Social Complexity in Southwest Asia: The Mesopotamian Advantage." Unpublished paper. Department of Anthropology, UCSD.

Algaze, Guillermo (2008) *Ancient Mesopotamia at the Dawn of Civilization: The Evolution of an Urban Landscape*. Chicago, IL: University of Chicago Press.

Allan, Tony (2001) *The Middle East Water Question: Hydropolitics and the Global Economy.* London: I.B. Tauris.

Allan, Tony (2011) *Virtual Water: Tackling the Threat to Our Planet's Most Precious Resource.* London: I.B. Tauris.

Alley, Richard B. (2000) *The Two-Mile Time Machine: Ice Cores, Abrupt Climate Change, and Our Future.* Princeton, NJ: Princeton University Press.

Al-Masri, Abdulrahman (2014) "Turkey's Control of the Euphrates Might Lead to Disaster." *Middle East Monitor,* June 23. Available at https://www.middleeastmonitor.com/20140623-turkeys-control-of-the-euphrates-might-lead-to-disaster/.

Amanou, Hajer, Mohsen Ben Sassi, Hatem Aocadi, Hicham Khemiri, Mohktar Majouachi, Yves Beckers, and Hedi Hammami (2018) "Climate Change-Related Risks and Adaptation Strategies as Perceived in Dairy Cattle Frmaing Systems in Tunisia." *Climate Risk* 20: 38–49.

Arranz-Otaegui, A., J.A. Lopez-Saez, J.L. Araus, M. Portillo, A. Balbo, E. Iriarte,. L. Gourichon, F. Braemer, L. Zapata, and J.J. Ibanez (2017) "Landscape Transformations at the Dawn of Agriculture in Southern Syria (10.7–9.9 Ka Cal BP); Plant-specific Responses to the Impact of Human Activities and Climate Change." *Quaternary Science Reviews* 158: 145–163.

Artzy, M. and D. Hillel (1988) "A Defence of the Theory of Progressive Soil Salinization in Ancient Southern Mesopotamia." *Geoarchaeology: An International Journal* 3, 3: 235–238.

Ash, K. and N. Obradovich (2019) "Climate Stress, Internal Migration, and Syrian Civil War Onset." *Journal of Conflict Resolution* 64, 1: 3–31.

Asouti, Eleni (2009) "The Relationship between Early Holocene Climate Change and Neolithic Settlement in Central Anatolia, Turkey: Current Issues and Prospects for Future Research." *Documenta Praehistorica* 36. Available at https://revije.ff.uni-lj.si/DocumentaPraehistorica/article/view/36.1.

Asseng, S.F., P. Ewart, R.P. Martre, D.B. Rotter, D. Lovell, B.A. Cammarano, M.J.. Kimball, G.W. Ottman, J.W. Wall, M.P. White, P.O. Reynolds, P.V.V. Alderman, P.K. Prasad, J. Aggarwal, B. Anothai, C. Basso, A.J. Biemath, G. Challiner, J. De Sanctis, E. Doltra, M. Fereres, S. Garcia-Vila, G. Gayler, L.A. Hogenboom, R.C. Hunt, M. Izauralde, C.D. Jabloon, K.C. Jones, A-K. Kersebaum, C. Koehler, S. Muller, C. Naresh Kumar, G. Nendal, J.E. O'Leary, T. Olesen, E. Pabsuo, E. Priesack, A.C. Eyshirezaei, M.A.. Ruane, I. Semenov, C. Seherbak, P. Stockle, T. Stratonovitch, I. Streck, F. Supit, P.J. Tao, K. Thorbun, E. Waha, D. Wang, T. Wallach, Z. Wolf, Z. Zhao, and Y. Zhu (2014) "Rising Temperatures Reduce Global Wheat Production." *Nature Climate Change* 5: 143–147.

Bai, Ying, and James Kai-Sing Kung (2011) "Climate Shocks and Sino-Nomadic Conflict." *Review of Economics and Statistics* 93, 3: 970–81.

Baines, J. and N. Yoffee (1998) "Order, Legitimacy, and Wealth in Ancient Egypt and Mesopotamia," in G.M. Feinman and J. Marcus, eds., *Archaic States.* Santa Fe, NM: School of American Research Press.

Bakker, J., E. Paulisen, D. Kaniewski, J. Pablome, V. DeLaet, G. Vestraeten, and M. Waelkens (2013) "Climate, People Fire and Vegetation: New Insights into Vegetation Dynamics in the Eastern Mediterranean Since the 1st Century AD." *Climate of the Past* 9: 57–87.

Balaraman, Kavya (2017) "A New Dam on the Nile Reveals Threats from Warming." *Scientific American.* Available at https://www.scientificamerican.com/article/a-new-dam-on-the-nile-reveals-threats-from-warming/.

Balter, Michael (2010) "The Tangled Roots of Agriculture." *Science* 327, 5964: 404–406.

Bard, Kathryn A. (1993) "State Collapse in Egypt in the Late Third Millennium BC." *Paper presented at the 57th Annual Meeting of the Society of American Archaeology*, St. Louis, MO, April.

Barich, B. (2014) "Northwest Libya from the Early to Late Holocene: New Data on Environment and Subsistence from the Jebel Gharbi." *Quaternary International* 320: 15–27.

Barjamovic, Gojko (2013) "Mesopotamian Empires," in Peter Fibiger Bang and Walter Scheidel, eds., *The Oxford Handbook of the State in the Ancient Near East and Mediterranean*. Oxford: Oxford University Press.

Bar-Matthews, Miryam and Avner Ayalon (1997) "Late Quaternary Paleoclimatic in the Eastern Mediterannean Region for Stable Isotope Analysis of Speleothems at Soreq Cave, Israel." *Quaternary Research* 47: 155–168.

Bar-Matthews, Miryam, Avner Ayalon, and Aaron Kaufman (1998) "Middle to Late Holocene Paleoclimatic in the Eastern Mediterranean Region from Stable Isotopic Composition of Speleothmens from Soreq Cave, Israel," in A.S. Issar and N. Brown, eds., *Water, Environment and Society in Times of Climatic Change*. Amsterdam: Kluwer.

Bar-Matthews, Miryam, Avner Ayalon, Aaron Kaufman, and Gerald Wasserburg (1999) "The Eastern Mediterranean Paleoclimate as a Reflection of Regional Events: Soreq Cave, Israel." *Earth and Planetary Science Letters* 166: 85–95.

Bar-Matthews, Miryam, J. Keinan, and Avner Ayalon (2019) "Hydro-climate Research of the Late Quaternary of the Eastern Mediterranean-Levant Region Based on Speleothems Research - A Review." *Quaternary Science Reviews* 221: 105872.

Barnard, Anne (2017) "For Syrian Refugees, There is No Going Home." *The New York Times*, February 23. Available at https://www.nytimes.ocm/2017/02/23/world/ middle east/lebanon-syria-refugees-geneva.html.

Barnett, Jon and W. Neil Adger (2007) "Climate Change, Human Security and Violent Conflict." *Political Geography* 26: 639–55.

Bar-Yosef, Ofer (2011) "Climatic Fluctuation and Early Farming in West and East Asia." *Current Anthropology* 52, 54: S175–S193.

Bar-Yosef, Ofer and Anna Belfer-Cohen (1989) "The Origins of Sedentism and Farming Communities in the Levant." *Journal of World Prehistory* 3, 4: 447–498.

Bastin, J.F., E. Clark, T. Elliott, S. Hart, J. van den Hoogen, I. Hordijk, H. Ma, A. Majunder, G. Manli, J. Maschler, L. Mo, D. Routh, K. Yu, C.M. Zohner, and T.W. Crowther (2019) "Understanding Climate Change From a Global Analysis of City Analogues." *PLoS ONE* 14, 10: e0224120. https://doi.org/101371/journalpone0224120.

Beck, Andrea (2014) "Drought, Dams, and Survival: Linking Water to Conflict and Cooperation in Syria's Civil War." *International Affairs Forum* 5(1): 11–22.

Beihl, Peter F., Ingmar Franz, Sonia Ostaptchook, David Orton, Jana Rogasch, Eva Rosenstock (2012) "One Community and Two Tells: The Phenomenon of Relocating Tell Settlements at the Turn of the 7th and 6th Millennia in Central Anatolia," in Robert Hofmann, Ferzi-Kemal Moetz, and Johannes Muller, eds., *Tells: Social and Environmental Space*, Vol. 3. Bonn: Verlag Dr. Rudolf Habelt GmbH.

Bell, Barbara (1971) "The Dark Ages in Ancient History I: The First Dark Age in Egypt." *American Journal of Archaeology* 75: 1–26.

Bell, Barbara (1975) "Climate and the History of Egypt: The Middle Kingdom." *American Journal of Archaeology* 79, 3: 223–269.

Bellwood, Peter (2005) *First Farmers: The Origins of Agricultural Societies*. Malden, MA: Blackwell.

Ben David, Amir (2016) "Dams Are Drying at the Jordan River." *Ynetnews.com*, June 25. Available at https://www.ynetnews.com/articles/0.7340.L-4820175.00.html.

Benedictow, Ole J. (2004) *The Black Death, 1346–1353: The Complete History*. Woodbridge, UK: Boydell Press.

Bentalab, Ilhem, Claude Caratini, Michel Fontugne, Marie-Therese Morzandec-Kerfourn, Jean-Pierre Pascal, and Colette Tissot (1997) "Monsoon Regional Variations during the late Holocene in Southwestern India," in H.N. Dalfes, G. Kukla, and H. Weiss, eds., *Third Millennium BC Climate Change and Old World Collapse*. Berlin: Springer.

Berger, Jean-Francois, Laurent Lespez, Catherine Kuzucuglu, Arthur Glais, Fuad Hourani, Adrien Barra, and Jean Guillaine (2016) "Interactions between Climate Change and Human Activities during the Early to Mid-Holocene in the Eastern Mediterranean Basins." *Climate of the Past* 12: 1847–1877.

Berkelhammer, M.A., L. Sinha, H. Stott, F.S. Heng, R. Pausat, and K. Yoshimura (2012) "An Abrupt Shift in the Indian Monsoon 4000 Years Ago," in Liviue Giosan, Dorian Q. Fuller, Kathleen Nicoll, Rowan K. Flad, and Peter D. Clift, eds., *Climates, Landscapes and Civilizations*. Washington, DC: American Geophysical Union.

Berking, Jonas, Janina Korper, Sebastian Wagner, Ulrich Cubasch, and Brigitta Schitt (2012) "Heavy Rainfalls in a Desert(ed) City: A Climate-Archaeological Case Study from Sudan," in Liviue Giosan, Dorian Q. Fuller, Kathleen Nicoll, Rowan K. Flad, and Peter D. Clift, eds., *Climates, Landscapes and Civilizations*. Washington, DC: American Geophysical Union.

Bernauer, Thomas, Tobias Bohmelt, and Vally Koubi (2012) "Environmental Changes and Violent Conflict." *Environmental Research Letters* 7: 1–8.

Biehl, Peter F. (2012) "Rapid Change Versus Long-Term Social Change during the Neolithic-Chalcolithic Transition in Central Anatolia." *Interdisciplinaria Archaeologica* 3, 1: 75–83.

Biehl, Peter F., Ingmar Franz, Sonia Ostaptchook, and Eva Rosenstock (2010) "Climate and Socio-economic Change during the Transition between the Late Neolithic and Early Chalcolithic in Central Anatolia." Paper presented at *The 8.2Ka Climate Event and Archaeology in the Ancient Near East*, Leiden University, March.

Bini, M., G. Zanchetta, A. Persouiu, R. Cartier, A. Catala, I. Cacho, J.R. Dean, F. DiRita, R.N. Drysdale, M. Finne, I. Isola, B. Jalai, F. Lirer, D. Magri, A. Masi, L. Marki, A.M. Mercuri, O. Peyron, L. Sadori, M.A. Sicre, F. Wek, C. Zielhofer, and E. Brisset (2019) "The 4.2ka BP Event in the Mediterranean Region: An Overview." *Climate of the Past* 15, 2: 555–577.

Bizri, Omar (2012) "Research, Innovation, Entrepreneurship and the Rentier Culture in the Arab Countries," in Thomas Anderson and Abdelkader Djeflat, eds., *The Real Issues of the Middle East and the Arab Spring: Addressing Research, Innovation and Entrepreneurship*. Berlin: Springer.

Blom, Philipp (2019) *Nature's Mutiny: How the Little Ice Age of the Long Seventeenth Century Transformed the West and Shaped the Present*. New York: W.W. Norton.

Borger, Julian (2007) "Darfur Conflict Heralds Era of Wars Triggered by Climate Change." *The Guardian* (June 23). Available at https://www.theguardian.com/environment/2007/jun/23/sudan.climatechange.

Borsch, Stuart J. (2005) *The Black Death in Egypt and England: A Comparative Study*. Austin, TX: University of Texas Press.

Bottema, S. (1995) "The Younger Dryas in the Eastern Mediterranean." *Quaternary Science Reviews* 14, 9: 883–891.

Bottema, S. (1997) "Third Millennium Climate in the Near East Based upon Pollen Evidence," in H.N. Dalfes, G. Kukla, and H. Weiss, eds., *Third Millennium BC Climate Change and Old World Collapse*. Berlin: Springer.

Boucek, Christopher (2010) "War in Saada: From Local Insurrection to National Challenger." *Carnegie Papers, Middle East Program no.* 110, April. Washington, DC: Carnegie Endowment for International Peace.

Bowden, M.J., R.W. Kates, P.A. Kay, W.E. Riebsame, R.A. Warrick, D.L. Johnson, H.A. Gould, and D. Weiner (1981) "The Effect of Climate Fluctuations on Human Populations: Two Hypotheses," in T.M. Wigley, M.J. Ingram, and G. Farmer, eds., *Climate and History: Studies in Past Climates and Their Impact on Man.* Cambridge: Cambridge University Press.

Bowersock, G.W. (2017) *The Crucible of Islam.* Cambridge, MA: Harvard University Press.

Breeze, P.S., H.S. Groucutt, N.A. Drake, T.S. White, R.P. Jennings, and M.D. Petraglia (2016) "Palaeohydrological Corridors for Hominin Dispersals in the Middle East −250–70,000 Years Ago." *Quaternary Science Reviews* 144: 155–185.

Brehony, Noel (2015) "Yemen and the Houthis: Genesis of the 2015 Crisis." *Asian Affairs* 46, 2: 232–250.

Britton, Bianca (2016) "Climate Change Could Render Sudan 'Uninhabitable'." *CNN.* Available at https://www.cnn.com/2016/12/07/africa/sudan-climate-change/index .html.

Bromberg, Gidon (2008) "Will the Jordan River Keep On Flowing?" *Yale Environment 360.* Yale School of Forestry and Environmental Studies. Available at https://e360 .yale.edu/features/will_the_jordan_river_keep_on_flowing.

Bronson, B. (1988) "The Role of Barbarians in the Fall of States," in N. Yoffee and C.L. Cowgill, eds., *The Collapse of Ancient States and Civilizations.* Tucson, AZ: University of Arizona Press.

Brooke, John L. (2014) *Climate Change and the Course of Global History: A Rough Journey.* Cambridge: Cambridge University Press.

Brown, Ian A. (2010) "Assessing Eco-scarcity as a Cause of the Outbreak of Conflict in Darfur: A Remote Sensing Approach." *International Journal of Remote Sensing* 31, 10: 2513–2520.

Brown, Lester (2011) *World on the Edge: How to Prevent Economic and Environmental Collapse.* New York: W. W. Norton and Company.

Brown, Oli and Alec Crawford (2009) *Rising Temperatures, Rising Tensions: Climate Change and the Risk of Violent Conflict in the Middle East.* Winnipeg, MB: International Institute for Sustainable Development.

Brown, Terence A., Martin K. Jones, Wayne Powell, and Robin G. Allaby (2009) "The Complex Origins of Domesticated Crops in the Fertile Crescent." *Trends in Ecology and Evolution* 24, 2: 103–109.

Buhaug, Halvard (2010) "Climate Not to Blame for African Civil Wars." *Proceedings of the National Academy of Sciences of the United States of America* 107, 38: 16477–16482.

Buhaug, Halvard (2015) "Climate-Conflict Research: Some Reflections on the Way Forward." *Wiley Interdisciplinary Reviews: Climate Change* 6, 3: 269–75.

Buhaug, Halvard, Havard Hegre, and Havard Strand (2010) "Sensitivity Analysis of Climate Variability and Civil War." *Peace Research Institute Oslo (PRIO)* 17 (November): 1–17.

Buhaug, Halvard, Jonas Nordkvelle, Thomas Bernauer, Tobias Böhmelt, Michael Brzoska, Joshua W. Busby, Antonio Ciccone, Hanne Fjelde; Erik Gartzke; Nils Petter Gleditsch; Jack A. Goldstone; Håvard Hegre; Helge Holtermann; Vally Koubi; Jasmin S.A. Link; Peter Michael Link; Päivi Lujala; John O'Loughlin; Clionadh Raleigh; Jürgen Scheffran; Janpeter Schilling; Todd G. Smith; Ole Magnus Theisen; Richard S.J. Tol; Henrik Urdal, and Nina von Uexkull (2014) "One Effect to Rule Them All? A Comment on Climate and Conflict." *Climatic Change* 127, 3: 391–397.

Buhaug, Halvard, and Ole M. Theisen (2012) "On Environmental Change and Armed Conflict," in H.G. Brauch, J. Scheffran, M. Brzoka, P.M. Link and J. Schilling, eds., *Climate Change, Human Security and Violent Conflict*. Berlin: Springer.

Bulliet, Richard (2011) *Cotton, Climate and Camels in Early Islamic Iran: A Moment in World History*. New York: Columbia University Press.

Buntgen, Ulf, Vladimir S. Myglan, Fredrik Charpentier Ljungqvist, Michael McCormick, Nicola Di Cosmo, Michael Sigl, Johann Jungclaus, Sebastian Wagner, Paul J. Krusic, Jan Esper, Jed O. Kaplan, Michiel A.C. de Vaan, Jurg Luterbacher, Lukas Wacker, Willy Tegel, and Alexander V. Kirdyanov (2016) "Cooling and Societal Change during the Late Antique Little Ice Age from 536 to around 660 AD." *Letters, Nature Geoscience* 9: 231–236.

Burke, Aaron A. (2017) "Amorites, Climate Change, and the Negotiation of Identity at the End of the Third Millennium B.C.," in Felix Hoflmayer, ed., *The Late Third Millennium in the Ancient Near East: Chronology, C14, and Climate Change*. Chicago, IL: Oriental Institute, University of Chicago.

Burke, Marshall, Solomon M. Hsiang, and Edward Miguel (2015) "Climate and Conflict." *Annual Review of Economics* 7: 577–617.

Burke, Marshall, Edward Miguel, S. Satyanath, J. Dykema, and D. Lobell (2009) "Warming Increases Risk of Civil War in Africa." *Proceedings of the National Academy of Sciences of the United States of America* 106, 49: 20670–20674.

Burroughs, William J. (2005) *Climate Change in Prehistory: The End of the Reign of Chaos*. Cambridge: Cambridge University Press.

Busby, Joshua and Nina von Uexkull (2018) "Climate Shocks and Humanitarian Crises: Which Countries Are Most at Risk?" *Foreign Affairs* 97, 4: 4–55.

Butler, Declan (2007) "Darfur's Climate Roots Challenged." *Nature* 447: 1038.

Butzer, Karl W. (1976) *Early Hydraulic Civilization in Egypt*. Chicago, IL: University of Chicago Press.

Butzer, Karl W. (1995) "Environmental Change in the Near East and Human Impact on the Land," in J.M. Sasson, ed., *Civilizations of the Ancient Near East*, Vol. 1. New York: Simon and Schuster.

Butzer, Karl W. (1997) "Sociopolitical Discontinuity in the Near East c. 2200 BCE: Scenarios from Palestine and Egypt," in H.N. Dalfes, G. Kukla and H. Weiss, eds., *Third Millennium BC Climate Change and Old World Collapse*. Berlin: Springer.

Butzer, Karl W. (2012) "Collapse, Environment, and Society." *Proceedings of the National Academy of Sciences of the United States of America (PNAS)* 109, 10: 3632–3639.

Byman, Daniel (2018) "Yemen's Disastrous War." *Survival* 60, 5: 141–158.

Campbell, B. (2016) *The Great Transition: Climate, Disease and Society in the Late-Medieval World*. Cambridge: Cambridge University Press.

Campbell, B.M., S.J. Vermeulen, P.K. Aggarwal, C. Carner-Dolloff, E. Giruetz, A.M. Loboguerreo, J. Ramirez-Villegas, T. Rosenstock, L. Sebastian, P.K. Thornton, and E. Wollenbarg (2016) "Reducing Risks to Food Security from Climate Change." *Global Food Security* 11: 33–43.

Carlson, A.E. (2013) "The Younger Dryas Climate Event," in S.A. Elias, ed., *The Encyclopedia of Quaternary Science*, Vol. 3. Amsterdam: Elsevier.

Carpenter, Stephen R. and William A. Brock (2008) "Adaptive Capacity and Traps." *Ecology and Society* 13, 2: art 40.

Carrington, Damian (2016) "Climate Change Will Stir 'Unimaginable' Refugee Crisis Says Military." *The Guardian*, December 1. Availalbe at https://www.theguardian.com/environment/2016/dec/01/climate-change-trigger-unimaginable-refugee-crisis-senior-military.

Cassis, M., O. Doonan, H. Elton, J. Newhard (2018) "Evaluating Archaeological Evidence for Demographics, Abandonment and Recovery in Late Antique and Byzantine Anatolia." *Human Ecology* 46, 3: 381–398.

Challinor, A.J., J. Watson, D.B. Lobell, S.N. Howden, D.R. Smith, and N. Chhetri (2014) "A Meta-analysis of Crop Yields Under Climate Change and Adaptation." *Nature Climate Change* 4: 287–291.

Charles River Editors (2016) *The Hyksos: The History of the Foreign Invaders Who Conquered Ancient Egypt and Established the Fifteenth Dynasty.* Cambridge, MA: Charles River Editors.

Chase, Kenneth (2003) *Firearms: A Global History to 1700.* Cambridge: Cambridge University Press.

Chavalas, M. (1992) "Ancient Syria: A Historical Sketch," in M.W. Chavalas and J.L. Hayes, eds., *New Horizons in the Study of Ancient Syria.* Malibu, CA: Udena Publications.

Cheddadi, Rachid, Alessio Palmisano, Jose Antonio Lopez-Saez, Madja Novrelbeit, Christoph Zielhofer, Jalal Tabel, Ali Rhoujjati, Carla Khater, Jessie Woodbridge, Gulio Lucarini, Cyrian Broodbank, William J. Fletcher, and C. Neil Roberts (2019) "Human Demography Changes in Morocco and Environmental Impact during the Holocene." *The Holocene* 29, 5: 816–829.

Chen, C., I. Noble, J. Hellman, J. Coffee, M. Murillo, and N. Chawla (2015) "University of Notre Dame Global Adaptation Index: Country Index Technical Report." Available at https://gain.nd.edu/assets/254377/nd_gain_technical_document_2015.pdf.

Chew, Sing (1999) "Ecological Relations and the Decline of Civilizations in the Bronze Age World System: Mesopotamia and Harappa 2500–1700 B.C.," in W.L. Goldfrank, D. Goodman, and A. Szasz, eds., *Ecology and the World-System.* Westport, CT: Greenwood Publishing Group.

Chew, Sing (2001) *World Ecological Degradation: Accumulation, Urbanization and Deforestation, 3000 BC-AD 2000.* Walnut Creek, CA: Altamira.

Chew, Sing (2006) *The Recurring Dark Ages: Ecological Stress, Climate Change, and System Transformation.* Oakland, CA: AltaMira Press.

Childe, V. Gordon (1936) *Man Makes Himself.* London: Collins.

Christensen, Peter (1993/2016) *The Decline of Iranshahr: Irrigation and Environment in the Middle East, 500 BC-AD 1500,* translated by Steven Sampson. London: I.B. Tauris.

Chulov, Marin (2018) "'We Can't Go Back.' Syria's Refugees Fear for their Future After War." *The Guardian,* August 30. Available at https://www.theguardian.com/world/2018/aug/30/we-cant-go-back-syrias-refugees-fear-for-their-future-after-war.

Clare, L. (2016) *Culture Change and Continuity in the Eastern Mediterranean during Rapid Climate Change Assessing Impacts of a Little Ice Age in the 7th Millennium calBC.* Rahden/Westf: Verlag Marie Leidorf.

Clarke, Joanne, Nick Brooks, Edward B. Banning, Miryam Bar-Matthews, Stuart Campbell, Lee Clare, Maruro Cremaschi, Savinodi Lernia, Nick Drake, Marina Gallinaro, Stuart Manning, Kathleen Nicole, Graham Philip, Steve Rosen, Ulf-Dietrich Schoop, MaryAnne Tafuri, Bernhard Weninger, and Andrea Zerboni (2016) "Climatic Changes and Social Transformations in the Near East and North Africa during the 'Long' 4th Millennium BC: A Comparative Study of Environmental and Archaeological Evidence." *Quaternary Science Reviews* 136: 96–121.

Cline, Eric H. (1994) *Sailing the Wine Dark Sea: International Trade and the Late Bronze Age Aegean.* Oxford: British Archaeological Reports.

Cline, Eric H. (2014) *1177 BC, The Year Civilization Collapsed.* Princeton, NJ: Princeton University Press.

CNA Military Advisory Board (2007) *National Security and the Threat of Climate Change.* Alexandria, VA: Center for Naval Analysis Corporation. Available at https://www.cna.org/cna_files/pdf/national%20security%20and%20the%20threat%20of%20climate%20change.pdf.

Committee on Abrupt Climate Change, National Research Council (2002) *Abrupt Climate Change: Inevitable Surprises.* Washington, DC: National Academies Press.

Conniff, Richard (2017) "The Vanishing Nile: A Great River Faces a Multitude of Threats." *Yale Environment 360.* Yale School of Forestry and Environmental Studies. Available at https://e360.yale.edu/features/vanishing-nile-a-great-river-faces-a-multitude-of-threats-egypt-dam.

Conolly, James, Sue Colledge, Keith Dobney, Jean-Denis Vilgne, Joris Peters, Barbara Stapp, Katie Manning, Stephen Shennan (2010) "Meta-analysis of Zooarchaeological Data for SW Asia and SE Europe Provides Insight into the Origins and Spread of Animal Husbandry." *Journal of Archaeological Science* 38, 3: 338–345.

Constantinidou, K., P. Hadjinicolaou, G. Zittis, and J. Lelieveld (2016) "Effects of Climate Change on the Yield of Winter Wheat in the Eastern Mediterranean and Middle East." *Climate Research* 69, 2: 129–141.

Contreras, D.A. and C. Makarwicz (2016) "Regional Climate, Local Paleoenvironment, and Early Cultivation in the Middle Wadi el Hasa, Jordan," in Daniel A. Contreras, ed., *The Archaeology of Human-Environment Interactions: Strategies for Investigating Anthropogenic Landscapes, Dynamic Environments, and Climate Change in the Human Past.* New York: Routledge.

Cope, Carol (1991) "Gazelle Hunting Strategies in the Natufian," in O. Bar-Yosef and F. Valla, eds., *The Natufian Culture in the Levant.* Ann Arbor, MI: International Monographs in Prehistory.

Coutsoukis, Photius (2001) "Total Populations by Country, 1950, 2000, 2015, 2025, 2050." Available at Photius.com/ranking/world2050_rank.html.

Crawford, H. (1992) "An Early Dynastic Trading Network in North Mesopotamia?," in D. Charpin and F. Joannes, eds., *La Circulation Des Biens, Des Personnes et Des Idees Dans Le Proche Orient Ancien.* Paris: Editions Recherche sur les Civilisations.

Culican, W. (1966) *The First Merchant Venturers: The Ancient Levant in History and Commerce.* London: Thames and Hudson.

Cullen, H.M., P.B. deMenocal, S. Hemming, G. Hemming, F.H. Brown, T. Guilderson, and F. Sirocko (2000) "Climate Change and the Collapse of the Akkadian Empire: Evidence from the Deep Sea." *Geology* 28, 4: 379–382.

Curtis, Kimberly (2017) "Climate Refugees 'Explained'." *Dispatch*, April 24. United Nations News and Commentary. Available at https://www.undispatch.com/climate-refugees-explained/.

Dales, G. (1976) "Shifting Trade Patterns between the Iranian Plateau and the Indus Valley in the Third Millennium BC," in J. Deshayes, ed., *Le Plateau Iranien et l'Asie Centrale des Origines au la Conquete Islamique.* Paris: Collogue Internationaux de CNRS.

Danti, Michael D. (2010) "Late Middle Holocene Climate and Northern Mesopotamia: Varying Cultural Responses to the 5.2 and 4.2 ka Aridification Events," in A. Bruce Mainwaring, Robert Geigengack, and Claudio Vita-Finzi, eds., *Climate Crises in Human History.* Philadelphia, PA: Amercan Philosophical Society.

De Châtel, Francesca (2014) "The Role of Drought and Climate Change in the Syrian Uprising: Untangling the Triggers of the Revolution." *Middle Eastern Studies* 50, 4: 521–35.

De Waal, Alex (2007) "Is Climate Change the Culprit for Darfur?" *African Arguments.* Available at https://africanarguments.org/2007/06/25/is-climate-change-the-culprit-for-darfur/.

Dee, Michael W. (2017) "Absolutely Dating Climatic Evidence and the Decline of Old Kingdom Egypt," in Felix Hoflmayer, ed., *The Late Third Millennium in the Ancient Near East: Chronology, C14, and Climate Change.* Chicago, IL: Oriental Institute, University of Chicago.

Dell, Melissa, Benjamin F. Jones, and Benjamin A. Olken (2008) "Climate Change and Economic Growth: Evidence for the Last Half Century." *Working Paper no.* 14132. Cambridge, MA: National Bureau of Economic Research.

Dell, Melissa, Benjamin F. Jones, and Benjamin A. Olken (2012) "Temperature Shocks and Economic Growth: Evidence for the Last Half Century." *American Economic Journal: Macroeconomics* 4, 3: 66–95.

DeMenocal, Peter, Joseph Ortiz, Tom Guilderson, Jess Adkins, Michael Sarnthein, Linda Baker, and Martha Yarusinsky (2000) "Abrupt Onset and Termination of the African Humid Period: Rapid Climate Responses to Gradual Insolation Forcing." *Quaternary Science Reviews* 19: 347–361.

DeNicola, Erica, Omar S. Abuirzaiza, Azhar Siddique, Haider Khwage, and David O. Carpenter (2015) "Climate Change and Water Scarcity: The Case of Saudi Arabia." *Annals of Global Health* 81, 3: 324–353.

Dennell, R. and M.D. Petraglia (2012) "The Dispersal of Homo Sapiens across Southern Asia: How Early, How Often, How Complex?" *Quaternary Science Reviews* 47: 15–22.

Diamond, Jared (1997) *Guns, Germs, and Steel: The Fates of Human Societies.* New York: W.W. Norton.

Diester-Haass, L. (1973) "Holocene Climate in the Persian Gulf as Deduced from Grain Size and Pteropod Distribution." *Marine Geology* 14: 207–23.

Diffenbaugh, Noah S. and Marshall Burke (2019) "Global Warming Has Increased Global Economic Inequality." *Proceedings of the National Academy of Sciences of the United States of America (PNAS)* 116, 20: 9808–9813.

Dobler, Sacha (2018) "Black Death and Abrupt Earth Changes in the 14th Century." Available at https://abruptearthchanges.files.wordpress.com/2018/02/01-02-2018 -updated-black-death-and-abrupt-earth-changes.pdf.

D'Odorico, Paolo, Joel Carr, Carole Davlin, Jampel Dell'Angelo, Megan Komar, Francisco Iaio, Luca Redolfi, Lorenzo Rosa, Samir Suweis, Stefanie Tanea, and Marta Tunivetti (2019) "Global Virtual Water Trade and the Hydrological Cycle: Patterns, Drivers, and Socio-environmental Impacts." *Environmental Research Letters* 14: 053001.

Dols, Michael W. (1979a) *The Black Death in the Middle East.* Princeton, NJ: Princeton University Press.

Dols, Michael W. (1979b) "The Second Plague Pandemic and Its Recurrences in the Middle East: 1347–1894." *Journal of the Economic and Social History of the Orient* 22, 2: 162–189.

Dominguez-Castro, Fernando, Jose Manuel Vaqero, Manuela Marin, Maria Cruz Gallego, and Ricardo Garcia-Herrera (2012) "How Useful Could Arabic Documentary Sources Be for Reconstructing Past Climate?" *Weather* 67, 3: 76–82.

Dornauer, Aron (2017) "Bioclimatic and Agroecological Properties of Crop Taxa: A Survey of the Cuneiform Evidence Concerning Climatic Change and the Early/Middle Bronze Age Transition," in Felix Hoflmayer, ed., *The Late Third Millennium in the Ancient Near East: Chronology, C14, and Climate Change.* Chicago, IL: Oriental Institute, University of Chicago.

Drews, Robert (1993) *The End of the Bronze Age: Changes in Warfare and the Catastrophe, ca. 1200.* Princeton, NJ: Princeton University Press.

Droubi, A. (2009) "Syria: Country Study." *Climate Change, Water, and the Policy-making Process in the Levant and North Africa Conference.* Issam Fares Institute for Public Policy and International Affairs, the American University of Beirut, August 4.

Durrell, Jack (2018) "Investing in Resilience: Addressing Climate-Induced Displacement in the MENA Region." *CGIAR-CCAFS Research Program on Climate Change, Agriculture and Food Security*. Beirut, Lebanon: International Center for Agricultural Research in the Dry Areas (ICARDA). Available at https://reliefweb.int/report/yemen/investing -resilience-addressing-climate-induced-displacement-mena-region.

Eastwood, W.J., N. Roberts, H.F. Lamb, and J.C. Tibby (1999) "Holocene Environmental Changes in Southwest Turkey: A Paleoecological Record of Lake and Catchement-Related Changes." *Quaternary Science Reviews* 18: 671–695.

The Economist Intelligence Unit (2018) "Global Food Security Index 2018: Building Resilience in the Face of Rising Food-Security Risks." Available at https:// foodsecurityindex.eiu.com/.

Edens, Christopher (1993) "Indus-Arabian Interaction during the Bronze Age: A Review of Evidcence," in G.L. Possehl, ed., *Harappan Civilization: A Recent Perspective*, 2nd rev. ed. New Delhi: Oxford University Press.

El Feki, S., B. Heilman, and G. Barker, eds., (2017) *Understanding Masculinities: Results from the International Men and Gender Equality Survey (Images) – Middle East and North Africa*. Cairo: UN Women and Promundo-US. Available at https://www.unwomen .org/en/digital-library/publications/2017/5/understanding-masculinities-results -from-the-images-in-the-middle-east-and-north-africa.

El-Khoury, Gabi (2017) "Socio-economic Developments in the Arab Countries: Selected Indicators." *Contemporary Arab Affairs* 10, 3: 471–480.

Ellenblum, Ronnie (2012) *The Collapse of the Eastern Mediterranean: Climate Change and the Decline of the East, 950–1072*. Cambridge: Cambridge University Press.

Elmi, Abdirashid A. (2017) "Risks to Critical Environmental Resource of Public Wellbeing from Climate Change in the Eyes of Public Opinion in Kuwait." *Environmental Progress & Sustainable Energy (online)*. Available at https://aiche .onlinelibrary.wiley.com/doi/abs/10.1002/ep.12662.

Erickson, Amanda (2017) "1 Million People Have Contracted Cholera in Yemen. You Should Be Outraged." *The Washington Post* (December 22).

Erinc, S. (1978) "Changes in the Physical Environment in Turkey Since the End of the Last Glacial," in W.C. Brice, ed., *The Environmental History of the Near and Middle East Since the Last Ice Age*. London: Academic Press.

European Political Strategy Centre (2017) "10 Trends Shaping Migration." Brussels: European Union. Available at https://ec.europa.eu/epsc/publications/other -publications/10-trends-shaping-migration_en.

Fader, Marienela, Dieter Gerten, Michael Krause, Wolfgang Lucht, and Wolfgang Cramer (2013) "Spatial Decoupling of Agricultural Production and Consumption: Quantifying Dependencies of Countries on Food Imports Due to Domestic Land and Water Constraints." *Environmental Research Letters* 8, 1: 014046.

Fagan, Brian (2004) *The Long Summer: How Climate Changed Civilization*. New York: Basic Books.

Fahim, M.A., M.K. Hassanein, A.A. Khalil, and A.F. AbouHadad (2013) "Climate Change Adaptation Needs for Food Security in Egypt." *Nature and Science* 11, 12: 68–74.

Fairbridge, R., O. Erol, M. Karaca, and Y. Yilmaz (1997) "Background to Mid-Holocene Climate Change in Anatolia and Adjacent Regimes," in H.N. Dalfes, G. Kukla, and H. Weiss, eds., *Third Millennium BC Climate Change and Old World Collapse*. Berlin: Springer.

Fairservis, W.A., Jr. (1992) "The Development of Civilization in Egypt and South Asia: A Hoffman-Fairservis Dialectic," in R. Friedman and B. Adams, eds., *The Followers of Horus: Studies Dedicated to Michael Allen Hoffman, 1944–1990*. Oxford: Oxbow Books.

Faris, Stephan (2007) "The Real Roots of Darfur." *Atlantic* (April). Available at https://www.theatlantic.com/magazine/archive/2007/04/the-real-roots-of-darfur/305701/. .

Femia, Francesco and Caitlin Werrell (2012) "Syria: Climate Change, Drought and Social Unrest." *The Center For Climate and Security*. Available at https://climateandsecurity.org/2012/02/29/syria-climate-change-drought-and-social-unrest/.

Finkelstein, Israel (1998) "Philistine Chronology: High, Middle or Low?" in S. Gitlin, A. Mazar and E. Stern, eds., *Mediterranean Peoples in Transition: Thirteenth to Early Tenth Centuries BCE*. Jerusalem: Israel Exploration Society.

Finkelstein, Israel (2000) "The Philistine Settlements: When, Where and How Many?" in E.D. Oren, ed., *The Sea Peoples and Their World: A Reassessment*. Philadelphia, PA: University of Pennsylvania Press.

Finkelstein, Israel (2007) "Is the Philistine Paradigm Still Viable?" in M. Bietak and E. Czerny, eds., *The Synchronization of Civilization in the Eastern Mediterranean in the Second Millennium BCE*, Proceedings of the SCIEM 2000- 2nd Euro Conference, Vienna: Verlagden Osterreichischen Academie der Wissenchaften.

Finne, M., J. Woodbridge, I. Labuhn, and C.N. Roberts (2019) "Holocene Hydro-Climatic Variability in the Mediterranean: A Synthetic Multi-Proxy Reconstruction." *The Holocene* 29, 5: 847–863.

Fischer, Hubertus, Katrin J. Meissner, Alan C. Mix, Nerilie J. Abram, Jacqueline Austermann, Victor Brovkin, Emilie Capron, Daniele Colombaroli, Anne-Laure Daniau, Kelsey A. Dyez, Thomas Felis, Sarah A. Finkelstein, Samuel L. Jaccard, Erin L. McClymont, Alessio Rovere, Johannes Sutter, Eric W. Wolff, Stephane Affolter, Pepijn Bakker, Juan Antonio Ballesteros-Canovas, Carlo Barbante, Thibaut Caley, Anders E. Carlson, Olga Churakova, et al. (2018) "Paleoclimate Constraints on the Impact of 2 Degrees C Anthropogenic Warming and Beyond." *Nature Geoscience* 11: 474–485.

Fleitmann, Dominik, Manfred J. Udelsee, Stephen J. Burns, Raymond S. Brindly, Jan Kramens, and Albert Matter (2008) "Evidence for a Widespread Climate Anomaly at around 9.2 ka Before Present." *Paleooceanography and Paleoclimatology* 23, 1: 1–6.

Flint, Julie and Alex de Waal (2008) *Darfur: A New History of a Long War*. London: Zed Books.

Flohr, Pascal, Dominik Fleitman, Roger Matthews, Wendy Matthews, and Stuart Black (2016) "Evidence of Resilience to past Climate Change in Southwest Asia: Early Farming Communities and the 9.2 and 8.2 ka Events." *Quaternary Science Reviews* 136, 15: 23–39.

Food and Agriculture Organization of the United Nations (2017) "The State of Food Security and Nutrition in the World." Available at http://www.fao.org/3/a-i7695e.pdf.

Food and Agriculture Organization of the United Nations (2019) "AQUASTAT Core Database." Available at http://www.fao.org/aquastat/en/.

Foresight (2011) "Migration and Global Environmental Change: Final Project Report." London: The Government Office for Science. Available at https://www.gov.uk/government/uploads/system/uploads/attachment_data/file/287717/11-1116-migration-and-global-environmental-change.pdf.

Foster, Benjamin R. (2016) *The Age of Agade: Inventing Empire in Ancient Mesopotamia*. New York: Routledge.

Fountain, Henry and Ben C. Solomon (2019) "Glaciers Are Retreating. Millions Rely on Their Water." *The New York Times*, January 16. Available at https://www.nytimes.com/interactive/2019/04/17/climate/melting-glaciers-globally.html.

Frank, Andre Gunder (1993) "Bronze Age World System Cycles." *Current Anthropology* 34: 383–405.

Freeman, Jack (2009) "The al Houthi Insurgency in the North of Yemen: An Analysis of the Shabab al Mouminean." *Studies in Conflict and Terrorism* 32, 11: 1008–1019.

Freyberg, Tom (2015) "Large Scale Solar Desalination Race Continues in the Middle East." *Water World*. Available at http://www.waterworld.com/articles/2015/01/large -scale-solar-desalination-race-continues-in-the-middle-east.html.

Fröhlich, Christiane (2016) "Climate Migrants as Protestors? Dispelling Misconceptions about Global Environmental Change in Pre-Revolutionary Syria." *Contemporary Levant* 1, 1: 38–50.

Frumkin, Amos, Israel Carmi, Israel Zak, and Mordeckai Margaritz (1994) "Middle Holocene Environmental Change Determined from Salt Caves of Mount Sedom, Israel," in O. Bar-Yosef and R. Kra, eds., *Late Quaternary Chronology and Paleoclimates of the Eastern Mediterranean*. Tucson, AZ: University of Arizona Press.

Fuller, Dorian Q., George Willcox, and Robin G. Allaby (2011) "Cultivation and Domestication Had Multiple Origins: Arguments against the Core Area Hypothesis for the Origins of Agriculture in the Near East." *World Archaeology* 43, 4: 628–652.

Fund for Peace (2019) *Fragile States Index: Annual Report 2019*. Washington, DC: Fund for Peace Organization. Available at Fundforpeace.org/wp-content/uploads/2019/04 /9511904-fragilestatesindex.pdf.

Gartzke, Erik (2012) "Could Climate Change Precipitate Peace?" *Journal of Peace Research* 49, 1: 177–192.

Gasche, H., J.A. Armstrong, S.W. Cole, and V.G. Gurzadyan (1998) *Dating the Fall of Babylon: A Reappraisal of Second-Millennium Chronology*. Ghent, Belgium/Chicago: University of Ghent and the Oriental Institute of the University of Chicago.

Gasse, F. (2000) "Hydrology and Changes in the African Tropics Since the Last Glacial Maxim." *Quaternary Science Reviews* 19: 189–212.

Gasse, F. and E. van Campo (1994) "Abrupt Post-glacial Events in West Asia and North Africa Monsoon Domains." *Earth and Planetary Science Letters* 126: 435–456.

Gebrechorkos, Solomon H., Christian Bernhofer, and Stephen Hulsmann (2019) "Impacts of Projected Change in Climate on Water Balance in Basins of East Africa." *Science of the Total Environment* 682: 160–170.

Gemenne, F., J. Barnett, W.N. Adger, and G.D. Dabelko (2014) "Climate and Security: Evidence, Emerging Risks, and a New Agenda." *Climatic Change* 123, 1: 1–9.

Genz, Hermann (2017) "The Transition from the Third to the Second Millennium B.C. in the Coastal Plain of Lebanon: Continuity or Break?," in Felix Hoflmayer, ed., *The Late Third Millennium in the Ancient Near East: Chronology, C14, and Climate Change*. Chicago, IL: Oriental Institute, University of Chicago.

Gerasimenko, N.P. (1997) "Environmental and Climatic Changes between 3 and 5 Ka BP in Southeastern Ukraine," in H.N. Dalfes, G. Kukla and H. Weiss, eds., *Third Millennium BC Climate Change and Old World Collapse*. Berlin: Springer.

Ghafar, Adel Abdel and Firas Masri (2016) "The Persistence of Poverty in the Arab World." *Brookings Institution*, February 28. Available at https://www.brookings.edu/ opinions/the-persistence-of-poverty-in-the-arab-world/.

Giosan, Liviu, Peter D. Clift, Mark G. Macklin, Dorian Q. Fuller, Stefan Constantinescu, Julie A. Durcan, Thomas Stevens, Geoff A.T. Duller, Ali R. Tabrez, Kavita Gangal, Ronojoy Adhikari, Anwar Alizai, Florin Filip, Sam VanLaningham, and James P.M. Syvitski (2012) "Fluvial Landscapes of the Harappan Civilization." *Proceedings of the National Academy of Science (PNAS)*. Available at http://www.pnas.org/cgi/doi/10 .1073/pnas.1112743109.

Girvetz, Evan, Julian Ramirez-Villegas, Lieven Claessens, Christine Iamamna, Carlos Navarro-Racines, Andrea Noowak, Phil Thornton, and Todd S. Roesnstock (2019) "Future Climate Projections in Africa: Where Are We Headed?" in T.S. Rosenstoock, A. Nowak, and E. Girvetz, eds., *The Climate-Smart Agriculture Papers.* Berlin: Springer.

Gleditsch, Nils Petter (2012) "Whither the Weather? Climate Change and Conflict." *Journal of Peace Research* 49: 3–9.

Gleick, Peter H. (2014) "Water, Drought, Climate Change, and Conflict in Syria." *Weather, Climate, and Society* 6, 3: 331–340.

Global Commission on the Economy and Climate (2018) *Unlocking the Inclusive Growth Story of the 21st Century: Accelerating Climate Action in Urgent Times.* Washington, DC: World Resources Institute. Available at https://newclimateeconomy.report/2018/.

Goder-Goldberger, M., N. Gubenk, and E. Hovers (2016) "Diffusion with Modifications: Nubian Assemblages in the Central Negev Highlands of Israel and Their Implications for Middle Paleolithic Inter-regional Interactions." *Quaternary Interactions* 408: 121–139.

Gohari, Alireza, Saeid Eslamanian, Jahangir Abedi-Koupaei, Alreza Massah Bavani, Dingbao Kang, and Kaveh Modeni (2013) "Climate Change Impacts on Crop Production in Iran's Zayandeh-Rud River Basin." *Science of the Total Environment* 442: 405–419.

Goodfriend, G.A. (1991) "Holocene Trends in ^{18}O in Land Snail Shells from the Negev Desert and Their Implications for Changes in Rainfall Source Areas." *Quaternary Research* 35, 3: 417–426.

Goodfriend, G.A. (1999) "Terrestrial Stable Isotopes in Late Quaternary Paleoclimates in the Eastern Mediterranean." *Quaternary Science Reviews* 18, 4-5: 501–513.

Gordon, Joel Adam (2017) "Mitigation of Climate Risk and Adaptation to Climate Security in Israel and the Middle East: Policy Measures toward Geopolitical Cooperation and Regional Transformation." *AEJI New Series on Climate Change Strategy, National Security, and Regional Sustainability.* CCS/NS/RS/2017.

Gornell, Jemma, Richard Betts, Eleanor Burke, Robin Clark, Joanne Camp, Kate Willett, and Andrew Wiltshire (2010) "Implications of Climate Change for Agricultural Productivity in the Early Twenty-first Century." *Philosophical Transactions of the Royal Society of London series B.* Available at https://royalsocietypublishing.org/doi/full/10.1098/rstb.2010.0158.

Grant, Audra K., Nicholas E. Burger, and Quentin Wodon (2014) "Climate-Induced Migration in the MENA Regions: Results from the Qualitative Fieldwork," in Quentin Wodon, Andrea Liverani, George Joseph, and Natalie Bougnoux, eds., *Climate Change and Migration: Evidence from the Middle East.* Washington, DC: World Bank.

Green, Monica H. (2018) "Climate and Disease in Medieval Eurasia," in David Ludden, ed., *Oxford Research Encyclopedia of Asian History.* New York: Oxford University Press.

Greenberg, Joel (2014) "Israel No Longer Worried about Its Water Supply Thanks to Desalination Plants." *The Bulletin,* May 2. Available at https://www.bendbulletin.com/nation/israel-no-longer-worried-about-its-water-supply-thanks-to-desalination-plants/article_a918ad90-895f-5b4b-bc73-0a84c30fb0dd.html.

Groucutt, H.S. and J. Blinkhorn (2013) "The Middle Paleolithic in the Desert and Its Implications for Understanding Hominin Adaptation and Dispersal." *Quaternary International* 300: 1–12.

The Guardian (2020) "Russia Announces Plan to 'Use the Advantages' of Climate Change." *The Guardian,* January 5. Available at http://Theguardian.com/world/2020/jan/05/Russia-announces-plan-to-use-the-advantages-of-climate-change.

Gunderson, Lance H., Craig R. Allen, and C.S. Holling, eds. (2010) *Foundations of Ecological Resilience*. Washington, DC: Island Press.

Gunderson, Lance H. and C.S. Holling, eds. (2002) *Panarchy: Understanding Transformation in Human and Natural Systems*. Washington, DC: Island Press.

Gunel, Gokce (2016) "The Infinity of Water: Climate Change Adaptation in the Arabian Peninsula." *Public Culture* 28, 2: 291–315.

Hadsall, Guy (2007) *Barbarian Migrations and the Roman West, 357–568*. Cambridge: Cambridge University Press.

Haldon, John (2013) "The Byzantine Successor State," in Peter Fibiger Bang and Walter Scheidel, eds., *The Oxford Handbook of the State in the Ancient Near East and Mediterranean*. Oxford: Oxford University Press.

Haldon, John and Arlene Rosen (2018) "Society and Environment in the East Mediterranean ca 300–1800 CE: Problems of Resilience, Adaptation and Transformation. Introductory Essay." *Human Ecology* 46, 3: 275–290.

Hallo, William W. and William K. Simpson (1971) *The Ancient Near East: A History*. New York: Harcourt, Brace Jovanovich.

Hallo, William W. and William K. Simpson (1998) *The Ancient Near East: A History*. New York: Harcourt, Brace Jovanovich.

Halverson, Nathan (2016) "Why World Leaders Are Terrified of Water Shortages: From Yemen to Syria to Arizona, Droughts Are a Growing Threat." *Mother Jones*, April 14. Available at http://www.motherjones.com/environment/2016/04/water-scarcity-wikileaks.

Hamblin, William J. (2006) *Warfare in the Ancient Near East to 1600 BC: Holy Warriors at the Dawn of History*. London: Routledge.

Hammond, S.T., J.H. Brown, J.R. Burger, T.P. Flanagan, T.S. Fristoe, N. Mercado-Silva, J.C. Nekola, and J.G. Okie (2014) "Food Spoilage, Storage, and Transport: Implications for a Sustainable Future." *Biosci* 65, 8: 758–768.

Harper, Kyle (2017) *The Fate of Rome: Climate, Disease and the End of an Empire*. Princeton, NJ: Princeton University Press.

Hartmann, Betsy (2010) "Rethinking Climate Refugees and Climate Conflict: Rhetoric, Reality and the Politics of Policy Discourse." *Journal of International Development* 22, 2: 233–246.

Hassan, Fekri A. (1994) "Population Ecology and Civilization in Ancient Egypt," in C.L. Crumley, ed., *Historical Ecology: Cultural Knowledge and Changing Landscapes*. Santa Fe, NM: School of American Research Press.

Hassan, Fekri A. (1997) "Nile Floods and Political Disorder in Early Egypt," in H. Nuzhet Dalfes, George Kukla, and Harvey Weiss, eds., *Third Millennium BC Climate Change and Old World Collapse*. Berlin: Springer.

Hassan, Fekri A. (2007) "Droughts, Famine, and the Collapse of the Old Kingdoms: Re-reading Power," in Zahi Hawass and Janet Richards, eds., *The Art and Archaeology of Ancient Egypt: Essays in Honour of David B. O'Connor*, Vol. 1. Cairo: Conseil Supremes des Antiquities de l'Egypte.

Hassan, Fekri A. (2011) "Nile Flood Discharge during the Medieval Climate Anomaly." *PAGES* 19, 1: 30–31.

Heather, Peter (2005) *The Fall of the Roman Empire: A New History*. London: Macmillan.

Henry, D.O. (2013) "The Natufian and the Younger Dryas," in Ofer Bar-Yosef and Francois R. Valla, eds., *Natufian Foragers in the Levant: Terminal Pleistocene Social Changes in Western Asia*. Cambridge, MA: International Monographs in Prehistory.

Hoffman, M.A. (1979) *Egypt before the Pharaohs: The Prehistoric Foundations of Egyptian Civilization*. New York: Alfred A. Knopf.

Hoflmayer, Felix (2017) "The Late Third Millennium B.C. in the Ancient Near East and Eastern Mediterranean: A Time of Collapse and Transformation," in Felix Hoflmayer, ed., *The Late Third Millennium in the Ancient Near East: Chronology, C14, and Climate Change.* Chicago, IL: Oriental Institute, University of Chicago.

Hokayem, Emile and David B. Roberts (2016) "The War in Yemen." *Survival* 58, 6: 157–186.

Hole, Frank (1994) "Environmental Instabilities and Urban Origins," in G. Stein and M.S. Rothman, eds., *Early States in the Near East: The Organizational Dynamics of Complexity.* Madison, WI: Prehistory Press.

Holthuijzen, Wieteke and Jacqueline Maxmilian (2011) "Dry, Hot, and Brutal; Climate Change and Desertification in the Sahel of Mali." *Journal of Sustainable Development in Africa* 13, 7245–7268.

Hsiang, Solomon M. and Marshall Burke (2014) "Climate, Conflict, and Social Stability: What Does the Evidence Say?" *Climatic Change* 123, 1: 39–55.

Hsiang, Solomon M., Marshall Burke, and Edward Miguel (2013) "Quantifying the Influence of Climate on Human Conflict." *Science* 341: 1–14.

Hsiang, Solomon M. and Kyle C. Meng (2014) "Reconciling Disagreement Over Climate-Conflict Results in Africa." *Proceedings of the National Academy of Sciences of the United States of America (PNAS)* 111, 6: 2100–2103.

Hsiang, Solomon M., Kyle C. Meng, and Mark Cane (2011) "Civil Conflicts Are Associated with the Global Climate." *Nature* 476: 438–41.

Ide, Tobias (2018) "Climate War in the Middle East? Drought, the Syrian Civil War and the State of Climate-Conflict Research. *Current Climate Change Reports* 4, 4: 347–354.

IEP (2020) *Global Peace Index 2020: Measuring Peace in a Complex World.* June 2020, Sydney, NSW: Institute for Economics & Peace. Available at Visionofhumanity.org/app/uploads/2020/06/GPI_2020_web.pdf.

ILC, CIRAD, CDE, GIGA, and GIZ (2014) "The Land Matrix Global Observatory." Available at http://landmatrix.org/en/.

Ilin-Tomich, Alexander (2016) "Second Intermediate Period," in Wolfram Grajetzki and Willeke Wendrich, eds., *UCLA Encyclopedia of Egyptology.* Los Angeles: Department of Near Eastern Languages, UCLA.

Im, Eun-Soon, Jeremy S. Pal, and Elfatih Eltahir (2017) "Deadly Heat Waves Projected in the Densely Populated Agricultural Regions of South Asia." *Science Advances* 3, 8: 1–7.

INRA (2015) "Addressing Agricultural Import Dependence in the Middle East-North African Region Through the Year 2050." Short Summary of the Study- October 2015. Institut National de la Recherche Agronomique. Available at: Inrae.fr/sites/default/files/pdf/addressing-agricultural-import-dependence-in-the-middle-east-north-africa-region-through-to-the-year-2050-doc.pdf.

International Monetary Fund (IMF) (2019) "World Economic Outlook." Available at https://www.imf.org/en/Publications/WEO/Issues/2019/10/01/world-economic-outlook-october-2019.

IPCC 2014 (2014) "Climate Change 2014: Synthesis Report Contribution of Working Groups I, II and III to the Fifth Assessment Report of the Intergovernmental Panel on Climate Change," in Core Writing Teams, R.K. Pachauri, and L.A. Megari, eds. Geneva, Switzerland: IPCC. Ippcc.ch/Sv15/.

IPCC 2018 (2018) *Global Warming of 1.5 Degrees C. An IPCC Special Report on the Impacts of Global Warming of 1.5 Degrees C above Pre-industrial Levels and Related Global Greenhouse Gas Emission Pathways, in the Context of Strengthening the Global Response to the Threat of Climate. V. Sustainable Development, and Efforts to Eradicate Poverty.* V. Masson-Delmotte, P. Zhai, H.O. Portner, D. Roberts, J. Skea, P.R. Shukla, A. Pirani, W. Moufouna-Okla, L. Pean, Pidcock, S. Connors, J.B.R. Matthews, Y. Chen, X. Shou, M.I. Gomos, E. Lonnoy, T. Maycock, M. Tignor, and T. Waterfield, eds., Geneva, Switzerland: IPCC. Available at Ipcc.ch/site/assets/uploads/sites/2/2019/06 /SR15_Full_Report_High, Res.pdf.

Issar, Arie S. and Mattanyah Zohar (2007) *Climate Change: Environment and History of the Near East,* 2nd ed. Berlin: Springer.

Izdebski, A., J. Mordechai, and S. White. (2018) "The Social Burden of Resilience: A Historical Perspective." *Human Ecology* 46, 3: 291–303.

Jacobson, Rowan (2016) "Israel Proves the Desalination Era Is Here." *Scientific American.* Available at https://www.scientificamerican.com/article/israel-proves-the -desalination-era-is-here/#.

Jacobson, T. and R.M. Adams (1958) "Salt and Silt in Ancient Mesopotamian Agriculture." *Science* 128: 1251–1258.

Jaouadi, S., V. Lebreton, V. Bout-Remazeilles, G. Slani, R. Lakhdar, R. Boussoffa, L. Dexileau, N. Kallel, B. Mannai-Tayech, and N. Combrourieu-Nebout (2016) "Environmental Changes, Climate and Anthropegenic Impact in South-East Tunisia during the Last 8 Kyr." *Climate of the Past* 12: 1339–1359.

Jobbins, G. and G. Henley (2015) *Food in an Uncertain Future: The Impacts of Climate Change on Food Security and Nutrition in the Middle East and North Africa.* London: Overseas Development Institute and Rome: World Food Programme.

Joffe, A.H. (2000) "Egypt and Syro-Mesopotamia in the 4th Millennium: Implications of the New Chronology." *Current Anthropology* 41: 113–23.

Johnson, Kyle (2013) "Surpassing Xerxes: The Advent of Ottoman Gunpowder Technology in the Fifteenth and Sixteenth Centuries," in Nicole Goetz, ed., *The Age of Gunpowder: An Era of Technological, Tactical, Strategic and Leadership Innovations.* Atlanta, GA: Department of History, Emory University.

Johnson, Thomas C. and Eric O. Odada, eds. (1996) *The Limnology, Climatology and Paleoclimatology of the East African Lakes.* New York: Gordon and Breach.

Jones, Clive (2011) "The Tribes that Bind: Yemen and the Paradox of Political Violence." *Studies in Conflict and Terrorism* 34, 12: 902–916.

Jones, M., N. Abu-Jaber, A. AlShdaifat, D. Baird, B.I. Cook, M.O. Cuthbert, J.R. Dean, M. Djamali, W. Eastwood, D. Fleitmann, A. Haywood, O. Kwiecien, J. Larsen, L.A. Maher, S.E. Metcalfe, A. Parker, C.A. Petrie, N. Primmer, T. Richter, N. Roberts, J. Roe, J.C. Tindall, E. Ünal-İmer, L. Weeks (2019) "20,000 Years of Societal Vulnerability and Adaptation to Climate Change in Southwest Asia." *WIREs Water.* e1330

Kaniewski, D., N. Marriner, R. Cheddadi, C. Morhange, J. Bretschneider, G. Jans, T. Otto, F. Luce, and E. Van Campo (2019) "Cold and Dry Outbreaks in the Eastern Mediterranean 3200 Years Ago." *Geology* 47, 10: 933–937.

Kaniewski, D., E. Paulissen, E. Van Campo, M. Al-Maqdissi, J. Bretschneider, and K. Van Leberghe (2008) "Middle East Coastal Ecosystem Response to Middle-to-Late Holocene Abrupt Climate Changes." *Proceedings of the National Academy of Sciences of the United States of America (PNAS)* 105, 37: 13941–13946.

Kaniewski, D., E. Paulissen, E. Van Campo, H. Weiss, T. Otto, J. Bretschneider, and K. Van Lerbergh (2010) "Late Second-Early First Millennium BC Abrupt Climate Changes in Coastal Syria and Their Possible Significance for the History of the Eastern Mediterranean." *Quaternary Research* 74: 207–215.

Kaniewski, D., E. Van Campo, J. Guiot, S. Le Burel, T. Otto, and C. Baetman (2013) "Environmental Roots of the Late Bronze Age Crisis." *PLoS ONE* 8, 8: 1–10.

Kaniewski, D., E. Van Campo, E. Paulissen, H. Weiss, J. Bakker, I. Rossignol, and K. van Lerberghe (2011) "The Medieval Climate Anomaly and the Little Ice Age in Coastal Syria Inferred from Pollen derived Paleoclimate Patterns." *Global and Planetary Change* 78: 178–187.

Kantor, H. (1992) "The Relative Chronology of Egypt and its Foreign Correlations Before the First Intermediate Period," in G. Stein and M.S. Rothman, eds., *Chronologies in Old World Archaeology*, Vol. 1, 3rd ed. Chicago, IL: University of Chicago Press.

Karimi, Vahid, Ezatollah Karami, and Marzieh Keshawarz (2018) "Climate Change and Agriculture: Impacts and Adaptive Responses in Iran." *Journal of Integrative Agriculture* 17, 1: 1–15.

Kathayat, Gayatri, Hai Cheng, Ashish Sinha, Liang Yum Xianglei Li, Haiwei Zhang, Hangying Li, Youfeng Ning, and R. Lawrence Edwards (2017) "The Indian Monsoon Variability and Civilization Changes in the Indian Subcontinent." *Science Advances* 3, 12. Available at https://advances.sciencemag.org/content/3/12/e1701296.

Kay, P.A. and D.L. Johnson (1981) "Estimated Tigris-Euphrates Stream Flow from Regional Paleoenvironmental Proxy Data." *Climatic Change* 3: 251–63.

Kelley, Colin, Shahrzad Mohtadi, Mark Cane, Richard Seager, Yochanan Kushnir (2015) "Climate Change in the Fertile Crescent and Implications of the Recent Syrian Drought." *PNAS* 112(11): 341–46. Available at http://www.pnas.org/content/112/11/3241.full.

Kelly-Buccelati, M. (1990) "Trade in Metals in the Third Millennium: Northeastern Syria and Eastern Anatolia," in P. Matthiae, M. Van Loon, and H. Weiss, eds., *Resurrecting the Past: A Joint Tribute to Adnan Bounni*. Istanbul: Nederlands Historisch Arahcaelogisch Instituut.

Kerr, R.A. (1998) "Sea-Floor Dust Shows Drought Felled Akkadian Empire." *Science* 279: 325–26.

Kerry, John (2015) "Climate Change Intensifies Conflicts." *Al Jazira*. Available at http://www.aljazeera.com/news/2015/11/kerry-political-crises-linked-climate-change-151110190932931.html.

Kevane, Michael and Leslie Gray (2008) "Darfur: Rainfall and Conflict." *Environmental Research Letters* 3, 3: 1–10.

Keys, David (1999) *Catastrophe: An Investigation into the Origins of the Modern World*. New York: Ballantine.

Ki Moon, Ban (2007) "A Climate Culprit in Darfur." *The Washington Post*, 16 June. Available at http://www.washingtonpost.com/wp-dyn/content/article/2007/06/15/ AR2007061501857.html?noredirect=on.

King, Andrew D. and Luke J. Harrington (2018) "The Inequality of Climate Change From 1.5 to 2 Degrees C of Global Warming." *Geophysical Research Letters* 45: 5030–5033.

Kirch, P.V. (1984) *The Evolution of the Polynesian Chiefdoms*. Cambridge: Cambridge University Press.

Kitoh, Akiu, Akiyo Yatagai, and Pinhas Alpert (2008) "First Super-High-Resolution Model Projection that the Ancient 'Fertile Crescent' Will Disappear in this Century." *Hydrological Research Letters*. Available at https://www.jstage.jst.go.jp/article/hrl/2/0 /2_0_1/_article.

Klengl, H. (1992) *Syria, 3000 to 300 BC: A Handbook of Political History*. Berlin: Akademie Verlag.

Knapp, A. Bernard and John F. Cherry (1994) *Provenance Studies and Bronze Age Cyprus: Production, Exchange and Politico-Economic Change*. Madison, WI: Prehistory Press.

Knapp, A. Bernard and Sturt W. Manning (2016) "Crisis in Context: The End of the Late Bronze Age in the Eastern Mediterranean." *American Journal of Archaeology* 120, 1: 99–149.

Kohl, P. (1987) "The Ancient Economy, Transferable Technology and the Bronze Age World-System: A View from the Northeastern Frontier of the Ancient Near East," in M. Rowlands, M. Larsen and K. Kristainsen, eds., *Centre and Periphery in the Ancient World*. Cambridge: Cambridge University Press.

Korotayev, Andrey, Vladimir Klimenko, and Dmitry Proussakov (1999) "Origins of Islam: Political-Anthropological and Environmental Context." *Acta Orientalia Academiae Scientiarum Hungaricae* 52, 3–4: 243–276.

Koubi, Vally, Thomas Bernauer, Anna Kalbhenn, and Gabriele Spilker (2012) "Climate Variability, Economic Growth, and Civil Conflict." *Journal of Peace Research* 49: 113–127.

Krementski, C.V. (1997) "The Late Holocene Environmental and Climate Shift in Russia and Surrounding Lands," in H.N. Dalfes, G. Kukla and H. Weiss, eds., *Third Millennium BC Climate Change and Old World Collapse*. Berlin: Springer.

Kropelin, S. (2005) "The Geomorphological and Palaeoclimatic Framework of Prehistoric Occupation in the Wad Bakht Area," in J. Linstadter, ed., *Wad Bakht Africa Praehistorica*. Kohn, Germany: 18 Heinrick-Barth Institut.

Kubursi, Atif, ed. (2012) *Food and Water Security in the Arab World: Proceedings of the First Arab Development Symposium*. Washington, DC: The International Bank for Reconstruction and Development, World Bank, and the Arab Fund for Economic and Social Development.

Kuhrt, A. (1995) *The Ancient Near East, c. 3000–330 BC*, Vol. 1. London: Routledge.

Kuper, R. and S. Kropelin (2006) "Climate-Controlled Holocene Occupation in the Sahara: Motor of Africa's Evolution." *Science* 313: 803–807.

Kuran, Timur (2011) *The Long Divergence: How Islamic Law Held Back the Middle East*. Princeton, NJ: Princeton University Press.

Lagi, M., K.Z. Bertrand, and Y. Bar-yam (2011) *The Food Crises and Political Instability in North Africa and the Middle East*. Cambridge, MA: New England Complex Systems Institute.

Langgut, D., A. Almogi-Labin, M. Bar-Matthews, N. Pickarski, and M. Weinstein-Evron (2018) "Evidence for a Humid Interval at ~ 56–44 Ka in the Levant and Its Potential Link to Modern Humans Dispersal Out of Africa." *Journal of Human Evolution*. https://doi.org/10.1016/jhevol.2018.08.002.

Langgut, Dafna, Israel Finkelstein, and Thomas Litt (2013) "Climate and the Late Bronze Collapse: New Evidence for the Southern Levant." *Tel Aviv* 40: 149–175.

Lawler, A. (2008) "Indus Collapse: The End or Beginning of an Asian Culture?" *Science* 320: 1281–1283.

LeBlanc, S.A. (1999) *Prehistoric Warfare in the American Southwest*. Salt Lake City, UT: University of Utah Press.

Lee, Harry F., David D. Zhang, Peter Brecke, and Jie Fei (2013) "Positive Correlation between the North Atlantic Oscillation and Violent Conflicts in Europe." *Climate Research* 56, 1: 1–10.

Leemans, W.F. (1960) *Foreign Trade in the Old Babylonian Period*. Leiden: E.J. Brill.

Lemcke, Gerry and Michael Sturm (1997) "δ18o and Trace Element Measurements as Proxy for the Reconstruction of Climate Changes at Lake Van (Turkey): Preliminary Results," in H.N. Dalfes, G. Kukla, and H. Weiss, eds., *Third Millennium BC Climate Change and Old World Collapse*. Berlin: Springer.

Lespez, L., A. Gleis, J-A. Lopez-Saez, Y. Le Drezen, Z. Tsirtsoni, R. Davidson, L. Birce, and D. Malamidov. (2016) "Middle Holocene Rapid Environmental Changes and Human Adaptation." *Quaternary Research* 85, 2: 227–244.

Lieberman, Amy (2015) "Where Will the Climate Refugees Go?" *Al Jazeera*, December 22. Available at https://www.aljazeera.com/indepth/features/2015/11/climate-refugees-151125093146088.html.

Liu, C., N. Hofstra, and E. Franz (2013) "Impacts of Climate Change on the Microbial Safety of Pre-harvest Leafy Green Vegetables as Indicated by Escherichia coli 0157 and Salmonella Spp." *International Journal of Food Microbiology* 163, 2–3: 119–128.

Liverani, Mario (1987) "The Collapse of the Near Eastern Regional System of the End of the Bronze Age: The Case of Syria," in M. Rowlands, M. Larsen and K. Kristiansen, eds., *Centre and Periphery in the Ancient World*. Cambridge: Cambridge University Press.

Luming, Sebastian, Mariusz Galka, and Fritz Vahoenholt (2017) "Warming and Cooling: The Medieval Climate Anomaly in Africa and Arabia." *Paleoceanography* 32: 1219–1235.

Luo, T., R. Young, and P. Reig (2015) "Aqueduct Projected Water Stress Rankings." Technical note. Washington, DC: World Resources Institute. Available at http://www.wri.org/publication/aqueduct-projected-water-stress-country-rankings.

Lupton, A. (1996) *Stability and Change: Socio-political Development in North Mesopotamia and Southeast Anatolia, 4000–2700 BC*. Oxford: British Archaeological Reports.

Madella, M. and D.Q. Fuller (2006) "Paleoecology and the Harappan Civilization of the South: A Reconsideration." *Quaternary Science Reviews* 25: 1283–1301.

Maher, L.A., E.B. Banning, and M. Chazan (2011) "Oasis or Mirage? Assessing the Role of Abrupt Climate Change in the Prehistory of the Southern Levant." *Cambridge Archaeological Journal* 21, 1: 1–29.

Maietta, Eildh Kennedy, Francois Bourse, Catherine Saumet, Tyler Rundel, Leonie Le Borgne, Jade Legrand, Marie-Jeanne Berger, Sterling Carter, and Samuel Carcanague (2017) *The Future of Aid: INGOs in 2030*. Paris: Inter-Agency Regional Analysts Network (IARAN). Available at http://futureofaid.iaran.org/The_Future_Of_Aid_INGOs_In_2030-36.pdf.

Makarewicz, C.A. (2012) "The Younger Dryas and Hunter-Gatherer Transitions to Food Production in the Near East," in Metin L. Eren, ed., *Hunter-Gatherer Behavior: Human Response during the Younger Dryas*. Walnut Creek, CA: Left Coast Press.

Mani, Muthukumana, Sushenjit Bandyopadyay, Shun Chonabayshi, Anil Markandya, and Thomas Mosier (2018) *South Asia Hotspots: Impacts of Temperature and Precipitants of Changes on Living Standards*. Washington, DC: World Bank.

Manning, Joseph G. (2013) "Egypt," in Peter Fibiger Ban and Walter Scheidel, eds., *The Oxford Handbook of the State in the Ancient Near East and Mediterranean*. Oxford: Oxford University Press.

Manning, Joseph G., Francis Ludlow, Alexander R. Stine, William R. Boos, Michael Sigl, and Jennifer R. Marlon (2017) "Volcanic Suppression of Nile Summer Flooding Triggers Revolt and Constrains Interstate Conflict in Ancient Egypt." *Nature Communications* 8, 900. https://doi.org/0.1038/541467-017-00957-y.

Manning, K. and A. Timpson (2014) "The Demographic Response to Holocene Climate Change in the Sahara." *Quaternary Science Reviews* 101: 28–35.

Marcinick, Arkadiusz and Lech Czerniak (2010) "Catalhoyuk East in the second half of the 7th Millennium." Paper presented at the *8.2Ka Climate Event and Archaeology in the Ancient Near East*, Leiden University, March.

Marcus, J. (1998) "The Peaks and Valleys of Ancient States: An Extension of the Dynamic Model," in G.M. Feinman and J. Marcus, eds., *Archaic States*. Santa Fe, NM: School of American Research Press.

Marfoe, Leon (1987) "Cedar Forest to Silver Mountain: Social Change and the Development of Long-Distance Trade in Early Near Eastern Societies," in Michael Rowlands, Mogens Larsen, and Kristian Kristiansen, eds., *Centre and Periphery in the Ancient World*. Cambridge: Cambridge University Press.

Mark, Samuel (1998) *From Egypt to Mesopotamia: A Study of Predynastic Trade Routes*. College Station: Texas A&M University Press.

Marks, A.E. and J.L. Rosa (2012) "Through a Prism of Paradigms: A Century of Research into the Origins of the Upper Paleolithic in the Levant," in Marcel Otte, ed., *Modes de Contacts e de Placements au Paleoloithique Eurasiatique*. Belgium: University of Liege.

Maron, Nimrod, Meirau Meiri, Yotan Tepper, Tali Erikson-Gini, Hagar Reshef, Leor Weisbrod, and Guy Bar-Oz (2019) "Zooarchaeology of the Social and Economic Upheaval in the Lat Antique-Early Islamic Sequence of the Negev Desert." *Scientific Reports* 9: 6702.

Marques, A., M.L. Nunes, S.K. Moore, and M.S. Strom (2010) "Climate Change and Seafood Safety: Human Health Implications." *Food Research International* 43: 1766–1779.

Matthews, R. (2003) *The Archaeology of Mesopotamia: Theories and Approaches*. London: Routledge.

Mayewski, Paul Andrew and Frank White (2002) *The Ice Chronicles: The Quest to Understand Global Climate Change*. Hanover, CT: University Press of New England.

Mazo, Jeffrey (2009) "Chapter Three: Darfur: The First Modern Climate-Change Conflict." *The Adelphi Papers* 49, 409: 73–86.

McCarl, Bruce A., Mark Musumba, Joel B. Smith, Paul Kirshan, Russel Jones, Akram El-Gamzori, Mohammed Bayoumi, and Riina Hynninen (2015) "Climatic Change Vulnerability and Adaptation Strategies in Egypt's Agricultural Sector." *Mitigation and Adaptation Strategies for Global Change* 20, 7: 1097–1105.

McEvedy, Colin and Richard Jones (1978) *Atlas of World Population History*. New York: Facts on File.

McMichael, P. (2009) "A Food Regime Genealogy." *Journal of Peasant Studies* 36, 1: 139–169.

McNeill, J.R. (2013) "The Eccentricity of the Middle East and North Africa's Environmental History," in Alan Mikhail, ed., *Water on Sand: Environmental Histories of the Middle East and North Africa*. Oxford: Oxford University Press.

McSweeney, Robert (2015) "Updated: The Climate Change Papers Most Featured in the Media." *CarbonBrief*, 29 July. Available at https://www.carbonbrief.org/updated-the-climate-change-papers-most-featured-in-the-media (accessed September 14, 2018).

Mekonnen, Mesfin M. and Hoekstra, Arjen Y. (2016) "Four Billion People Facing Severe Water Scarcity." *Science Advances* 2, 2, e1500323.

Merrillees, R. (1998) "Egypt and the Aegean," in E.H. Cline and D. Harris-Cline, eds., *The Aegean and the Orient in the Second Millennium*. Liege: Annales d'archeologie Egeenne de l'Universite de Liege et UT-PASP.

Miguel, Edward, Shanker Satyanath, and Ernest Sergenti (2004) "Economic Shocks and Civil Conflict: An Instrumental Variables Approach." *Journal of Political Economy* 112, 4: 725–753.

Mikhail, Alan (2016) "Climate and the Chronology of Iranian History." *Iranian Studies* 49: 963–972.

Missirian, Anouch and Wolfram Schlenker (2017) "Asylum Applicants Respond to Temperature Fluctuations." *Science* 358, 6370: 1610–1614.

Mithen, Steven (2003) *After the Ice: A Global History, 20,000–5000 BC*. Cambridge, MA: Harvard University Press.

Modelski, George (2003) *World Cities, −3000 to 2000*. Washington, DC: Faros 2000.

Moeller, N. (2005) "The First Intermediate Period: A Time of Famine and Climate Change?" *Agypten und Levante* 15: 153–167.

Monroe, Christopher M. (2009) *Scales of Fate: Trade, Tradition and Transformation in the Eastern Mediterranean, c. 1350–1175 BCE*. Munster: Ugarit-Verlag.

Moorey, P.R.S. (1987) "On Tracking Cultural Transfers in Prehistory: The Case of Egypt and Lower Mesopotamia in the Fourth Millennium BC," in M. Rowlands, M. Larsen, and K. Kristiansen, eds., *Centre and Periphery in the Ancient World*. Cambridge: Cambridge University Press.

Moorey, P.R.S. (1994) *Ancient Mesopotamian Materials and Industries: The Archaeological Evidence*. Oxford: Clarendon Press.

Mordechai, L. (2018) "Short-Term Cataclysmic Events in Premodern Complex Societies." *Human Ecology* 46, 3: 323–333.

Muhly, J. (1969) *Copper and Tin: The Distribution of Mineral Resources and the Nature of the Metals Trade in the Bronze Age*. PhD dissertation: Yale University.

Muhly, J. (1992) "The Crisis Years in the Mediterranean World: Transition or Cultural Disintegration?," in W.A. Ward and M.S. Joukowsky, eds., *The Crisis Years: The 12th Century BC: From Beyond the Danube to the Tigris*. Dubuque, IO: Kendall/Hunt.

Müller-Wollermann, Renate (2014) "End of the Old Kingdom," in Wolfram Grajeteki and Willeke Wendrich, eds., *UCLA Encyclopedia of Egyptology*. Los Angles, CA: UCLA Department of Near Eastern Languages. Available at https://escholarship.org/uc/item/2ns3652b.

Murphy, Denis J. (2007) *Peoples, Plants and Genes: The Story of Crops and Humanity*. Oxford: Oxford University Press.

Nelson, Gerald C., Mark W. Rosegrant, Jawoo Koo, Richard Robertson, Timothy Sulser, Tingju Zhu, Claudia Ringler, Siwa Msangi, Amanda Palazzo, Miroslav Batka, Marilia Magalhaes, Rowena Vlamonte-Santos, Mandy Ewing, and David Lee (2009) *Climate Change Impact on Agriculture and Costs of Adaptation*. Washington, DC: International Food Policy Research Institute.

Nesbitt, Mark (1999) "When and Where Did Domesticated Cereals First Occur in Southwest Asia?" in R.T.J. Cappers and S. Bottema, eds., *The Dawn of Farming in the Near East*. Berlin: Springer.

Neumann, J. and R.M. Sigrist (1978) "Harvest Dates in Ancient Mesopotamia as Possible Indicators of Climate Variations." *Climate Change* 1: 239–52.

Niang, Isabelle, Oliver C. Ruppel, Mohammed A. Adrabo, Ama Essel, Christopher Lennard, Jonathan Padgham, and Penny Urquhart (2014) "Africa," in V.R. Barros, C.B. Field, D.J. Dokken, M.D. Mastrandrea, K.J. Mach, T.E. Bilir, M. Chatterjee, K.L. Ebi, Y.O. Estrada, R.C. Genova, B. Firma, E.S. Kissel, A.N. Levy, S. MacCraken, P.R. Mastrandrea, and L.L. White, eds., *Report of the Intergovernmental Panel on Climate Change (IPCC)*. Cambridge: Cambridge University Press.

Nibbi, A. (1975) *The Sea Peoples and Egypt*. Park Ridge, NJ: Royes Press.

Nicoll, Kathleen (2004) "Records of Recent Environmental Change and Prehistoric Human Activity in Egypt and Northern Sudan." *Quaternary Science Reviews* 23: 561–580.

Nicoll, Kathleen (2012) "Geoarchaeological Perspectives on Holocene Climate Change as a Civilizing Factor in the Egyptian Sahara," in Liviue Giosan, Dorain Q. Fuller, Kathleen Nicoll, Rowan K. Flad, and Peter D. Clift, eds., *Climates, Landscapes and Civilizations*. Washington, DC: American Geophysical Union.

Nieuwenhuyse, Olivier and Peter Akkermans (2010) "Cultural Transformations in Upper Mesopotamia: The Case of Tell Sabi Ayad." Paper presented at *The 8.2Ka Climate Event and Archaeology in the Ancient Near East*, Leiden University, March.

Nissen, H.J. (1988) *The Early History of the Ancient Near East, 9000–2000 BC*, translated by E. Lutzeier with K.J. Northcott. Chicago, IL: University of Chicago Press.

Oates, D. and J. Oates (1977) "Early Irrigation Agriculture in Mesopotamia," in G. DeG. Siveking, I.H. Longworth, and K.E. Wilson, eds., *Problems in Economic and Social Archaeology*. London: Duckworth.

Oates, Joan (1993) "Trade and Power in the Fifth and Fourth Millennium B.C.: New Evidence from Northern Mesopotamia." *World Archaeology* 24, 3: 403–22.

Oates, T. (2014) "Mesopotamia: The Historical Periods," in C. Renfrew and P. Bahn, eds., *The Cambridge World Prehistory*. Cambridge: Cambridge University Press.

O'Connor, D. (1983) "New Kingdom and Third Intermediate period, 1552–664 BC," in B.G. Trigger, B.J. Kemp, D. O'Connor, and A.B. Lloyd, eds., *Ancient Egypt: A Social History*. Cambridge: Cambridge University Press.

Ohlsson, Leif (2000) "Water Conflicts and Social Resource Scarcity." *Physics and Chemistry of the Earth, Part B: Hydrology, Oceans and Atmosphere* 25, 3: 213–220.

Ollivier, Guillaume, Daniele Magda, Armelle Maze, Gael Pluecocq, and Claire Lamine (2018) "Agroecological Transitions: What Can Sustainability Transition Frameworks Teach Us? An Ontological and Empirical Analysis." *Ecology and Science* 23, 2: art 5.

Olson, Mancur (1982) *The Rise and Decline of Nations: Economic Growth, Stagflation, and Social Rigidities*. New Haven, CT: Yale University Press.

Oualkacha, L. Soutr, A. Agoumi, and A. Kettab (2017) "Climate Change Impacts in the Mahgreb Region: Status and Prospects of the Water Resources," in M. Ouessar, D. Gabriels, A. Tsunekawa, and S. Evett, eds., *Water and Land Security in Drylands*. Berlin: Springer.

Pal, Jeremy S. and Elfatih A.B. Eltahir (2016) "Future Temperature in Southwest Asia Projected to exceed a Threshold for Human Adaptability." *Nature Climate Change* 6: 197–200.

Palmisano, A., J. Woodbridge, N. Roberts, A. Bevan, R. Fyfe, S. Shennan, R. Cheddadi, R. Greenberg, D. Kaniewski, D. Langutt, S.A.G. Leroy, T. Litt, and A. Miebach (2019) "Holocene Landscape Dynamics and Long-term Population Trends in the Levant." *The Holocene* 29, 5: 708–727.

Parker, A. and A.S. Goudie (2008) "Geomorphological and Paleoenvironmental Investigations in the Southeastern Arabian Gulf Region and the Implication for the Archaeology of the Region." *Geomorphology* 1010: 458–470.

Parker, Geoffrey (1988) *The Military Revolution: Military Innovation and the Rise of the West, 1500–1800*. New York: Cambridge University Press.

Parker, Geoffrey (2013) *Global Crisis: War, Climate Change and Catastrophe in the Seventeenth Century*. New Haven, CT: Yale University Press.

Pearce, Fred (2009) "Fertile Crescent Will Disappear this Century." *New Scientist*, July 29. Available at https://www.newscientiest.com/article/mg20327194.200-fertile -crescent-will-disappear-this-century/.

Pearce, Fred (2014) "Mideast Water Wars in Iraq, a Battle for Control of Water." *Yale Environment 360*, August 25. Available at https://e360.yale.edu/features/ mideast_water_wars_in_iraq_a_battle_for_control_of_water#:~:text=Conflicts %20over%20water%20have%20long%20haunted%20the%20Middle%20East.&text =Behind%20the%20headline%20stories%20of,that%20sustain%20these%20desert %20nations.

Pearson, Nakia and Camille Niaufre (2013) "Desertification and Drought Related Migrations in the Sahel- The Cases of Mali and Burkina Faso." *The State of Environmental Migration* 3: 79–98.

Pei, Qing, Harry F. Lee, and David D. Zhang (2018) "Long-Term Association between Climate Change and Agriculturalists' Migration in Historical China." *The Holocene* 28, 2: 208–216.

Pei, Qing and David D. Zhang (2014) "Long-Term Relationship between Climate Change and Nomadic Migration in Historical China." *Ecology and Society* 19, 2: 67–83.

Pei, Qing, David D. Zhang, Harry F. Lee, and Guodong Li (2014) "Climate Change and Macro-economic Cycles in Pre-Industrial Europe." *PLoS ONE* 9, 2: e88155. Available at https://doi.org/10.1371/journal.pone.0088155.

Pei, Qing, David D. Zhang, Guodong Li, Bruce Winterhalder, and Harry F. Lee (2015) "Epidemics in Ming and Qing China: Impacts of Changes in Climate and Economic Well-Being." *Social Science & Medicine* 136–137: 73–80.

Perkins, Brian M. (2017) "Yemen: between Revolution and Regression." *Studies in Conflict and Terrorism* 40, 4: 300–317.

Peters, Joris, Klaus Schmidt, Oliver Dietrich, and Nadja Pollath (2014) "Gobekli Tepe: Agriculture and Domestication," in Claire Smith, ed., *Encyclopedia of Global Archaeology*. Berlin: Springer.

Petraglia, M.D., H.S. Groucutt, M. Guagnin, P.S. Breeze, and N. Boivin (2020) "Human Response to Climate Ecosystem Change in Ancient Arabia." *Proceedings of the National Academy of Sciences of the United States of America (PNAS)* 117, 15: 8263–8270.

Petraglia, M.D., A. Parton, H.S. Groucutt, and A. Alsharekh (2015) "Green Arabia: Human Prehistory at the Crossroads of Continents." *Quaternary International* 382: 1–7.

Petrie, Cameron A., N. Ravinda, Jennifer Bates Singh, Dixit Yama, A.I. Charly, David A. French, et al. (2017) "Adaptation to Variable Environments, Resilience to Climate Change: Investigating *Land, Water and Settlement* in Indus Northwest India." *Current Anthropology* 58, 1: 1–30.

Phadtare, Netajirao R. (2000) "Sharp Decrease in Summer Monsoon Strength 4000–3500 Cal Yr Bp in the Central Higher Himalaya of India Based on Pollen Evidence from Alpine Peat." *Quaternary Research* 51: 39–53.

Piesinger, C. (1983) *Legacy of Dilmun: The Roots of Ancient Maritime Trade in Eastern Coastal Arabia in the 4th/3rd Millennium BC*. PhD dissertation, Madison, WI: University of Wisconsin.

Pomeranz, Kenneth (2000) *The Great Divergence: China, Europe and the Making of the Modern Economy*. Princeton, NJ: Princeton University Press.

Porada, E., D. Hansen, S. Dunham, and S. Babcock (1992) "The Chronology of Mesopotamia, ca. 7000–1600 BC," in Robert W. Ehrlich, ed., *Chronologies in Old World Archaeology*, Vol. 1, 3rd ed. Chicago, IL: University of Chicago Press.

Porkka, Mina, Matti Kummu, Stefan Siebert, and Olli Varis (2013) "From Food Insufficiency towards Trade Dependency: A Historical Analysis of Global Food Availability." *PLoS ONE* 8, 12: e8214.

Porter, J.R., L. Xie, A.J. Challenor, K. Cohrane, M. Howden, M.M. Iqbal, D.B. Lobell, and M.I. Travasso (2014) "Food Security and Food Production Systems in Climate Change 2014: Impacts, Adaptation and Vulnerability." *Working Group II Contribution to IPCC fifth Assessment Report*. Geneva, Switzerland.

Possehl, Gregory L. (1997a) "The Transformation of the Indus Civilization." *Journal of World Prehistory* 11, 4: 425–472.

Possehl, Gregory L. (1997b) "Climate and the Eclipse of the Ancient Cities of the Indus," in H.N. Dalfes, G. Kukla, and H. Weiss, eds., *Third Millennium BC Climate Change and Old World Collapse*. Berlin: Springer.

Postgate, N. (1986) "The Transition from Uruk to Early Dynastic: Continuities and Discontinuities in the Record of Settlement," in U. Finkbeiner and W. Rollig, eds., *Gamdat Nasr: Period or Regional Style?* Weisbaden: Dr. Ludwig Reichert Verlag.

Potts, Malcom, Eliya Zulu, Michael Wehner, Federico Castillo, and Courtney Henderson (2013) *Crisis in the Sahel: Possible Solutions and the Consequences of Inaction.* Berkeley, CA: University of California. Available at https://nature.berkeley.edu/release/oasis_monograph_final.pdf.

Potts, T.F. (1990) *The Arabian Gulf in Antiquity: From Prehistory to the Fall of the Achaemenid Empire.* Oxford: Clarendon Press.

Potts, T.F. (1993) "Patterns of Trade in Third-Millennium BC Mesopotamia and Iran." *World Archaeology* 24, 3: 379–402.

Potts, T.F. (1994) *Mesopotamia and the East: An Archaeological and Historical Study of Foreign Relations, ca. 3400–2000 BC.* Oxford: Oxbow Books.

Potts, T.F. (1997) *Mesopotamian Civilization: The Material Foundations.* Ithaca, NY: Cornell University Press.

Powell, M. (1985) "Salt, Seed, and Yields in Sumerian Agriculture: A Critique of the Theory of Progressive Salinization." *Zeitschrift fur Assyrioloogie* 75: 7–38.

Preiser-Kapeller, Johannes (2015) "A Collapse of the Eastern Mediterranean? New Results and Theories in the Interplay between Climate and Societies in Byzantium and the Near East, ca. 1000–1200 AD." *Jahrbuch der Osterreichischen Byzantinistik* 65: 195–242.

Price, T. Douglas and Ofer Bar-Yosef (2011) "The Origins of Agriculture: New Data, New Ideas: An Introduction to Supplement 4." *Current Anthropology* 52, 54: S163–S174.

Pringle, Heather (1998) "The Slow Birth of Agriculture." *Science* 282, 5393: 1446.

Psima, Sarah, Victor S. Indasi, Modathir Zaroug, Husen Seid Endris, Masilin Gudoshva, Herbert O. Misiani, Alex Niuslima, Richard O. Anyah, George Otiemo, Bob A. Ogwang, Soman Jain, Alfred I. Kondowe, Emmah Mwangi, Chris Lennard, Grigory Nikulin, and Alessandro Dosio (2018) "Projected Climate Over the Greater Horn of Africa under 1.5 Degrees C and 2 Degrees C Global Warming." *Environmental Research Letters* 13: 1–10.

Pumpelly, Raphael (1908) "Explorations in Turkestan: Expedition of 1904," in Raphael Pumpelly, ed., *Civilizations of Anau: Origins, Growth and Influence of Environment.* Washington, DC: Carnegie Institution.

Rabil, Robert G. (2016) *The Syrian Refugee Crisis in Leabanon: The Double Tragedy of Refugees and Impacted Host Communities.* Lanham, MD: Lexington Books.

Rabinowitz, D. (2020) *The Power of Deserts: Climate Change, the Middle East, and the Promise of a Post-Oil Era.* Stanford, CA: Stanford University Press.

Rambeau, Claire and Stuart Black (2011) "Paleoenvironments of the Southern Levant 5,000 BP tyo Present: Linking he Geological and Archaeological Records," in Steven Mithen and Emily Black, eds., *Water, Life and Civilisation: Climate, Environments and Society in the Jordan Valley.* Cambridge: Cambridge University Press.

Raphael, Sarah Kate (2013) *Climate and Political Climate: Environmental Disasters in the Medieval Levant.* Leiden: Brill.

Rasmussen, Simon, Eske Willerslev, and Kristian Kristiansen (2015) "Early Divergent Strains of Yersina Pestis in Eurasia 5,000 Years Ago." *Cell* 163: 571–582.

Rattenagar, S. (1981) *Encounters: The Westerly Trade of the Harappa Civilization.* Delhi: Oxford University Press.

Redford, D. (1992) *Egypt, Canaan, and Israel in Ancient Times.* Princeton, NJ: Princeton University Press.

Reich, David (2018) *Who We Are and How We Got Here: Ancient DNA and the New Science of the Human Past*. New York: Pantheon.

Reihl, Simone (2017) "Regional Environments and Human Perception: The Two Most Important Variables in Adaptation to Climate Change," in Felix Hoflmayer, ed., *The Late Third Millennium in the Ancient Near East: Chronology, C14, and Climate Change*. Chicago, IL: Oriental Institute, University of Chicago.

Reliefweb (2017) "A Treasured Coastline: Addressing Climate Change Vulnerabilities and Risks in Vulnerable Coastal Areas of Tunisia." Available at https://reliefweb.int /report/tunisia/treasured-coastline-addressing-climate-change-vulnerabilities-and -risks-vulnerable.

Renssen, Hans (2010) "The Nature of the 8.2Ka event in the Middle East: Climate Model Results." Paper presented at the *8.2Ka Climate Event and Archaeology in the Ancient Near East*, Leiden University, March.

Rezai, Armon, Lance Taylor, and Duncan Foley (2017) "Economic Growth, Income Distribution, and Climate Change." *Working Paper Series no. 17*. Vienna: Institute for Ecological Economics, Vienna University.

Ricketts, R.D. and T.C. Johnson (1996) "Climate Change in the Turkana Basin as Deduced from a 4000 Year Long d18O Record." *Earth and Planetary Science Letters* 142: 7–17.

Rigaud, Kanta Kumari, Alex de Sherbinin, Bryan Jones, Jonas Bergman, Vivanne Clement, Kayly Ober, Jacob Schere, Susanna Adamo, Brend McCusker, Silke Hauser, and Amelia Midgley (2018) *Groundswell: Preparing for International Climate Migration*. Washington, DC: World Bank.

Roaf, Michael and Jane Galbraith (1994) "Pottery and P-Values: Seafaring Merchants of Ur? Re-examined." *Antiquity* 68: 770–83.

Roberts, N., M. Cassis, O. Doonan, W. Eastwood, H. Elton, J. Haldon, A. Izdebski, and J. Newhard (2018a) "Not the End of the World? Post-Classical Decline and Recovery in Rural Anatolia." *Human Ecology* 46, 3: 305–322.

Roberts, N., W. Eastwood, C. Kuzucuoglu, G. Fiorentino, and V. Caracuta (2011) "Climatic, Vegetation and Cultural Change in the Eastern Mediterranean during the Mid-Holocene Environmental Transition." *The Holocene* 21, 1: 147–162.

Roberts, N., A. Moreno, B. L. Valerio-Garces, J.P. Corella, M. Jones, S. Allcock, J. Woodbridge, M.Morellon, J. Luterbacher, E. Xoplaki, and M. Turkes (2012) "Palaeolimnological Evidence for an East-West Climate See-Saw in the Mediterranean Since AD 900." *Global and Planetary Change* 84085: 23–24.

Roberts, N., J. Woodbridge, A. Bevan, A. Palmisano, S. Shennan, and E. Asouti (2018b) "Human Responses and Non-responses to Climatic Variations during the Last Glacial-Interglacial Transition in the Eastern Mediterranean." *Quaternary Science Reviews* 184: 47–67.

Roberts, N., J. Woodbridge, A. Palmisano, A. Bevan, R. Fyfe, and S. Shennan (2019) "Mediterranean Landscape Change during the Holocene: Synthesis, Comparison and Regional Trends in Population, Land Cover, and Climate." *The Holocene* 29, 5: 923–937.

Robinson, S.A., S. Black, B. Sellwood, and P.J. Valdes. (2006) "A Review of Palaeoclimates and Palaeoenvironments in the Levant and Eastern Mediterranean from 25000 to 5000 Years BP: Setting the Environmental Background for the Evolution of Human Civilization." *Quaternary Science Reviews* 25: 1517–1541.

Rockstrom, Johan, Owen Gaffney, Joeri Rogelj, Malte Meinshausen, Nebojsa Nakicenovic, and Hans Joachim Schellnhuber (2017) "A Roadmap for Rapid Decarbonization." *Science* 355, 6331: 1269–1271.

Roessler, Philip (2016) *Ethnic Politics and State Power in Africa: The Logic of the Coup-Civil War Trap*. Cambridge: Cambridge University Press.

Rogers Archaeology Lab (2013) "Panarchy: Out with the Old, in with the New." 2/12/2013. Available at http://nmnh.typepad.com/rogers.archaeology_lab/2013/02 /panarchy.html.

Rohling, Eelco J., Angela Hayes, Paul A. Mayewski, and Michael Kucera (2009) "Holocene Climate Variability in the Eastern Mediterranean and the End of the Bronze Age," in Christoph Bachhuber and R. Gareth Roberts, eds., *Forces of Transformation: The End of the Bronze Age in the Mediterranean*. Oxford: Oxbow Books.

Rollefson, Gary and Ilse Kohler-Rollefson (1993) "PPNC Adaptations in the First Half of the 6th Millennium BC." *Paleorient* 19, 1: 33–42.

Rollefson, Gary, Yorke Rowan, and Alexander Wasse (2014) "The Late Neolithic Colonization of the Eastern Badia of Jordan." *Levant* 46, 2: 1–17.

Rosen, Arlene M. and Isabel Rivera Collazo (2012) "Climate Change, Adaptive Cycles and the Persistence of Foraging Economies during the Late Pleistocene/Holocene Transition in the Levant." *Proceedings of the National Academy of Sciences of the United States of America* 109, 10: 3540–3645.

Rosen, Steven A. (2011) "The Desert and the Pastoralist: An Archaeological Perspective on Human Landscape Interaction in the Negev over the Millennia." *Annals of Arid Zone* 50, 3–4: 1–15.

Rosen, William (2007) *Justinian's Flea: Plague, Empire and the Birth of Empire*. New York: Viking.

Rothman, Mitchell S., ed. (2001) *Uruk Mesopotamia & Its Neighbors: Cross-Cultural Interactions in the Era of State Formation*. Santa Fe, NM: SAR Press.

Russell, Anna, Hijlke Buitenhuis, and Rene Cappers (2010) "Cooking Up a Storm: Did the 8.2 Ka Event Play a Role in the Changing Patterns of Subsistence at Tell Sabi Aybad?" Paper presented at the *8.2Ka Climate Event and Archaeology in the Ancient Near East*, Leiden University, March.

Ryan, Philippa and Arlene Rosen (2010) "Diversification as a Neolithic Adaptive Strategy in Central Anatolia during the 9th Millennium BP: Phytolith Perspectives for Catalhoyuk." Paper presented at *The 8.2Ka Climate Event and Archaeology in the Ancient Near East*, Leiden University, March.

Sahoune, F., M. Belhamel, M. Zelmat, and R. Kerbachi (2013) "Climate Change in Algeria: Vulnerability and Strategy of Mitigation and Adaptation." *Energy Procedia* 36: 1286–1294.

Sakaguchi, Kendra, Anil Varughese, and Graeme Auld (2017) "Climate Wars? A Systematic Review of Empirical Analyses on the Links between Climate Change and Violent Conflict." *International Studies Review* 19, 4: 622–645.

Sandars, N.K. (1978) *The Sea Peoples: Warriors of the Ancient Mediterranean, 1250–1150 BC*. London: Thames and Hudson.

Sarker, A., A.D. Mukerjee, M.K. Bera, B. Das, N. Juyal, P. Morthekai, R.D. Deshpande, V.S. Shinde, and L.S. Rao (2016) "Open Isotype in Archaeological Bioapatite from India: Implications to Climate Change and Decline of Bronze Age Harappan Civilization." *Scientific Reports* 6, 26555.

Sayce, A.H. (2014) *The First Chaldean Empire and the Hyksos in Egypt*. Scotts Valley, CA: CreateSpace Independent Publishing Platform.

Sazak, Selim Can, and Lauren R. Sukin (2015) "The Other Liquid Gold: Nuclear Power and Desalination in Saudi Arabia." *Foreign Affairs* (Nov).

Scerri, E.M.L., H.S. Groucutt, R.P. Jennings, and M.D. Petraglia (2014) "Unexpected Technological Heterogeneity in Northern Arabia Indicates Complex Late Pleistocene Demography at the Gateway to Asia." *Journal of Human Evolution* 75: 125–142.

Scheelbeek, Pauline F.D., Frances A. Bird, Hanna L. Tuomisto, Rosemary Green, Francesca B. Harris, Edward J.M. Joy, Zaid Chalabi, Elizabeth Allen, Andy Haines, and Alan D. Dangour (2018) "Effect of Environmental Changes on Vegetable and Legume Yields and Nutritional Quality." *Proceedings of the National Academy of Sciences of the United States of America (PNAS)* 115, 26: 6804–6809.

Scheffran, Jurgen, Elina Marmer, and Papa Sow (2012) "Migration as a Contribution to Resilience and Innovation in Climate Adaptation: Social Networks and Co-development in Northwest Africa." *Applied Geography* 33 (April): 119–127.

Schilling, Janpeter, Korbinian Freier, Elke Hertig, and Jurgen Scheffran (2012) "Climate Change, Vulnerability and Adaptation in North Africa with Focus on Morocco." *Agriculture, Ecosystems and Environment* 156: 12–26.

Schloen, J. David (2017) "Economic and Political Implications of Raising the Date for the Disappearance of Walled Towns in the Early Bronze Age Southern Levant," in Felix Hoflmayer, ed., *The Late Third Millennium in the Ancient Near East: Chronology, C14, and Climate Change.* Chicago, IL: Oriental Institute, University of Chicago.

Schmid, Boris V., Ulf Buntgen, W. Ryan Easterday, Christian Ginzler, Lars Walloe, Barbara Bramanti, and Nils Chr. Stenseth (2015) "Climate-Driven Introduction of the Black Death and Successive Plague Reintroductions into Europe." *Proceedings of the National Academy of Sciences of the United States of America (PNAS)* 113, 10: 3020–3025.

Schoell, M. (1978) "Oxygen Isotope Analysis on Authigenic Carbonates from lake Van Sediments and Their Possible Bearing on the Climate of the past 10000 Years," in E.T. Degens and F. Kurtman, eds., *The Geology of Lake Van.* Ankara: The Mineral Research and Exploration Institute of Turkey.

Schumpeter, Joseph (1939) *Business Cycles: A Theoretical, Historical and Statistical Analysis of the Capitalist Process,* 2 vols. New York: McGraw-Hill.

Schwartz, Glenn M. (2017) "Western Syria and the Third-to Second-Millennium B.C. Transition," in Felix Hoflmayer, ed., *The Late Third Millennium in the Ancient Near East: Chronology, C14, and Climate Change.* Chicago, IL: Oriental Institute, University of Chicago.

Schwartz, John (2018) "Will We Survive Climate Change?" *The New York Times,* November 19. Available at https://www.nytimes.com/2018/11/19/science/climate -change-doom.html.

Scotti, Ariel (2017) "Two Billion People May become Refugees from Climate Change by the End of the Century." *Daily News,* June 27. Available at http://www.dailynews .com/news/world/billion-people-refugees-climate-change-article-1328594.

Seidlmayer, S. (2000) "The First Intermediate Period (c. 2160–2055 BC)," in I. Shaw, ed., *The Oxford History of Ancient Egypt.* Oxford: Oxford University Press.

Selby, Jan (2019) "Climate Change and the Syrian Civil War, Part II: The Jazira's agrarian Crisis." *Geoforum* 101, May: 260–274.

Selby, Jan, Omar S. Dahi, Christiane Frohlich, and Mike Hulme (2017) "Climate Change and the Syrian Civil War Revisited." *Political Geography* 60: 232–244.

Selby, Jan and Clemens Hoffmann (2014a) "Introduction: Rethinking Climate Change, Conflict and Security." *Geopolitics* 19: 747–756.

Selby, Jan and Clemens Hoffmann (2014b) "Beyond Scarcity: Rethinking Water, Climate Change and Conflict in the Sudans." *Global Environmental Change* 29: 360–370.

Seneviratne, Sonia I., Markus G. Donat, Andy J. Pitman, Reto Knutti, and Robert L. Wilby (2016) "Allowable CO_2 Emissions Based on Regional and Impact-Related Climate Targets." *Nature* 529, 7587: 477–483.

Serdeczny, Olivia, Sophie Adams, Florent Baarsch, Dim Coumou, Alexander Robinson, Bill Hare, Michiel Schaeffer, Mahe Perrette, and Julia Reinhardt (2016) "Climate

Change Impacts in Sub-Saharan Africa: From Physical Changes to Their Social Repercussions." *Regional Environmental Change* 17: 1585–1600.

Shaffer, J. (1992) "The Indus Valley, Baluchistan, and Helmand Traditions: Neolithic Through Bronze Age," in R.W. Ehrlich, ed., *Chronologies in Old World Archaeology*, Vol. 1, 3rd ed. Chicago, IL: University of Chicago Press.

Sharafi, Arash, Ali Pourmand, Elizabeth A. Canuel, Erin Ferer Tyler, larry C. Peterson, Bernhard Aaichner, Sarh J. Feakins, Raraj Daryaee, Morteza Djamali, Abdolmagid Naderi Beni, Hamid A.K. LOahijani, and Peter K. Swart (2015) "Abrupt Climate Variablility Since the Last Deglaciation Based on a High-Resolution, Multi-proxy Past Record from NW Iran: The Hand that Rocked the Cradle of Civilization?" *Quaternary Science Reviews* 123: 2115–230.

Shea, J.J. (2008) "Transitions or Turnovers? Climatically-forced Extinctions of Homo Sapiens and Neanderthals in the East Mediterranean Levant." *Quaternary Science Reviews* 27, 23: 2253–2270.

Sherratt, Andrew (1996) "Sedentary Agricultural and Nomadic Pastoral Populations (3000–700 BC)," in A.H. Dani and J.-P. Mohen, eds., *History of Humanity: From the Third Millennium to the Seventh Century BC*. London: Routledge.

Sherratt, Andrew (2007) "Diverse Origins: Regional Contributions to the Genesis of Farming," in Sue Colledge and James Conolly, eds., *The Origins and Spread of Domestic Plants in Southwest Asia and Europe*. Walnut Creek, CA: Left Coast Press.

Sherratt, Susan and Andrew Sherratt (1993) "The Growth of the Mediterranean Economy in the Early First Millennium BC." *World Archaeology* 24, 3: 361–378.

Shongwe, Mxolisi, Geert Jan van Oldenborgh, Bart Van den Hurk, and Maarten van Alast (2011) "Projected Changes in Mean and Extreme Precipitation in Africa under Global Warming – Part II: East Africa." *Journal of Climate* 24: 3718–3733.

Simmons, A.H. (2007) *The Neolithic Revolution in the Near East*. Tucson, AZ: University of Arizona Press.

Singer, Itamar (1999) "A Political History of Ugarit," in W.G.E. Watson and N. Wyatt, eds. *Handbook of Ugaritic Studies*. Leiden: Brill.

Sinha, A., G. Kathayat, H. Weiss, H. Li, H. Cheng, J. Reuter, A.W. Schneider, M. Berkelhammer, S.F. Adali, L.D. Stott, and R.L. Edwards (2019) "Role of Climate in the Rise and Fall of the New-Assyrian Empire." *Science Advances* 5, 11: eaax6656.

Slettebak, Rune T. (2012) "Don't Blame the Weather! Climate-Related Natural Disasters and Civil Conflict." *Journal of Peace Research* 49, 1: 163–76.

Soon, Willie and Sallie Balinunas (2003) "Proxy Climate and Environmental Changes of the Past 1000 Years." *Climate Research* 23: 9–110.

Sowers, Jeannie, Avner Vengosh, and Erika Weinthal (2011) "Climate Change, Water Resources, and the Politics of Adaptation in the Middle East and North Africa." *Climatic Change* 104, 3–4: 599–627.

Spier, Fred (1996) *The Structure of Big History: From the Big Bang until Today*. Amsterdam: Amsterdam University Press.

Spiess, A. (2012) "Environmental Degradation, Climate Uncertainties and Human Vulnerabilities," in J. Scheffran, M. Brzoska, H.G. Brauch, P.M. Link, and J. Schilling, eds., *Climate Change, Human Security and Violent Conflict. Challenges for Social Stability*. Berlin: Springer.

Stager, L. (1992) "The Periodization of Palestine from Neolithic Through Early Bronze Times," in R.W. Ehrlich, ed., *Chronologies in Old World Archaeology*, Vol. 1, 3rd ed. Chicago, IL: University of Chicago Press.

Stanley, Steven M. (1996) *Children of the Ice Age: How a Global Catastrophe Allowed Humans to Evolve*. New York: W.H. Freeman and Co.

Staubwasser, M., F. Sirocko, P. M. Grootes, and M. Segl (2003) "Climate Change at the 4.2 ka BP Termination of the Indus Valley Civilization and Holocene South Asian Monsoon Variability." *Geophysical Research Letters* 30, 8: 1425.

Stein, Gil (1999) *Rethinking World-Systems: Diasporas, Colonies and Interaction in Uruk Mesopotamia*. Tucson, AZ: University of Arizona Press.

Stein, Gil (2010) "Local Identities and Interaction Spheres: Modeling Regional Variation in the Ubaid Horizon," in Robert A. Carter and Graham Philip, eds., *Beyond the Ubaid: Transformation and Integration in the Late Prehistoric Societies of the Middle East*. Chicago, IL: Oriental Institute, University of Chicago.

Steinkeller, P. (1993) "Early Political Development in Mesopotamia and the Origins of the Sargonic Empire," in M. Liverani, ed., *Akkad, The First World Empire: Structure, Ideology, Traditions*. Padua, Italy: Sargon.

Sternberg, Troy (2012) "Chinese Drought, Bread and the Arab Spring," *Applied Geography* 34: 519–524.

Sternberg, Troy (2013) "Chinese Drought, Wheat and the Egyptian Uprising: How a Localized Hazard Became Globalized," in C.E. Werrell and F. Femia, eds., *The Arab Spring and Climate Change: A Climate and Security Correlations Series*. Washington, DC: Center for American Progress.

Sterzel, T., M. Lüdeke, M. Kok, C. Walther, D. Sietz , I. de Soysa, P. Lucas, and P. Janssen (2014) "Armed Conflict Distribution in Global Drylands Through the Lens of a Typology of Socio-Ecological Vulnerability." *Regional Environmental Change* 14, 4: 1419–1435.

Swain, Ashok and Anders Jagerskog (2016) *Emerging Security Threats in the Middle East: The Impact of Climate Change and Globalization*. Lanham, MD: Rowman and Littlefield.

Sweatman, Martin B. and Dimirios Tsikritsis (2017) "Decoding Gobekli Tepe with Archaeoastronomy: What Does the Fox Say?" *Mediterranean Archaeology and Archaeometry* 17, 1: 233–250.

Sylla, Mohamadou Bamba, Jeremy S. Pal, Aissatou Faye, Kangberi Dimobe, and Harald Kunstmann (2018) "Climate Change to Severely Impact West African Basin Scale Irrigation in 2 Degrees C and 1.5 Degrees C Global Warming Scenarios." *Scientific Reports* 8: 15395.

Tainter, J.A. (1988) *The Collapse of Complex Societies*. Cambridge: Cambridge University Press.

Tanarhte,M., P. Hadjinicolaou, and J. Lelieveld (2012) "Intercomparison of Temperature and Precipitation Datasets Based on Observations in the Mediterranean and the Middle East." *Journal of Geophysical Research* 117 (June): 1–24.

Tayebi, Zahra (2014) *Agricultural Productivity in the Greater Middle East*. MA thesis. Lincoln, NE: Agricultural Economics, University of Nebraska.

Theisen, Ole Magnus (2012) "Climate Clashes? Weather Variability, Land Pressure, and Organized Violence in Kenya, 1989–2004." *Journal of Peace Research* 49, 1: 81–96.

Theisen, Ole Magnus, Nils Petter Gleditsch, and Halvard Buhaug (2013) "Is Climate Change a Driver of Armed Conflict?" *Climatic Change* 117, 3: 613–625.

Thomas, Natalie and Sumant Nigam (2018) "Twentieth Century Climate Change Over Africa: Seasonal Hydroclimate Trends and Sahara Desert Expansion." *Journal of Climate* 31: 3349–3370.

Thompson, William R. (2001a) "Trade, Collapse, and Diffusion." Paper delivered at the *Annual Meeting of the International Studies Association*, Chicago, February.

Thompson, William R. (2001b) "The Globalization of Ancient Near Eastern Trade." Paper delivered at the *Annual Meeting of the American Schools of Oriental Research*, Boulder, CO, November.

Thompson, William R. (2002) "Testing a Cyclical Instability Theory in the Ancient Near East." *Comparative Civilizations Review* 46: 34–78.

Thompson, William R. (2004) "Complexity, Diminishing Marginal Returns, and Fragmentation in Ancient Mesopotamia." *Journal of World Systems Research* 10, 3: 613–652.

Thompson, William R. (2006a) "Early Globalization, Trade Crises and Reorientations in the Ancient Near East," in Oystein S. LaBianca and Sandra Schram, eds., *Connectivity in Antiquity: Globalization as Long-Term Historical Process*. London: Equinox.

Thompson, William R. (2006b) "Crises in the Southwest Asian Bronze Age." *Nature and Culture* 1: 88–131.

Thompson, William R. (2006c) "Climate, Water and Political-economic Crises in Ancient Mesopotamia and Egypt," in Alf Hornburg and Carole Cromley, eds., *The World System and the Earth System: Global Socio-environmental Change and Sustainability Since the Neolithic*. Oakland, CA: Left Coast Books.

Thompson, William R. (2018) "Incursions, Climate Change, and Early Globalization Patterns," in Christopher Chase-Dunn and Hiroko Inoue, eds., *Systemic Boundaries*. Berlin: Springer.

Thompson, William R. and Leila Zakhirova (2019) *Racing to the Top: How Energy Fuels System Leadership in World Politics*. New York: Oxford University Press.

Tigchelaar, Michelle, David S. Battisti, Rosamund L. Naylor, and Deepak K. Ray (2018) "Future Warming Increases Probability of Globally Synchronized Maize Production Shocks." *Proceedings of the National Academy of Sciences of the United States of America (PNAS)* 115, 26: 6644–6649.

Toker, E., D. Sivan, E. Stern, B. Shirman, M. Tsinplis, and G. Spada (2012) "Evidence for Centennial Scale Sea Level Variability during the Medieval Climatic Optimum (Crusader Period) in Israel and the Eastern Mediterranean." *Earth and Planetary Science Letters* 315–316: 51–61.

Tosi, M. (1993) "The Harappan Civilization Beyond the Indian Subcontinent," in G.L. Possehl, ed., *Harappan Civilization: A Recent Perspective*, 2nd rev. ed. New Delhi: Oxford University Press.

Traerup, Sara and Jean Stephen (2015) "Technologies for Adaptation to Climate Change: Examples from the Agricultural and Water Sectors in Lebanon." *Climatic Change* 131, 3: 435–449.

Transfeld, Mareike (2016) "Political Bargaining and Violent Conflict: Shifting Elite Alliances as the Decision Factor in Yemen's Transformation." *Mediterranean Politics* 21, 1: 150–169.

Trigger, B.G. (1983) "The Rise of Egyptian Civilization," in B.G. Trigger, B.J. Kemp, D. O'Connor, and A.B. Lloyd, eds., *Ancient Egypt: A Social History*. Cambridge: Cambridge University Press.

Trigger, B.G. (1993) *Early Civilizations: Ancient Egypt in Context*. Cairo: American University in Cairo Press.

Tull, K. (2020) "The Projected Impacts of Climate Change on Food Security in the Middle East and North Africa." Brighton, UK: Institute of Development Studies. Available at http://www.gov.uk/dfid-research-ouputs/the-projected-impacts-of-climate-change-on-food-security-in-the-middle-east-and-north-africa-mena

Tyyska, Vappu, Tenna Blower, Samantha De Boer, Shunya Kawai, and Ashley Walcott (2017) "The Syrian Refugee Crisis: A Short Orientation." Toronto, ON: Ryerson Centre for Immigration and Settlement, Ryerson University, Working Paper No. 2017/2, April. Available at https://www.ryerson.ca/content/dam/rcis/documents/RCIS%20Working%20Paper%202017_2%20Tyyska%20et%20al_Final.pdf.

UAE Ministry of Climate Change & Environment (2017) "National Climate Change Plan of the United Arab Emirates, 2017–2050." Dubai: UAE Ministry of Climate Change & Environment.

UNCHR (2018) "Syria Refugee Crisis." Available at https://www.unrefugees.org/emergencies/syria/.

UNEP (2007) *Sudan: Post-Conflict Environmental Analysis*. Nairobi: United Nations Environment Programme. Available at https://postconflict.unep.ch/publications/UNEP_Sudan.pdf (accessed September 9, 2018).

UNEP (2018) *The Emissions Gap Report*. Nairobi: United Nations Environment Programme. Available at http://wedocs.unep.org/bitstream/handle/20.500.11822/22070/EGR_2017.pdf?sequence+1&isAllowed=y.

United Nations (2017) "World Population Prospects: 2017 Revision," pp. 17–21. Available at https://www.un.org/development/desa/publications/world-population-prospects-the-2017-revision.html.

United Nations (n.d). "UN Sustainable Development Goal 6.4.2 – Water Stress Indicator." Available at http://www.fao.org/sustainable-development-goals/indicators/642/en/.

Ur, Jason A. (2010) "Cycles of Civilization in Northern Mesopotamia, 4400–2000 BC." *Journal of Archaeological Research* 18: 387–431.

Ur, Jason A. (2015) "Urban Adaptations to Climate Change in Northern Mesopotamia," in Susanne Kemer, Rachael Dann, and Pernille Langsgaard, eds., *Climate and Ancient Societies*. Copenhagen: Museum Tusculanum Press.

USAID (2017) "Climate Change Risk Profile: West Africa Sahel." Available at https://www.climatelinks.org/resources/climate-risk-profile-west-africa-sahel.

Van De Mieroop, Marc (2004) *A History of the Ancient Near East, ca. 3000–323 BC*. Malden, MA: Blackwell.

Van De Mieroop, Marc (2007) *A History of the Ancient Near East, ca. 3000–323 BC*, 2nd ed. Blackwell.

Van De Mieroop, Marc (2011) *A History of Ancient Egypt*. London: Blackwell.

Van Seters, John (2010) *The Hyksos: A New Investigation*. Eugene, OR: Wipf and Stock.

Van Zeist, W. and S. Bottema (1977) "Palynological Investigations in Western Iran." *Palaeohistoria* 17: 19–85.

Van Zeist, W. and H. Woldring (1978) "A Postglacial Pollen Diagram from Lake Van in East Anatolia." *Review of Palaeobotany and Palynology* 26: 249–76.

Vercoutter, J. (1967) "Archaic Egypt," in J. Bottero, E. Cassin, J. Vercoutter, eds., *The Near East: The Early Civilizations*, translated by R.F. Tannenbaum. New York: Delacorte Press.

Verhoeven, Marc (2004) "Beyond Boundaries: Nature, Culture and a Holistic Approach to Domestication in the Levant." *Journal of World Prehistory* 18, 3: 179–298.

Verner, Dorte, ed. (2012) *Adaptation to a Changing Climate in the Arab Countries: A Case for Adaptation Governance and Leadership in Building Climate Resilience*. Washington, DC: World Bank.

Verner, Dorte (2013) *Tunisia in a Changing Climate: Assessment and Actions for Increased Resilience and Development*. Washington, DC: World Bank.

Von Rad, U., M. Schaaf, K.H. Michels, H. Schulz, W.H. Berger, and F. Sirocko (1999) "A 5000-Year Record of Climate Change." *Quaternary Research* 51: 39–53.

von Uexkull, Nina, Mihai Croicu, Hanna Fjelde, and Halvard Buhaug (2016) "Civil Conflict Sensitivity to Growing-Season Drought." *Proceedings of the National Academy of Sciences of the United States of America (PNAS)* Available at http://www.pnas.org/cgi/doi/10.1073/pnas.1607542113.

Voss, Katalyn A., James S. Famiglietti, MinHui Lo, Caroline de Linage, Matthew Rodell, and Sean C. Swenson (2013) "Groundwater Depletion in the Middle East from GRACE with Implications for Transboundary Water Management in the Tigris-Euphrates-Western Iran Region." *Water Resources Research* 49, 2: 904–914.

Waha, Katharina, Linda Krummenauer, Sophie Adams, Valentin Aich, Florent Baarsch, Dim Couou, Marianela Fader, Holger Hoff, Guiy Jobbins, Rachel Marcus, Matthias Mengel, Ilona M. Otto, Mahe Perrette, Marcia Rocha, Alexander Robinson, and Carl-Friedrich Schleussner (2017) "Climate Change Impacts in the Middle East and Northern Africa (MENA) Region and Their Implications for Vulnerable Population Groups." *Regional Environmental Change.* doi:10.1007/s10113-0017-1144-2.

Wanda, Yoshihide, Tom Gleeson, and Laurent Esnault (2014) "Wedge Approach to Water Stress." *Nature Geoscience* 7: 615–617.

Wang, Shaowu, Quangsheng Ge, Fang Wang, Xinyu Wen, and Jianbin Huang (2013) "Abrupt Climate Changes of Holocene." *Chinese Geographical Science* 23, 1: 1–12.

Ward, W. (1971) *Egypt and the East Mediterranean World, 2200–1900 BC.* Beirut, Lebanon: Heidelberg Press.

Ward-Perkins, Bryan (2005) *The Fall of Rome and the End of Civilization.* Oxford: Oxford University Press.

Warren, P. (1995) "Minoan Crete and Pharaonic Egypt," in W. Vivian Davies and L. Schofield, eds., *Egypt, the Aegean and the Levant: Interconnections in the Second Millennium BC.* London: British Museum Press.

Water Resources Institute (2019) "Aqueduct." Available at https://wri.org/applications/aqueduct/country-rankings/.

Waterbury, John (2013) "The Political Economy of Climate Change in the Arab Region." UNDP: Arab Human Development Report, Research Paper Series. Available at http://citeseerx.ist.psu.edu/viewdoc/download?doi=10.1.1.363.5511&rep=rep1&type=pdf.

Waterbury, John (2017) "Water and Water Supply in the MENA: Less of the Same," in Adnan Badran, Sohail Murad, Elias Baydou, and Nuhad Daghir, eds., *Water, Energy & Food Sustainability in the Middle East: The Sustainability Triangle.* Berlin: Springer.

Watrous, L. (1998) "Egypt and Crete in the Early Middle Bronze Age: A Case of Trade and Cultural Diffusion," in *The Aegean and the Orient in the Second Millennium*, E.H. Cline and D. Harris-Cline, eds. Liège, Belgium: Annales d'Archeologie Egéenne de l'Université de Liège et UT-PASP.

Watson, P.J., S. LeBlanc, and C.L. Redman (1984) *Archaeological Explanation: The Scientific Method in Archaeology.* New York: Columbia University Press.

Weber, T., A. Haensler, D. Rechid, S. Pfeifer, B. Eggert, and D. Jacob (2018) "Analyzing Regional Climate Change in Africa in a 1.5, 2, and 3 Degree C Global Warming World." *Earth's Future* 6: 643–655.

Weiner, Malcolm (2016) "The Interaction of Climate Change and Agency in the Collapse of Civilization, c. 2300–2000 BC." *Radiocarbon* 56, 4: S1–S16.

Weinstein, J.M. (1998) "Egyptian Relations with the Eastern Mediterranean World at the End of the Second Millennium BCE," in S. Gitin, A. Mazar, and E. Stern, eds., *Mediterranean Peoples in Transition: Thirteenth to Early Tenth Centuries BCE.* Jerusalem: Israel Exploration Society.

Weis, Kenneth R. (2015) "The Making of a Climate Refugee." *Foreign Policy.* Available at https://foreignpolicy.com/2015/01/28/the-making-of-a-climate-refugee-kiribati-tarawa-teitiota/.

Weiss, Barry (1982) "The Decline of Late Bronze Age Civilizations as a Possible Response to Climate Change." *Climate Change* 4, 2: 173–198.

Weiss, Harvey (1997) "Beyond the Younger Dryas: Collapse as Adaptation to Abrupt Climate Change in Ancient West Asia and the Eastern Mediterranean," in G. Bawden and R.M. Reycraft, eds., *Environmental Disaster and the Archaeology of Human Response*. Albuquerque, NM: Maxwell Museum of Anthropology, University of New Mexico.

Weiss, Harvey (2000) "Beyond the Younger Dryas: Collapse as Adaptation to Abrupt Climate Change in Ancient West Asia and the Eastern Mediterranean," in G. Bawden and R.M. Reycraft, eds., *Environmental Disaster and the Archaeology of Human Response*. Albuquerque, NM: Maxwell Museum of Anthropology, University of New Mexico.

Weiss, Harvey (2014) "Seventeen Kings Who Lived in Tents," in Felix Hoflmayer, ed., *The Late Third Millennium in the Ancient Near East: Chronology, C14, and Climate Change*. Chicago, IL: Oriental Institute, University of Chicago.

Weiss, Harvey (2016) "Global Megadrought, Societal Collapse and Resilience at 4.2-3.9ka BP Across the Mediterranean and West Asia." *Pages Magazine* 24, 2: 62–63.

Weiss, Harvey and M.-A. Courty (1993) "The Genesis and Collapse of the Akkadian Empire," in M. Liverani, ed., *Akkad: The First World Empire*. Padua, Italy: Sargon.

Weiss, Harvey, M.-A. Courty, W W Etterstron, F. Guichard, L. Senior, R. Meadow and A. Curnow (1993) "The Genesis and Collapse of Third Millennium North Mesopotamian Civilization." *Science* 261, August 20: 995–1004.

Weninger, Bernhard (2009) "Yarmoukian Rubble Slides. Evidence for Early Holocene Rapid Climate Change in Southern Jordan." *Neo-Lithics: The Newsletter of Southwest Asian Neolithic Research*. 1, 09: 5–11

Weninger, Bernhard, Lee Clare, Eelco J. Rohling, and Ofer Bar-Yosef (2009) "The Impact of Rapid Climate Change on Prehistoric Societies during the Holocene in the Eastern Mediterranean." *Documenta Praehistorica* 36: 7–59.

Weninger, Bernhard and Thomas Harper (2015) "The Geographic Corridor for Rapid Climate Change in Southeast Europe and Ukraine," in E. Schultze, ed., *Neolithic and Copper Age between the Carpathians and the Aegean Sea*. Berlin: Deutsches Archaologist Institut.

Werz, Michael and Max Hoffman (2013) "Climate Change, Migration, and Conflict," in Caitlin E. Werrell and Francesco Femia, eds., *The Arab Spring and Climate Change*. Available at https://climateandsecurity.files.wordpress.com/2012/04/climatechan gearabspring-ccs-cap-stimson.pdf.

White, Sam (2011) *The Climate of Rebellion in the Early Modern Ottoman Empire*. Cambridge: Cambridge University Press.

White, Sam (2013) "The Little Ice Age Crisis of the Ottoman Empire: A Conjuncture in Middle East Environmental History," in Alan Mikhail, ed., *Water on Sand: Environmental Histories of the Middle East and North Africa*. Oxford: Oxford University Press.

Wick, Lucia, Gerry Lemcke, and Michael Sturm (2003) "Evidence of Lateglacial and Holocene Climatic Change and Human Impact in Eastern Anatolia: High-resolution Pollen, Charcoal, Isotopic and Geochemical Records from the Laminated Sediments of Lake Van, Turkey." *The Holocene* 13, 5: 665–675.

Wilcox, George (2013) "The Roots of Cultivation in Southwestern Asia." *Science* July 5, 341: 6141.

Wilkinson, T.J. (1994) "The Structure and Dynamics of Dry-Farming States in Upper Mesopotamia." *Current Anthropology* 35, 5: 483–520.

Wilkinson, T.J. (1997) "Environmental Fluctuations, Agricultural Production and Collapse: A View from Bronze Age Mesopotamia," in H. Nuzhet Dalfes, George Kukla, and Harvey Weiss, eds., *Third Millennium BC Climate Change and Old World Collapse*. Berlin: Springer.

Wilkinson, Toby A.H. (1996) *State Formation in Egypt: Chronology and Society*. Oxford: Basingstoke Press.

Wilkinson, Toby A.H. (2010) *The Rise and Fall of Ancient Egypt*. London: Bloomsbury.

Wilkinson, Toby A.H. (2011) *The Rise and Fall of Ancient Egypt*. New York: Random House.

Woertz, Eckart (2013) *Oil for Food: The Global Foods Crisis and The Middle East*. Oxford: Oxford University Press.

Woertz, Eckart (2017) "Agricultural Development in the Wake of the Arab Spring," in Giacomi Luciani, ed., *Combining Economic and Political Development: The Experience of MENA*. Leiden: Brill.

World Bank Group (2014) *Turn Down the Heat: Confronting the New Climate Normal*. Washington, DC: World Bank.

World Bank Group (2016) "Climate Action Plan, 2016–2020." Available at https://openknowledge.worldbank.org/bitstream/handle/10986/24451/K8860.pdf.

Worth, Robert F. (2010) "Earth is Parched Where Syrian Farms Thrived." *New York Times*, October 13. Available at https://www.nytimes.com/2010/10/14/world/middleeast/14syria.html.

Worth, Robert F. (2017) "The Man Who Danced on the Heads of Snakes." *New York Times*, December 7.

Wright, G. (1968) *Obsidian Analyses and Early Trade in the Near East: 7500 to 3500 BC*. PhD dissertation, University of Michigan.

Wright, Katherine I. (1994) "Ground-Stone Tools and Hunter Gatherer Subsistence in Southwest Asia: Implications for the Transition to Farming." *American Antiquity* 59, 2: 238–263.

Xoplaki, Elena, Dominik Fleitmann, Juerg Luterbacher, Sebastien Wagner, John F. Haldon, Eduardo Zorita, Ioannis Telelis, Andrea Toreti, and Adam Izdebski (2016) "The Medieval Climate Anomaly and Byzantium: A Review of the Evidence on Climatic Fluctuations, Economic Performance and Societal Change." *Quaternary Science Reviews* 136: 229–252.

Xoplaki, Elena, Juerg Luterbacher, Sebastian Wagner, Eduardo Zorita, Dominik Fleitmann, Johannes Preiser-Kapeller, Abigail M. Sargent, Sam White, Andrea Toreti, John Haldon, Lee Mordechai, Deniz Bozkurt, Sena Akcer-On, and Adam Izdebski (2018) "Modelling Climate and Societal Resilience in the Eastern Mediterranean in the Last Millennium." *Human Ecology* 46, 3: 363–379.

Xu, Chi, Timothy, A. Kohler, Timothy M. Lenton, Jens-Christian Svenning, and Marten Scheffer (2019) "Future of the Human Climate Niche." *Proceedings of the National Academy of Sciences (PNAS)*. Available at http://www.pnas.org/cgi/doi/10;1073/pnas.191014117.

Yassin, Nasser (2018) "101 Facts & Figures on the Syrian Refugee Crisis." Issam Fares Institute for Public Policy and International Affairs, American University of Beirut, Lebanon. Available at https://www.unhcr.org/lb/wp-content/uploads/sites/16/2018/06/AUB-IFI-101-fact-Sheet-2018_-English.pdf.

Yayboke, Erolk and Aaron N. Milner (2018) *Confronting the Global Forced Migration Crisis*. Lanham, MD: Rowman and Littlefield and Washington, DC: Center for Strategic and International Studies, Georgetown University.

Yoffee, Norman (1991) "The Collapse of Ancient Mesopotamian States and Civilization," in Norman Yoffee and George L. Cowgill, eds., *The Collapse of Ancient States and Civilizations*. Tucson, AZ: University of Arizona Press.

Yoffee, Norman (2010) "Collapse in Ancient Mesopotamia: What Happened, What Didn't," in P.A. McAnamy and Norman Yoffee, eds., *Questioning Collapse: Human Resilience, Ecological Vulnerability, and the Aftermath of Empire*. Cambridge: Cambridge University Press.

Yue, Ricci P.H. and Harry F. Lee (2018) "Climate Change and Plague History in Europe." *Science China Earth Sciences* 61, 2: 163–177.

Zarins, Juris (1989) "Ancient Egypt and the Red Sea Trade: The Case for Obsidian in the Predynastic and Archaic Periods," in A. Leonard, Jr. and B.B. Williams, eds., *Essays in Ancient Civilization Presented to Helene J. Kantor*. Chicago, IL: University of Chicago Oriental Institute.

Zarins, Juris (1992) "Pastoral Nomadism in Arabia: Ethnoarchaeology and the Archeological Record," in O.B. Yosef and A. Zhazanov, eds., *Pastoralism in the Levant*. Monographs in World Archeology, no. 10. Madison, WI: Prehistory Press.

Zeder, Melinda A. (2008) "Domestication and Early Agriculture in the Mediterranean Basin: Origins, Diffusion and Impact." *Proceedings of the National Academy of Sciences of the United States of America (PNAS)*, August 19, 105, 3: 11597–11604.

Zeder, Melinda A. (2011) "The Origins of Agriculture in the Near East." *Current Anthropology* 52, 54: S221–S235.

Zerboni, Andrea and Savino di Lernia (2010) "The 8.2Ka Event and the Transition Towards Food Production in the Central Sahara." Paper presented at *The 8.2Ka Climate Event and Archaeology in the Ancient Near East*, Leiden University, March.

Zettler, R. (1992) "12th Century BC Babylonia: Continuity and Change," in W.A. Ward and M.S. Joukowsky, eds., *The Crisis Years: The 12th Century BC, From beyond the Danube to the Tigris*. Dubuque, IA: Kendall/Hunt.

Zhang, David D., Peter Brecke, Harry F. Lee, Yuan-qing He, and Jane Zhang (2007) "Global Climate Change, War, and Population Decline in Recent Human History." *Proceedings of the National Academy of Sciences of the United States of America (PNAS)* 104, 49: 19214–19219.

Zhang, David D., C.Y. Jim, George .C-S. Lin, Yuan-Qing He, James J. Wang, and Harry F. Lee (2006) "Climate Change, Wars, and Dynastic Cycles in China over the Last Millennium." *Climatic Change* 76: 459–477.

Zhang, David D., Henry F. Lee, Cong Wang, Baosheng Li, Jane Zhang, Qing Pei, and Jingan Chen (2011a) "Climate Change and Large-Scale Human Population Collapses in the Pre-industrial Era." *Global Ecology and Biogeography* 20, 4: 520–531.

Zhang, David D., Harry F. Lee, Cong Wang, Baosheng Li, Qing Pei, Jane Zhang, and Yulun An (2011b) "The Causality Analysis of Climate Change and Large-scale Human Crisis." *Proceedings of the National Academy of Sciences of the United States of America (PNAS)* 108, 42: 17296–17301.

Zhang, David D., Jane Zhang, Harry F. Lee and Yuan-qing He (2007) "Climate Change and War Frequency in Eastern China over the Last Millennium." *Human Ecology* 35, 4: 403–414.

Zhao, Chuang, Bing Liu, Shilong Piao, Huhui Wang, David B. Lobell, Yao Huang, Mengtian Huang, Yitong Yao, Simon Bassu, Phillippe Ciais, Jean-Louis Durand, Joshua Elliott, Frank Ewert, Ivan A. Janssens, Tao Li, Erde Lin, Qiang Liu, Pierre Martre, Chirtopher Muller, Shushi Peng, Josep Penuclas, Alex C. Ruane, Daniel Wallad, Tao Wang, Donghai Wu, Zhuo Lie, Yan Zhu, Zaichun Zhu, and Senthold Asseng (2017) "Temperature Increase Reduces Global Yields of Major Crops in Four Independent Estimates." *Proceedings of the National Academy of Sciences of the United States of America (PNAS)* 114, 35: 9326–9331.

Zimmer, C. (2016) "How the First Farmers Changed History." *New York Times*, October 17. Available at https://www.nytimes.com/2016/10/18/science/ancient-farmers-archaeology-dna.html.

Zuckerman, Sharon (2007) "Anatomy of Destruction: Crisis Architecture, Termination Rituals, and the Fall of Canaanite Hazor." *Journal of Mediterranean Archaeology* 20, 1: 3–32.

INDEX